Introduction to Multivariate Analysis

Linear and Nonlinear Modeling

CHAPMAN & HALL/CRC
Texts in Statistical Science Series

Series Editors
Francesca Dominici, *Harvard School of Public Health, USA*
Julian J. Faraway, *University of Bath, UK*
Martin Tanner, *Northwestern University, USA*
Jim Zidek, *University of British Columbia, Canada*

Statistical Theory: A Concise Introduction
F. Abramovich and Y. Ritov

Practical Multivariate Analysis, Fifth Edition
A. Afifi, S. May, and V.A. Clark

Practical Statistics for Medical Research
D.G. Altman

Interpreting Data: A First Course in Statistics
A.J.B. Anderson

Introduction to Probability with R
K. Baclawski

Linear Algebra and Matrix Analysis for Statistics
S. Banerjee and A. Roy

Statistical Methods for SPC and TQM
D. Bissell

Bayesian Methods for Data Analysis, Third Edition
B.P. Carlin and T.A. Louis

Second Edition
R. Caulcutt

The Analysis of Time Series: An Introduction, Sixth Edition
C. Chatfield

Introduction to Multivariate Analysis
C. Chatfield and A.J. Collins

Problem Solving: A Statistician's Guide, Second Edition
C. Chatfield

Statistics for Technology: A Course in Applied Statistics, Third Edition
C. Chatfield

Bayesian Ideas and Data Analysis: An Introduction for Scientists and Statisticians
R. Christensen, W. Johnson, A. Branscum, and T.E. Hanson

Modelling Binary Data, Second Edition
D. Collett

Modelling Survival Data in Medical Research, Second Edition
D. Collett

Introduction to Statistical Methods for Clinical Trials
T.D. Cook and D.L. DeMets

Applied Statistics: Principles and Examples
D.R. Cox and E.J. Snell

Multivariate Survival Analysis and Competing Risks
M. Crowder

Statistical Analysis of Reliability Data
M.J. Crowder, A.C. Kimber, T.J. Sweeting, and R.L. Smith

An Introduction to Generalized Linear Models, Third Edition
A.J. Dobson and A.G. Barnett

Nonlinear Time Series: Theory, Methods, and Applications with R Examples
R. Douc, E. Moulines, and D.S. Stoffer

Introduction to Optimization Methods and Their Applications in Statistics
B.S. Everitt

Extending the Linear Model with R: Generalized Linear, Mixed Effects and Nonparametric Regression Models
J.J. Faraway

A Course in Large Sample Theory
T.S. Ferguson

Multivariate Statistics: A Practical Approach
B. Flury and H. Riedwyl

Readings in Decision Analysis
S. French

Markov Chain Monte Carlo: Stochastic Simulation for Bayesian Inference, Second Edition
D. Gamerman and H.F. Lopes

Bayesian Data Analysis, Third Edition
A. Gelman, J.B. Carlin, H.S. Stern, D.B. Dunson, A. Vehtari, and D.B. Rubin

Multivariate Analysis of Variance and Repeated Measures: A Practical Approach for Behavioural Scientists
D.J. Hand and C.C. Taylor

Practical Data Analysis for Designed Practical Longitudinal Data Analysis
D.J. Hand and M. Crowder

Logistic Regression Models
J.M. Hilbe

Richly Parameterized Linear Models: Additive, Time Series, and Spatial Models Using Random Effects
J.S. Hodges

Statistics for Epidemiology
N.P. Jewell

Stochastic Processes: An Introduction, Second Edition
P.W. Jones and P. Smith

The Theory of Linear Models
B. Jørgensen

Principles of Uncertainty
J.B. Kadane

Graphics for Statistics and Data Analysis with R
K.J. Keen

Mathematical Statistics
K. Knight

Introduction to Multivariate Analysis: Linear and Nonlinear Modeling
S. Konishi

Nonparametric Methods in Statistics with SAS Applications
O. Korosteleva

Modeling and Analysis of Stochastic Systems, Second Edition
V.G. Kulkarni

Exercises and Solutions in Biostatistical Theory
L.L. Kupper, B.H. Neelon, and S.M. O'Brien

Exercises and Solutions in Statistical Theory
L.L. Kupper, B.H. Neelon, and S.M. O'Brien

Design and Analysis of Experiments with SAS
J. Lawson

A Course in Categorical Data Analysis
T. Leonard

Statistics for Accountants
S. Letchford

Introduction to the Theory of Statistical Inference
H. Liero and S. Zwanzig

Statistical Theory, Fourth Edition
B.W. Lindgren

Stationary Stochastic Processes: Theory and Applications
G. Lindgren

The BUGS Book: A Practical Introduction to Bayesian Analysis
D. Lunn, C. Jackson, N. Best, A. Thomas, and D. Spiegelhalter

Introduction to General and Generalized Linear Models
H. Madsen and P. Thyregod

Time Series Analysis
H. Madsen

Pólya Urn Models
H. Mahmoud

Randomization, Bootstrap and Monte Carlo Methods in Biology, Third Edition
B.F.J. Manly

Introduction to Randomized Controlled Clinical Trials, Second Edition
J.N.S. Matthews

Statistical Methods in Agriculture and Experimental Biology, Second Edition
R. Mead, R.N. Curnow, and A.M. Hasted

Statistics in Engineering: A Practical Approach
A.V. Metcalfe

Beyond ANOVA: Basics of Applied Statistics
R.G. Miller, Jr.

A Primer on Linear Models
J.F. Monahan

Applied Stochastic Modelling, Second Edition
B.J.T. Morgan

Elements of Simulation
B.J.T. Morgan

Probability: Methods and Measurement
A. O'Hagan

Introduction to Statistical Limit Theory
A.M. Polansky

Applied Bayesian Forecasting and Time Series Analysis
A. Pole, M. West, and J. Harrison

Statistics in Research and Development, Time Series: Modeling, Computation, and Inference
R. Prado and M. West

Introduction to Statistical Process Control
P. Qiu

Sampling Methodologies with Applications
P.S.R.S. Rao

A First Course in Linear Model Theory
N. Ravishanker and D.K. Dey

Essential Statistics, Fourth Edition
D.A.G. Rees

Stochastic Modeling and Mathematical Statistics: A Text for Statisticians and Quantitative
F.J. Samaniego

Statistical Methods for Spatial Data Analysis
O. Schabenberger and C.A. Gotway

Large Sample Methods in Statistics
P.K. Sen and J. da Motta Singer

Decision Analysis: A Bayesian Approach
J.Q. Smith

Analysis of Failure and Survival Data
P. J. Smith

Applied Statistics: Handbook of GENSTAT Analyses
E.J. Snell and H. Simpson

Applied Nonparametric Statistical Methods, Fourth Edition
P. Sprent and N.C. Smeeton

Data Driven Statistical Methods
P. Sprent

Generalized Linear Mixed Models: Modern Concepts, Methods and Applications
W. W. Stroup

Survival Analysis Using S: Analysis of Time-to-Event Data
M. Tableman and J.S. Kim

Applied Categorical and Count Data Analysis
W. Tang, H. He, and X.M. Tu

Elementary Applications of Probability Theory, Second Edition
H.C. Tuckwell

Introduction to Statistical Inference and Its Applications with R
M.W. Trosset

Understanding Advanced Statistical Methods
P.H. Westfall and K.S.S. Henning

Statistical Process Control: Theory and Practice, Third Edition
G.B. Wetherill and D.W. Brown

Generalized Additive Models: An Introduction with R
S. Wood

Epidemiology: Study Design and Data Analysis, Third Edition
M. Woodward

Experiments
B.S. Yandell

Texts in Statistical Science

Introduction to Multivariate Analysis

Linear and Nonlinear Modeling

Sadanori Konishi
Chuo University
Tokyo, Japan

CRC Press
Taylor & Francis Group
Boca Raton London New York

CRC Press is an imprint of the
Taylor & Francis Group, an **informa** business
A CHAPMAN & HALL BOOK

Originally published in Japanese by Iwanami Shoten, Publishers, Tokyo, 2010. This English language edition published in 2014 by Chapman & Hall/CRC, Boca Raton, FL, U.S.A., by arrangement with the author c/o Iwanami Shoten, Publishers, Tokyo.

CRC Press
Taylor & Francis Group
6000 Broken Sound Parkway NW, Suite 300
Boca Raton, FL 33487-2742

ISBN 13: 978-1-4665-6728-3 (hbk)

Library of Congress Cataloging-in-Publication Data

Konishi, Sadanori, author.
Introduction to multivariate analysis : linear and nonlinear modeling / Sadanori Konishi.
pages cm. -- (Chapman & Hall/CRC Texts in Statistical Science series.)
Includes bibliographical references and index.
Summary: "Multivariate techniques are used to analyze data that arise from more than one variable in which there are relationships between the variables. Mainly based on the linearity of observed variables, these techniques are useful for extracting information and patterns from multivariate data as well as for the understanding the structure of random phenomena. This book describes the concepts of linear and nonlinear multivariate techniques, including regression modeling, classification, discrimination, dimension reduction, and clustering"-- Provided by publisher.
ISBN 978-1-4665-6728-3 (hardback)
1. Multivariate analysis. I. Title.

QA278.K597 2014
519.5'35--dc23 2014013362

Visit the Taylor & Francis Web site at
http://www.taylorandfrancis.com

and the CRC Press Web site at
http://www.crcpress.com

Contents

List of Figures xiii

List of Tables xxi

Preface xxiii

1 Introduction 1
- 1.1 Regression Modeling 1
  - 1.1.1 Regression Models 2
  - 1.1.2 Risk Models 4
  - 1.1.3 Model Evaluation and Selection 5
- 1.2 Classification and Discrimination 7
  - 1.2.1 Discriminant Analysis 7
  - 1.2.2 Bayesian Classification 8
  - 1.2.3 Support Vector Machines 9
- 1.3 Dimension Reduction 11
- 1.4 Clustering 11
  - 1.4.1 Hierarchical Clustering Methods 12
  - 1.4.2 Nonhierarchical Clustering Methods 12

2 Linear Regression Models 15
- 2.1 Relationship between Two Variables 15
  - 2.1.1 Data and Modeling 16
  - 2.1.2 Model Estimation by Least Squares 18
  - 2.1.3 Model Estimation by Maximum Likelihood 19
- 2.2 Relationships Involving Multiple Variables 22
  - 2.2.1 Data and Models 23
  - 2.2.2 Model Estimation 24
  - 2.2.3 Notes 29
  - 2.2.4 Model Selection 31
  - 2.2.5 Geometric Interpretation 34
- 2.3 Regularization 36

2.3.1 Ridge Regression 37
2.3.2 Lasso 40
2.3.3 L_1 Norm Regularization 44

3 Nonlinear Regression Models 55
3.1 Modeling Phenomena 55
3.1.1 Real Data Examples 57
3.2 Modeling by Basis Functions 58
3.2.1 Splines 59
3.2.2 *B*-splines 63
3.2.3 Radial Basis Functions 65
3.3 Basis Expansions 67
3.3.1 Basis Function Expansions 68
3.3.2 Model Estimation 68
3.3.3 Model Evaluation and Selection 72
3.4 Regularization 76
3.4.1 Regularized Least Squares 77
3.4.2 Regularized Maximum Likelihood Method 79
3.4.3 Model Evaluation and Selection 81

4 Logistic Regression Models 87
4.1 Risk Prediction Models 87
4.1.1 Modeling for Proportional Data 87
4.1.2 Binary Response Data 91
4.2 Multiple Risk Factor Models 94
4.2.1 Model Estimation 95
4.2.2 Model Evaluation and Selection 98
4.3 Nonlinear Logistic Regression Models 98
4.3.1 Model Estimation 100
4.3.2 Model Evaluation and Selection 101

5 Model Evaluation and Selection 105
5.1 Criteria Based on Prediction Errors 105
5.1.1 Prediction Errors 106
5.1.2 Cross-Validation 108
5.1.3 Mallows' C_p 110
5.2 Information Criteria 112
5.2.1 Kullback-Leibler Information 113
5.2.2 Information Criterion AIC 115
5.2.3 Derivation of Information Criteria 121
5.2.4 Multimodel Inference 127

5.3 Bayesian Model Evaluation Criterion 128
5.3.1 Posterior Probability and BIC 128
5.3.2 Derivation of the BIC 130
5.3.3 Bayesian Inference and Model Averaging 132

6 Discriminant Analysis 137
6.1 Fisher's Linear Discriminant Analysis 137
6.1.1 Basic Concept 137
6.1.2 Linear Discriminant Function 141
6.1.3 Summary of Fisher's Linear Discriminant Analysis 144
6.1.4 Prior Probability and Loss 146
6.2 Classification Based on Mahalanobis Distance 148
6.2.1 Two-Class Classification 148
6.2.2 Multiclass Classification 149
6.2.3 Example: Diagnosis of Diabetes 151
6.3 Variable Selection 154
6.3.1 Prediction Errors 154
6.3.2 Bootstrap Estimates of Prediction Errors 156
6.3.3 The .632 Estimator 158
6.3.4 Example: Calcium Oxalate Crystals 160
6.3.5 Stepwise Procedures 162
6.4 Canonical Discriminant Analysis 164
6.4.1 Dimension Reduction by Canonical Discriminant Analysis 164

7 Bayesian Classification 173
7.1 Bayes' Theorem 173
7.2 Classification with Gaussian Distributions 175
7.2.1 Probability Distributions and Likelihood 175
7.2.2 Discriminant Functions 176
7.3 Logistic Regression for Classification 179
7.3.1 Linear Logistic Regression Classifier 179
7.3.2 Nonlinear Logistic Regression Classifier 183
7.3.3 Multiclass Nonlinear Logistic Regression Classifier 187

8 Support Vector Machines 193
8.1 Separating Hyperplane 193
8.1.1 Linear Separability 193
8.1.2 Margin Maximization 196

8.1.3 Quadratic Programming and Dual Problem 198
8.2 Linearly Nonseparable Case 203
8.2.1 Soft Margins 204
8.2.2 From Primal Problem to Dual Problem 208
8.3 From Linear to Nonlinear 212
8.3.1 Mapping to Higher-Dimensional Feature Space 213
8.3.2 Kernel Methods 216
8.3.3 Nonlinear Classification 218

9 Principal Component Analysis 225
9.1 Principal Components 225
9.1.1 Basic Concept 225
9.1.2 Process of Deriving Principal Components and Properties 230
9.1.3 Dimension Reduction and Information Loss 234
9.1.4 Examples 235
9.2 Image Compression and Decompression 239
9.3 Singular Value Decomposition 243
9.4 Kernel Principal Component Analysis 246
9.4.1 Data Centering and Eigenvalue Problem 246
9.4.2 Mapping to a Higher-Dimensional Space 249
9.4.3 Kernel Methods 252

10 Clustering 259
10.1 Hierarchical Clustering 259
10.1.1 Interobject Similarity 260
10.1.2 Intercluster Distance 261
10.1.3 Cluster Formation Process 263
10.1.4 Ward's Method 267
10.2 Nonhierarchical Clustering 270
10.2.1 *K*-Means Clustering 271
10.2.2 Self-Organizing Map Clustering 273
10.3 Mixture Models for Clustering 275
10.3.1 Mixture Models 275
10.3.2 Model Estimation by EM Algorithm 277

A Bootstrap Methods 283
A.1 Bootstrap Error Estimation 283
A.2 Regression Models 285
A.3 Bootstrap Model Selection Probability 285

B Lagrange Multipliers **287**
B.1 Equality-Constrained Optimization Problem 287
B.2 Inequality-Constrained Optimization Problem 288
B.3 Equality/Inequality-Constrained Optimization 289

C EM Algorithm **293**
C.1 General EM Algorithm 293
C.2 EM Algorithm for Mixture Model 294

Bibliography **299**

Index **309**

List of Figures

1.1 The relation between falling time (x sec) and falling distance (y m) of a body. 3

1.2 The measured impact y (in acceleration, g) on the head of a dummy in repeated experimental crashes of a motorcycle with a time lapse of x (msec). 4

1.3 Binary data {0, 1} expressing the presence or absence of response in an individual on exposure to various levels of stimulus. 5

1.4 Regression modeling; the specification of models that approximates the structure of a phenomenon, the estimation of their parameters, and the evaluation and selection of estimated models. 6

1.5 The training data of the two classes are completely separable by a hyperplane (left) and the overlapping data of the two classes may not be separable by a hyperplane (right). 10

1.6 Mapping the observed data to a high-dimensional feature space and obtaining a hyperplane that separates the two classes. 10

1.7 72 chemical substances with 6 attached features, classified by clustering on the basis of mutual similarity in substance qualities. 13

2.1 Data obtained by measuring the length of a spring (y cm) under different weights (x g). 17

2.2 The relationship between the spring length (y) and the weight (x). 18

2.3 Linear regression and the predicted values and residuals. 20

2.4 (a) Histogram of 80 measured values obtained while repeatedly suspending a load of 25 g and its approximated probability model. (b) The errors (i.e., noise) contained in these measurements in the form of a histogram having its origin at the mean value of the measurements and its approximated error distribution. 21

2.5 Geometrical interpretation of the linear regression model $\boldsymbol{y} = X\boldsymbol{\beta} + \boldsymbol{\varepsilon}$. $M(X)$ denotes the $(p+1)$-dimensional linear subspace spanned by the $(p+1)$ n-dimensional column vectors of the design matrix X. 35

2.6 Ridge estimate (left panel) and lasso estimate (right panel): Ridge estimation shrinks the regression coefficients β_1, β_2 toward but not exactly to 0 relative to the corresponding least squares estimates $\hat{\boldsymbol{\beta}}$, whereas lasso estimates the regression coefficient β_1 at exactly 0. 41

2.7 The profiles of estimated regression coefficients for different values of the L_1 norm $= \sum_{i=1}^{13} |\beta_i(\lambda)|$ with λ varying from 6.78 to 0. The axis above indicates the number of nonzero coefficients. 45

2.8 The function $p_\lambda(|\beta_j|)$ (solid line) and its quadratic approximation (dotted line) with the values of β_j along the x axis, together with the quadratic approximation for a β_{j0} value of 0.15. 48

2.9 The relationship between the least squares estimator (dotted line) and three shrinkage estimators (solid lines): (a) hard thresholding, (b) lasso, and (c) SCAD. 50

3.1 Left panel: The plot of 104 tree data obtained by measurement of tree trunk girth (inch) and tree weight above ground (kg). Right panel: Fitting a polynomial of degree 2 (solid curve) and a growth curve model (dashed curve). 57

3.2 Motorcycle crash trial data ($n = 133$). 59

3.3 Fitting third-degree polynomials to the data in the subintervals $[a, t_1]$, $[t_1, t_2]$, $\cdots$, $[t_m, b]$ and smoothly connecting adjacent polynomials at each knot. 60

3.4 Functions $(x - t_i)_+ = \max\{0, x - t_i\}$ and $(x - t_i)_+^3$ included in the cubic spline given by (3.10). 61

3.5 Basis functions: (a) $\{1, x\}$; linear regression, (b) polynomial regression; $\{1, x, x^2, x^3\}$, (c) cubic splines, (d) natural cubic splines. 62

3.6 A cubic B-spline basis function connected four different third-order polynomials smoothly at the knots 2, 3, and 4. 63
3.7 Plots of the first-, second-, and third-order B-spline functions. As may be seen in the subintervals bounded by dotted lines, each subinterval is covered (piecewise) by the polynomial order plus one basis function. 65
3.8 A third-order B-spline regression model is fitted to a set of data, generated from $u(x) = \exp\{-x\sin(2\pi x)\} + 0.5 + \varepsilon$ with Gaussian noise. The fitted curve and the true structure are, respectively, represented by the solid line and the dotted line with cubic B-spline bases. 66
3.9 Curve fitting; a nonlinear regression model based on a natural cubic spline basis function and a Gaussian basis function. 70
3.10 Cubic B-spline nonlinear regression models, each with a different number of basis functions (a) 10, (b) 20, (c) 30, (d) 40, fitted to the motorcycle crash experiment data. 73
3.11 The cubic B-spline nonlinear regression model $y = \sum_{j=1}^{13} \hat{w}_j b_j(x)$. The model is estimated by maximum likelihood and selected the number of basis functions by AIC. 75
3.12 The role of the penalty term: Changing the weight in the second term by the regularization parameter γ changes $S_\gamma(\boldsymbol{w})$ continuously, thus enabling continuous adjustment of the model complexity. 78
3.13 The effect of a smoothing parameter λ: The curves are estimated by the regularized maximum likelihood method for various values of λ. 82

4.1 Plot of the graduated stimulus levels shown in Table 4.1 along the x axis and the response rate along the y axis. 89
4.2 Logistic functions. 90
4.3 Fitting the logistic regression model to the observed data shown in Table 4.1 for the relation between the stimulus level x and the response rate y. 90
4.4 The data on presence and non-presence of the crystals are plotted along the vertical axis as $y = 0$ for the 44 individuals exhibiting their non-presence and $y = 1$ for the 33 exhibiting their presence. The x axis takes the values of their urine specific gravity. 92

4.5 The fitted logistic regression model for the 77 set of data expressing observed urine specific gravity and presence or non-presence of calcium oxalate crystals. 93

4.6 Plot of post-operative kyphosis occurrence along $Y = 1$ and non-occurrence along $Y = 0$ versus the age (x; in months) of 83 patients. 99

4.7 Fitting the polynomial-based nonlinear logisitic regression model to the kyphosis data. 103

5.1 Fitting of 3rd-, 8th-, and 12th-order polynomial models to 15 data points. 107

5.2 Fitting a linear model (dashed line), a 2nd-order polynomial model (solid line), and an 8th-order polynomial model (dotted line) to 20 data. 119

6.1 Projecting the two-dimensional data in Table 6.1 onto the axes $y = x_1$, $y = x_2$ and $y = w_1x_1 + w_2x_2$. 139

6.2 Three projection axes (a), (b), and (c) and the distributions of the class G_1 and class G_2 data when projected on each one. 140

6.3 Fisher's linear discriminant function. 143

6.4 Mahalanobis distance and Euclidean distance. 151

6.5 Plot of 145 training data for a normal class G_1 (○), a chemical diabetes class G_2 (▲), and clinical diabetes class G_3 (×). 152

6.6 Linear decision boundaries that separate the normal class G_1, the chemical diabetes class G_2, and the clinical diabetes class G_3. 154

6.7 Plot of the values obtained by projecting the 145 observed data from three classes onto the first two discriminant variables (y_1, y_2) in (6.92). 170

7.1 Likelihood of the data: The relative level of occurrence of males 178 cm in height can be determined as $f(178|170, 6^2)$. 176

7.2 The conditional probability $P(x|G_i)$ that gives the relative level of occurrence of data $\boldsymbol{x}$ in each class. 178

7.3 Decision boundary generated by the linear function. 184

7.4 Classification of phenomena exhibiting complex class structures requires a nonlinear discriminant function. 185

7.5 Decision boundary that separates the two classes in the nonlinear logistic regression model based on the Gaussian basis functions. 187

8.1 The training data are completely separable into two classes by a hyperplane (left panel), and in contrast, separation into two classes cannot be obtained by any such linear hyperplane (right panel). 194
8.2 Distance from $\boldsymbol{x}_0 = (x_{01}\ x_{02})^T$ to the hyperplane $w_1x_1 + w_2x_2 + b = \boldsymbol{w}^T\boldsymbol{x} + b = 0$. 196
8.3 Hyperplane (H) that separates the two classes, together with two equidistant parallel hyperplanes (H_+ and H_-) on opposite sides. 197
8.4 Separating hyperplanes with different margins. 198
8.5 Optimum separating hyperplane and support vectors represented by the black solid dots and triangle on the hyperplanes H_+ and H_-. 202
8.6 No matter where we draw the hyperplane for separation of the two classes and the accompanying hyperplanes for the margin, some of the data (the black solid dots and triangles) do not satisfy the inequality constraint. 205
8.7 The class G_1 data at (0, 0) and (0, 1) do not satisfy the original constraint $x_1 + x_2 - 1 \geq 1$. We soften this constraint to $x_1 + x_2 - 1 \geq 1 - 2$ for data (0, 0) and $x_1 + x_2 - 1 \geq 1 - 1$ for (0, 1) by subtracting 2 and 1, respectively; each of these data can then satisfy its new inequality constraint equation. 205
8.8 The class G_2 data (1, 1) and (0, 1) are unable to satisfy the constraint, but if the restraint is softened to $-(x_1 + x_2 - 1) \geq 1 - 2$ and $-(x_1 + x_2 - 1) \geq 1 - 1$ by subtracting 2 and 1, respectively, each of these data can then satisfy its new inequality constraint equation. 206
8.9 A large margin tends to increase the number of data that intrude into the other class region or into the region between hyperplanes H_+ and H_-. 207
8.10 A small margin tends to decrease the number of data that intrude into the other class region or into the region between hyperplanes H_+ and H_-. 207
8.11 Support vectors in a linearly nonseparable case: Data corresponding to the Lagrange multipliers such that $0 < \hat{\alpha}_i \leq \lambda$ (the black solid dots and triangles). 211

8.12 Mapping the data of an input space into a higher-dimensional feature space with a nonlinear function. 214
8.13 The separating hyperplane obtained by mapping the two-dimensional data of the input space to the higher-dimensional feature space yields a nonlinear discriminant function in the input space. The black solid data indicate support vectors. 216
8.14 Nonlinear decision boundaries in the input space vary with different values σ in the Gaussian kernel; (a) $\sigma = 10$, (b) $\sigma = 1$, (c) $\sigma = 0.1$, and (d) $\sigma = 0.01$. 221

9.1 Projection onto three different axes, (a), (b), and (c) and the spread of the data. 226
9.2 Eigenvalue problem and the first and second principal components. 230
9.3 Principal components based on the sample correlation matrix and their contributions: The contribution of the first principal component increases with increasing correlation between the two variables. 237
9.4 Two-dimensional view of the 21-dimensional data set, projected onto the first (x) and second (y) principal components. 239
9.5 Image digitization of a handwritten character. 240
9.6 The images obtained by first digitizing and compressing the leftmost image 7 and then decompressing transmitted data using a successively increasing number of principal components. The number in parentheses shows the cumulative contribution rate in each case. 242
9.7 Mapping the observed data with nonlinear structure to a higher-dimensional feature space, where PCA is performed with linear combinations of variables z_1, z_2, z_3. 250

10.1 Intercluster distances: Single linkage (minimum distance), complete linkage (maximum distance), average linkage, centroid linkage. 262
10.2 Cluster formation process and the corresponding dendrogram based on single linkage when starting from the distance matrix in (10.7). 265

10.3 The dendrograms obtained for a single set of 72 six-dimensional data using three different linkage techniques: single, complete, and centroid linkages. The circled portion of the dendrogram shows a chaining effect. 266

10.4 Fusion-distance monotonicity (left) and fusion-distance inversion (right). 267

10.5 Stepwise cluster formation procedure by Ward's method and the related dendrogram. 271

10.6 Stepwise cluster formation process by k-means. 272

10.7 The competitive layer comprises an array of m nodes. Each node is assigned a different weight vector $\boldsymbol{w}_j = (w_{j1}, w_{j2}, \cdots, w_{jp})^T$ $(j = 1, 2, \cdots, m)$, and the Euclidean distance of each p-dimensional data to the weight vector is computed. 274

10.8 Histogram based on observed data on the speed of recession from Earth of 82 galaxies scattered in space. 276

10.9 Recession-speed data observed for 82 galaxies are shown on the upper left and in a histogram on the upper right. The lower left and lower right show the models obtained by fitting with two and three normal distributions, respectively. 279

List of Tables

2.1 The length of a spring under different weights. 16

2.2 The n observed data. 17

2.3 Four factors: temperature (x_1), pressure (x_2), PH (x_3), and catalyst quantity (x_4), which affect the quantity of product (y). 23

2.4 The response y representing the results in n trials, each with a different combination of p predictor variables x_1, $x_2, \cdots, x_p$. 23

2.5 Comparison of the sum of squared residuals ($\hat{\sigma}^2$) divided by the number of observations, maximum log-likelihood $\ell(\hat{\beta})$, and AIC for each combination of predictor variables. 33

2.6 Comparison of the estimates of regression coefficients by least squares (LS) and lasso L_1. 44

4.1 Stimulus levels and the proportion of individuals responded. 88

5.1 Comparison of the values of RSS, CV, and AIC for fitting the polynomial models of order 1 through 9. 119

6.1 The 23 two-dimensional observed data from the varieties A and B. 138

6.2 Comparison of prediction error estimates for the classification rule constructed by the linear discriminant function. 161

6.3 Variable selection via the apparent error rates (APE). 161

Preface

The aim of statistical science is to develop the methodology and the theory for extracting useful information from data and for reasonable inference to elucidate phenomena with uncertainty in various fields of the natural and social sciences. The data contain information about the random phenomenon under consideration and the objective of statistical analysis is to express this information in an understandable form using statistical procedures. We also make inferences about the unknown aspects of random phenomena and seek an understanding of causal relationships.

Multivariate analysis refers to techniques used to analyze data that arise from multiple variables between which there are some relationships. Multivariate analysis has been widely used for extracting useful information and patterns from multivariate data and for understanding the structure of random phenomena. Techniques would include regression, discriminant analysis, principal component analysis, clustering, etc., and are mainly based on the linearity of observed variables.

In recent years, the wide availability of fast and inexpensive computers enables us to accumulate a huge amount of data with complex structure and/or high-dimensional data. Such data accumulation is also accelerated by the development and proliferation of electronic measurement and instrumentation technologies. Such data sets arise in various fields of science and industry, including bioinformatics, medicine, pharmaceuticals, systems engineering, pattern recognition, earth and environmental sciences, economics, and marketing. Therefore, the effective use of these data sets requires both linear and nonlinear modeling strategies based on the complex structure and/or high-dimensionality of the data in order to perform extraction of useful information, knowledge discovery, prediction, and control of nonlinear phenomena and complex systems.

The aim of this book is to present the basic concepts of various procedures in traditional multivariate analysis and also nonlinear techniques for elucidation of phenomena behind observed multivariate data, focusing primarily on regression modeling, classification and discrimination, dimension reduction, and clustering. Each chapter includes many figures and illustrative examples to promote a deeper understanding of various techniques in multivariate analysis.

In practice, the need always arises to search through and evaluate a large number of models and from among them select an appropriate model that will work effectively for elucidation of the target phenomena. This book provides comprehensive explanations of the concepts and derivations of the AIC, BIC, and related criteria, together with a wide range of practical examples of model selection and evaluation criteria. In estimating and evaluating models having a large number of predictor variables, the usual methods of separating model estimation and evaluation are inefficient for the selection of factors affecting the outcome of the phenomena. The book also reflects these aspects, providing various regularization methods, including the L_1 norm regularization that gives simultaneous model estimation and variable selection.

The book is written in the hope that, through its fusion of knowledge gained in leading-edge research in statistical multivariate analysis, machine learning, and computer science, it may contribute to the understanding and resolution of problems and challenges in this field of research, and to its further advancement.

This book might be useful as a text for advanced undergraduate and graduate students in statistical sciences, providing a systematic description of both traditional and newer techniques in multivariate analysis and machine learning. In addition, it introduces linear and nonlinear statistical modeling for researchers and practitioners in various scientific disciplines such as industrial and systems engineering, information science, and life science. The basic prerequisites for reading this textbook are knowledge of multivariate calculus and linear algebra, though they are not essential as it includes a self-contained introduction to theoretical results.

This book is basically a translation of a book published in Japanese by Iwanami Publishing Company in 2010. I would like to thank Uichi Yoshida and Nozomi Tsujimura of the Iwanami Publishing Company for giving me the opportunity to translate and publish in English.

I would like to acknowledge with my sincere thanks Yasunori Fujikoshi, Genshiro Kitagawa, and Nariaki Sugiura, from whom I have learned so much about the seminal ideas of statistical modeling. I have been greatly influenced through discussions with Tomohiro Ando, Yuko Araki, Toru Fujii, Seiya Imoto, Mitsunori Kayano, Yoshihiko Maesono, Hiroki Masuda, Nagatomo Nakamura, Yoshiyuki Ninomiya, Ryuei Nishii, Heewon Park, Fumitake Sakaori, Shohei Tateishi, Takahiro Tsuchiya, Masayuki Uchida, Takashi Yanagawa, and Nakahiro Yoshida.

I would also like to express my sincere thanks to Kei Hirose, Shuichi Kawano, Hidetoshi Matsui, and Toshihiro Misumi for reading the manuscript and offering helpful suggestions. David Grubbs patiently encouraged and supported me throughout the final preparation of this book. I express my sincere gratitude to all of these people.

Sadanori Konishi

Tokyo, January 2014

Chapter 1

Introduction

The highly advanced computer systems and progress in electronic measurements and instrumentation technologies have together facilitated the acquisition and accumulation of data with complex structure and/or high-dimensional data in various fields of science and industry. Data sets arise in such areas as genome databases in life science, remote-sensing data from earth-observing satellites, real-time recorded data of motion process in system engineering, high-dimensional data in character recognition, speech recognition, image analysis, etc. Hence, it is desirable to research and develop new statistical data analysis techniques to efficiently extract useful information as well as elucidate patterns behind the data in order to analyze various phenomena and to yield knowledge discovery. Under the circumstances linear and nonlinear multivariate techniques are rapidly developing by fusing the knowledge in statistical science, machine learning, information science, and mathematical science.

The objective of this book is to present the basic concepts of various procedures in the traditional multivariate analysis and also nonlinear techniques for elucidation of phenomena behind the observed multivariate data, using many illustrative examples and figures. In each chapter, starting from an understanding of the traditional multivariate analysis based on the linearity of multivariate observed data, we describe nonlinear techniques, focusing primarily on regression modeling, classification and discrimination, dimension reduction, and clustering.

1.1 Regression Modeling

Regression analysis is used to model the relationship between a response variable and several predictor (explanatory) variables. Once a model has been identified, various forms of inferences such as prediction, control, information extraction, knowledge discovery, and risk evaluation can be done within the framework of deductive argument. Thus, the key to solving various real-world problems lies in the development and construction of suitable linear and nonlinear regression modeling.

1.1.1 Regression Models

Housing prices vary with land area and floor space, but also with proximity to stations, schools, and supermarkets. The quantity of chemical products is sensitive to temperature, pressure, catalysts, and other factors. In Chapter 2, using *linear regression* models, which provide a method for relating multiple factors to the outcomes of such phenomena, we describe the basic concept of *regression modeling*, including model specification based on data reflecting the phenomena, model estimation of the specified model by least squares or maximum likelihood methods, and model evaluation of the estimated model. Throughout this modeling process, we select a suitable one among competing models.

The volume of extremely high-dimensional data that are observed and entered into databases in biological, genomic, and many other fields of science has grown rapidly in recent years. For such data, the usual methods of separating model estimation and evaluation are ineffectual for the selection of factors affecting the outcome of the phenomena, and thus effective techniques are required to construct models with high reliability and prediction. This created a need for work on modeling and has led, in particular, to the proposal of various regularization methods with an L_1 penalty term (the sum of absolute values of regression coefficients), in addition to the sum of squared errors and log-likelihood functions. A distinctive feature of the proposed methods is their capability for simultaneous model estimation and variable selection. Chapter 2 also describes various regularization methods, including *ridge* regression (Hoerl and Kennard, 1970) and the least absolute shrinkage and selection operator (*lasso*) proposed by Tibshirani (1996), within the framework of linear regression models.

Figure 1.1 shows the results of an experiment performed to investigate the relation between falling time (x sec) and falling distance (y m) of a body. The figure suggests that it should be possible to model the relation using a polynomial. There are many phenomena that can be modeled in this way, using polynomial equations, exponential functions, or other specific nonlinear functions to relate the outcome of the phenomenon and the factors influencing that outcome.

Figure 1.2, however, poses new difficulties. It shows the measured impact y (in acceleration, g) on the head of a dummy in repeated experimental crashes of a motorcycle into a wall, with a time lapse of x (msec) as measured from the instant of collision (Härdle, 1990). For phenomena with this type of apparently complex nonlinear structure, it is quite difficult to effectively capture the structure by modeling with specific

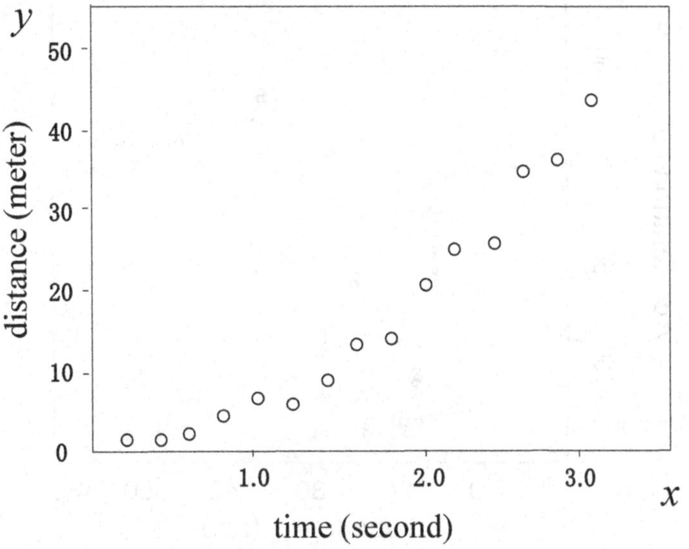

Figure 1.1 *The relation between falling time (x sec) and falling distance (y m) of a body.*

nonlinear functions such as polynomial equations and exponential functions.

Chapter 3 discusses *nonlinear regression* modeling for extracting useful information from data containing complex nonlinear structures. It introduces models based on more flexible splines, *B*-splines, and radial basis functions for modeling complex nonlinear structures. These models often serve to ascertain complex nonlinear structures, but their flexibility often prevents their effective function in the estimation of models with the traditional least squares and maximum likelihood methods. In such cases, these estimation methods are replaced by regularized least squares and regularized maximum likelihood methods.

The latter two techniques, which are generally referred to as *regularization* methods, are effectively used to reduce over-fitting of models to data and thus prevent excessive model complexity, and are known to contribute for reducing the variability of the estimated models. This chapter also describes regularization methods within the framework of nonlinear regression modeling.

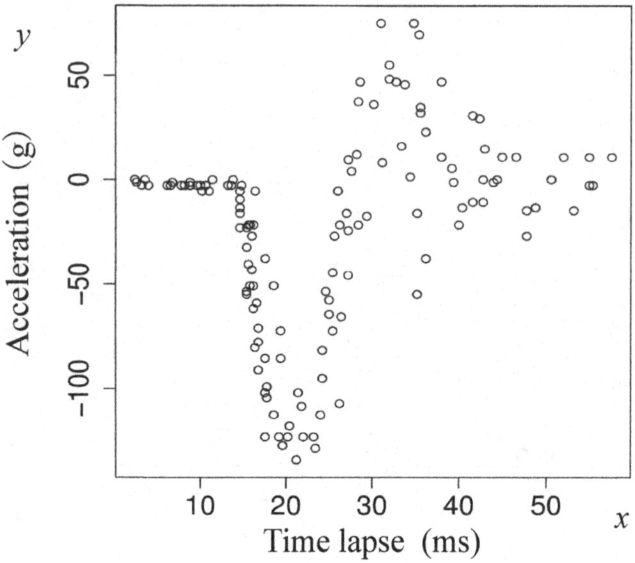

Figure 1.2 *The measured impact y (in acceleration, g) on the head of a dummy in repeated experimental crashes of a motorcycle with a time lapse of x (msec).*

1.1.2 Risk Models

In today's society marked by complexity and uncertainty, we live in a world exposed by various types of risks. The risk may be associated with occurrences such as traffic accidents, natural disasters such as earthquakes, tsunamis, or typhoons, or development of a lifestyle disease, with transactions such as credit card issuance, or with many other occurrences too numerous to enumerate. It is possible to gauge the magnitude of risk in terms of probability based on past experience and information gained in life in society, but often with only a limited accuracy.

All of this poses the question of how to probabilistically assess unknown risks for a phenomenon using information obtained from data. For example, in searching for the factors that induce a certain disease, the problem is in how to construct a model for assessing the probability of its occurrence based on observed data. The effective probabilistic model for assessing the risk may lead to its future prevention. Through such risk modeling, moreover, it may also be possible to identify important disease-related factors.

Chapter 4 presents an answer to this question, in the form of modeling for the risk evaluation, and in particular describes the basic concept of *logistic regression modeling*, together with its extension from linear to nonlinear modeling. This includes models to assess risks based on binary data {0, 1} expressing the presence or absence of response in an individual or object on exposure to various levels of stimulus, as shown in Figure 1.3.

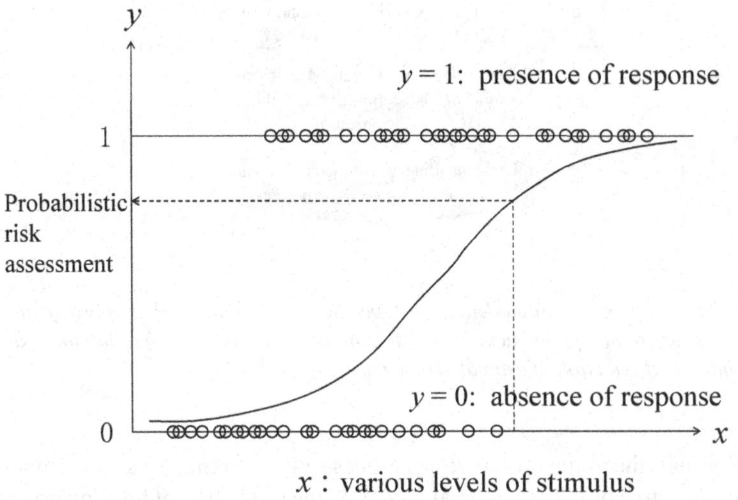

Figure 1.3 *Binary data {0, 1} expressing the presence or absence of response in an individual on exposure to various levels of stimulus.*

1.1.3 Model Evaluation and Selection

Figure 1.4 shows a process consisting essentially of the conceptualization of *regression modeling*; the specification of models that approximates the structure of a phenomenon, the estimation of their parameters, and the evaluation and selection of estimated models.

In relation to the data shown in Figure 1.1 for a body dropped from a high position, for example, it is quite natural to consider a polynomial model for the relation between the falling time and falling distance and to carry out polynomial model fitting. This represents the processes of model specification and parameter estimation. For elucidation of this

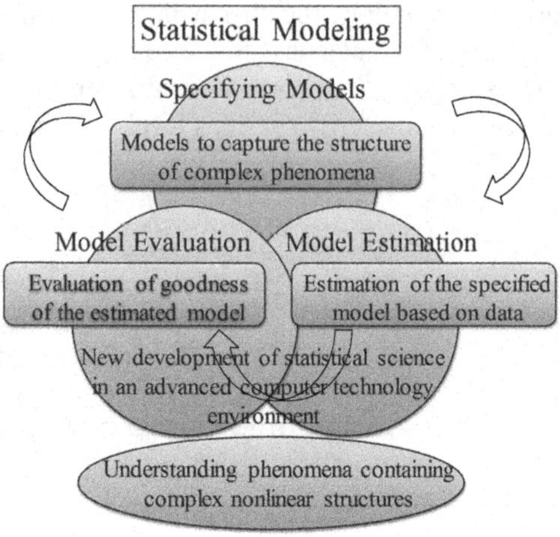

Figure 1.4 *Regression modeling; the specification of models that approximates the structure of a phenomenon, the estimation of their parameters, and the evaluation and selection of estimated models.*

physical phenomenon, however, a question may remain as to the optimum degree of the polynomial model. In the prediction of housing prices with linear regression models, moreover, a key question is what factors to include in the model. Furthermore, in considering nonlinear regression models, one is confronted by the availability of infinite candidate models for complex nonlinear phenomena controlled by smoothing parameters, and the need for selection of models that will appropriately approximate the structures of the phenomena, which is essential for their elucidation.

In this way, the need always arises to search through and evaluate a large number of models and from among them select one that will work effectively for elucidation of the target phenomena, based on the information provided by the data. This is commonly referred to as the *model evaluation and selection* problem.

Chapter 5 focuses on the model evaluation and selection problems, and presents various model selection criteria that are widely used as indicators in the assessment of the *goodness* of a model. It begins with a description of evaluation criteria proposed as estimators of prediction error, and then discusses the AIC (Akaike information criterion) based on Kullback-Leibler information and the BIC (Bayesian information criterion) derived from a Bayesian view point, together with fundamental concepts that serve as the bases for derivation of these criteria.

The AIC, proposed in 1973 by Hirotugu Akaike, is widely used in various fields of natural and social sciences and has contributed greatly to elucidation, prediction, and control of phenomena. The BIC was proposed in 1978 by Gideon E. Schwarz and is derived based on a Bayesian approach rather than on information theory as with the AIC, but like the AIC it is utilized throughout the world of science and has played a central role in the advancement of modeling. Chapters 2 to 4 of this book show the various forms of expression of the AIC for linear, nonlinear, logistic, and other models, and give examples for model evaluation and selection problems based on the AIC.

Model selection from among candidate models constructed on the basis of data is essentially the selection of a single model that best approximates the data-generated probability structure. In Chapter 5, the discussion is further extended to include the concept of *multimodel inference* (Burnham and Anderson, 2002) in which the inferences are based on model aggregation and utilization of the relative importance of constructed models in terms of their weighted values.

1.2 Classification and Discrimination

Classification and discrimination techniques are some of the most widely used statistical tools in various fields of natural and social sciences. The primary aim in discriminant analysis is to assign an individual to one of two or more classes (groups) on the basis of measurements on feature variables. It is designed to construct linear and nonlinear decision boundaries based on a set of training data.

1.2.1 Discriminant Analysis

When a preliminary diagnosis concerning the presence or absence of a disease is made on the basis of data from blood chemistry analysis, information contained in the blood relating to the disease is measured, assessed, and acquired in the form of qualitative data. The diagnosis of normality or abnormality is based on multivariate data from several test results. In other words, it is an assessment of whether the person examined is included in a group consisting of normal individuals or a group consisting of individuals who exhibit a disease-related abnormality.

This kind of assessment can be made only if information from test results relating to relevant groups is understood in advance. In other words, because the patterns shown by test data for normal individuals and for individuals with relevant abnormalities are known in advance, it is possible to judge which group a new individual belongs to. Depending on the type of disease, the group of individuals with abnormalities may be further divided into two or more categories, depending on factors such as age and progression, and diagnosis may therefore involve assignment of a new individual to three or more target groups. In the analytical method referred to as *discriminant analysis*, a statistical formulation is derived for this type of problem and statistical techniques are applied to provide a diagnostic formula.

The objective of discriminant analysis is essentially to find an effective rule for classifying previously unassigned individuals to two or more predetermined groups or classes based on several measurements. The discriminant analysis has been widely applied in many fields of science, including medicine, life sciences, earth and environmental science, biology, agriculture, engineering, and economics, and its application to new problems in these and other fields is currently under investigation.

The basic concept of discriminant analysis was introduced by R. A. Fisher in the 1930s. It has taken its present form as a result of subsequent research and refinements by P. C. Mahalanobis, C. R. Rao, and others, centering in particular on linear discriminant analysis. Chapter 6 begins with discussion and examples relating to the basic concept of Fisher and Mahalanobis for two-class linear and quadratic discrimination. It next proceeds to the basic concepts and concrete examples of multiclass discrimination, and *canonical discriminant analysis* of multiclass data in higher-dimensional space, which enables visualization by projection to lower-dimensional space.

1.2.2 Bayesian Classification

The advance in measurement and instrumentation technologies has enabled rapid growth in the acquisition and accumulation of various types of data in science and industry. It has been accompanied by rapidly developing research on *nonlinear* discriminant analysis for extraction of information from data with complex structures.

Chapter 7 provides a bridge from linear to nonlinear classification based on *Bayes' theorem*, incorporating prior information into a modeling process. It discusses the concept of a likelihood of observed data using the probability distribution, and then describes Bayesian procedures for linear and quadratic classification based on a *Bayes factor*. The discussion then proceeds to construct *linear and nonlinear logistic* discriminant procedures, which utilizes the Bayesian classification and links it to the logistic regression model.

1.2.3 Support Vector Machines

Research in character recognition, speech recognition, image analysis, and other forms of pattern recognition is advancing rapidly, through the fusion of machine learning, statistics, and computer science, leading to the proposal of new analytical methods and applications (e.g., Hastie, Tibshirani and Friedman, 2009; Bishop, 2006). One of these is an analytical method employing the *support vector machine* described in Chapter 8, which constitutes a classification method that is conceptually quite different from the statistical methods. It has therefore come to be used in many fields as a method that can be applied to classification problems that are difficult to analyze effectively by previous classification methods, such as those based on high-dimensional data.

The essential feature of the classification method with the support vector machine is, first, establishment of the basic theory in a context of two perfectly separated classes, followed by its extension to linear and nonlinear methods for the analysis of actual data. Figure 1.5 (left) shows that the training data of the two classes are completely separable by a hyperplane (in the case of two dimensions, a straight line segment). In an actual context requiring the use of the classification method, as shown in Figure 1.5 (right), the overlapping data of the two classes may not be separable by a hyperplane.

With the support vector machine, mapping of the observed data to high-dimensional space is used for the extension from linear to nonlinear analysis. In diagnoses for the presence or absence of a given disease, for example, increasing the number of test items may have the effect of increasing the separation between the normal and abnormal groups and thus facilitate the diagnoses. The basic concept is the utilization of such a tendency, where possible, to map the observed data to a high-dimensional feature space and thus obtain a hyperplane that separates the two classes, as illustrated in Figure 1.6. An extreme increase in computational complexity in the high-dimensional space can also be surmounted by utilization of a *kernel* method. Chapter 8 provides a step-by-step description of this process of extending the support vector machine from linear to nonlinear analysis.

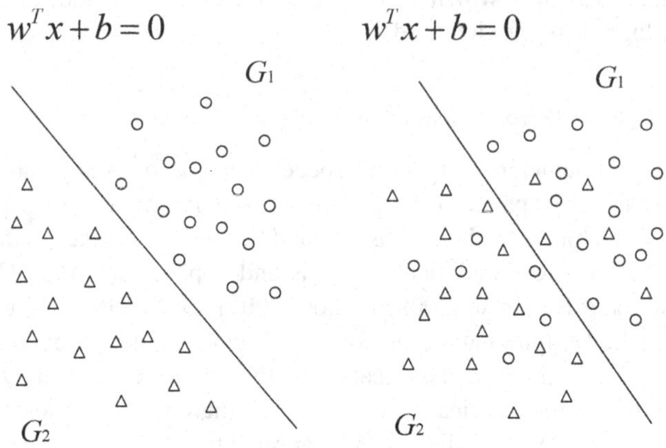

Figure 1.5 *The training data of the two classes are completely separable by a hyperplane (left) and the overlapping data of the two classes may not be separable by a hyperplane (right).*

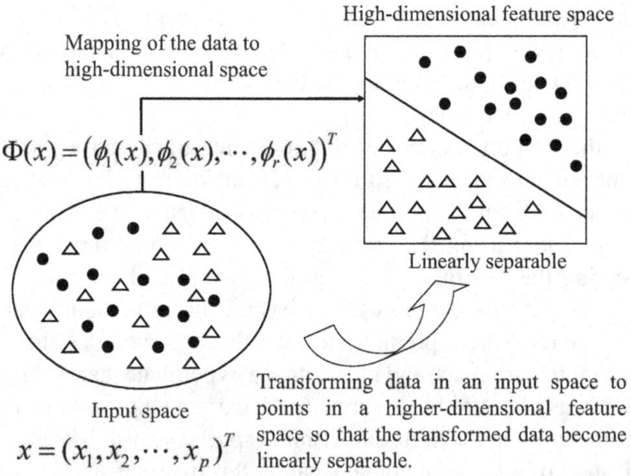

Figure 1.6 *Mapping the observed data to a high-dimensional feature space and obtaining a hyperplane that separates the two classes.*

1.3 Dimension Reduction

Multivariate analysis generally consists of ascertaining the features of individuals as a number of variables and constructing new variables by their linear combination. In this process the procedures in multivariate analysis have been proposed on the basis of the different type of criterion employed. *Principal component analysis* can be regarded as a procedure of capturing the information contained in the data by the system variability, and of defining a smaller set of variables with the minimum possible loss of information. This reduction is achieved by employing an orthogonal transformation to convert a large number of correlated variables into a smaller number of uncorrelated variables, called *principal components*. Principal components enable the extraction of information of interest from the data.

Principal component analysis can also be used as a technique for performing dimension compression (i.e., dimension reduction), thus reducing a large number of variables to a smaller number and enabling 1D (line), 2D (plane), and 3D (space) projections amenable to intuitive understanding and visual discernment of data structures. In fields such as pattern recognition, image analysis, and signal processing, the technique is referred to as *Karhunen-Loève expansion* and is also utilized for dimension compression.

Chapter 9 begins with a discussion of the basic concept of principal component analysis based on linearity. It next provides an example of the application of principal component analysis to the performance of dimension compression in transmitted image data reproduction. It also discusses *nonlinear principal component analysis* for multivariate data with complex structure dispersed in a high-dimensional space, using a kernel method to perform structure searching and information extraction through dimension reduction.

1.4 Clustering

The main purpose of discriminant analysis is to construct a discriminant rule and predict the membership of future data among multiple classes, based on the membership of known data (i.e., "training data"). In *cluster analysis*, in contrast, as described in Chapter 10, the purpose is to divide data into aggregates ("data clusters") with their similarity as the criterion in cases involved a mixture of data of uncertain class membership.

Cluster analysis is useful, for example, for gaining an understanding of complex relations among objects, through its grouping by the similarity criterion of objects with attached features representing multidimensional properties. Its range of applications is extremely wide, extending from ecology, genetics, psychology, and cognitive science to pattern recognition in document classification, speech and image processing, and throughout the natural and social sciences. In life sciences, in particular, it serves as a key method for elucidation of complex networks of genetic information.

1.4.1 Hierarchical Clustering Methods

Hierarchical clustering essentially consists of linking target data that are mutually similar, proceeding in the stepwise formation from small to large clusters in units of the smallest cluster. It is characterized by the generation of readily visible tree diagrams called *dendrograms* throughout the clustering process. Figure 1.7 shows 72 chemical substances with 6 attached features, classified by clustering on the basis of mutual similarity in substance qualities. The lower portions of the interconnected tree represent higher degrees of similarity in qualities, and as may be seen from Figure 1.7, small clusters interlink to form larger clusters proceeding up the dendrogram. The utilization of cluster analysis enables classification of large quantities of data dispersed throughout higher-dimension space, which is not intuitively obvious, into collections of mutually similar objects.

Chapter 10 provides discussion and illustrative examples of representative *hierarchical clustering* methods, including the nearest-neighbor, farthest-neighbor, group-average, centroid, median, and Ward's methods, together with the process and characteristics of their implementation and the basic prerequisites for their application.

1.4.2 Nonhierarchical Clustering Methods

In contrast to hierarchical clustering, with its stepwise formation of increasingly large clusters finally ending in the formation of a single large cluster, nonhierarchical clustering methods essentially consist of dividing the objects into a predetermined number of clusters. One representative method is *k-means* clustering, which is used in cases where large-scale data classification and hierarchical structure elucidation are unnecessary. Another is the *self-organizing map*, which is a type of neural network proposed by T. Kohonen. It is characterized by the projection of high-dimensional objects to a two-dimensional plane and visualization of collections of similar objects by coloring, and its application is under investigation in many fields.

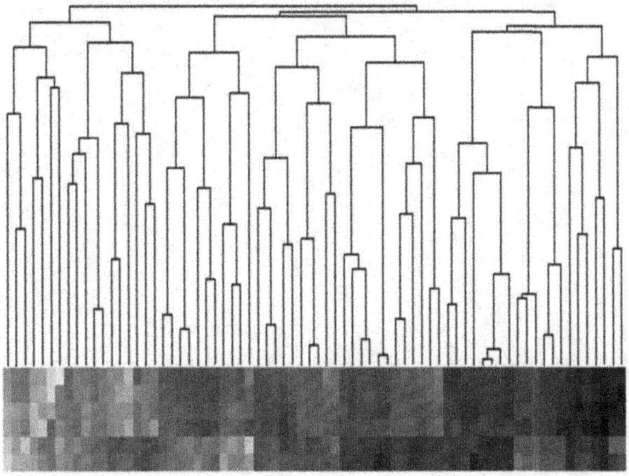

Figure 1.7 *72 chemical substances with 6 attached features, classified by clustering on the basis of mutual similarity in substance qualities.*

Chapter 10 describes clustering by *k*-means and self-organizing map techniques, and the processes of their implementation. It also shows a clustering method that utilizes a *mixture distribution model* formed by combining several probability distributions, together with examples of its application and model estimation.

The appendix contains an outline of the *bootstrap methods*, *Lagrange's method of undetermined multipliers*, and the *EM algorithm*, all of which are used in this book.

Chapter 2

Linear Regression Models

The modeling of natural and social phenomena from related data plays a fundamental role in their explication, prediction, and control, and in new discoveries, knowledge, and understanding of these phenomena. Models that link the outcomes of phenomena to multiple factors that affect them are generally referred to as *regression models*.

In regression modeling, the model construction essentially proceeds through a series of processes consisting of: (1) assuming models based on observed data thought to affect the phenomenon; (2) estimating the parameters of the specified models; and (3) evaluating the estimated models for selection of the optimum model. In this chapter, we consider the fundamental concepts of modeling as embodied in linear regression modeling, the most basic form of modeling for the explication of relationships between variables.

Regression models are usually estimated by least squares or maximum likelihood methods. In cases involving multicollinearity among predictor variables, the ridge regression is used to prevent instability in estimated linear regression models. In estimating and evaluating models having a large number of predictor variables, the usual methods of separating model estimation and evaluation are inefficient for the selection of factors affecting the outcome of the phenomena. In such cases, regularization methods with L_1 norm penalty, in addition to the sum of squared errors and log-likelihood functions, provide a useful tool for effective regression modeling based on high-dimensional data. This chapter also describes ridge regression, lasso, and various regularization methods with L_1 norm penalty.

2.1 Relationship between Two Variables

In this section, we describe the basic method of explicating the relationship between a variable that represents the outcome of a phenomenon and a variable suspected of affecting this outcome, based on observed data. The relationship used in our example is actually Hooke's well-known

Table 2.1 *The length of a spring under different weights.*

x_g	5	10	15	20	25	30	35	40	45	50
y_{cm}	5.4	5.7	6.9	6.4	8.2	7.7	8.4	10.1	9.9	10.5

law of elasticity, which states, essentially, that a spring changes shape under an applied force and that within the spring's limit of elasticity the change is proportional to the force.

2.1.1 Data and Modeling

Table 2.1 shows ten observations obtained by measuring the length of a spring (y cm) under different weights (x g). The data are plotted in Figure 2.1. The plot suggests a straight-line relationship between the two variables of spring length and suspended weight. If the measurements were completely free from error, all of the data points might actually lie in a straight line. As shown in Figure 2.1, measurement data generally include errors commonly referred to as noise, and modeling is therefore required to explicate the relationship between variables. To find the relationship between the two variables of spring length (y) and weight (x) from the data including the measurement errors, let us therefore attempt the modeling based on an initially unknown function $y = u(x)$.

We first consider a more specific expression for the unknown function $u(x)$ that represents the true structure of the spring phenomenon. The data plot, as well as our a priori knowledge that the function should be linear, suggests that the function should describe a straight line. We therefore adopt a *linear model* as our specified model, so that

$$y = u(x) = \beta_0 + \beta_1 x. \tag{2.1}$$

We then attempt to apply this linear model in order to explicate the relationship between the spring length (y) and the weight (x) as a physical phenomenon.

If there were no errors in the data shown in Table 2.1, then all 10 data points would lie on a straight line with an appropriately selected intercept (β_0) and slope (β_1). Because of measurement errors, however, many of the actual data points will depart from any straight line. To include consideration for this departure (ε) from a straight line by data points obtained with different weights, we therefore assume that they satisfy

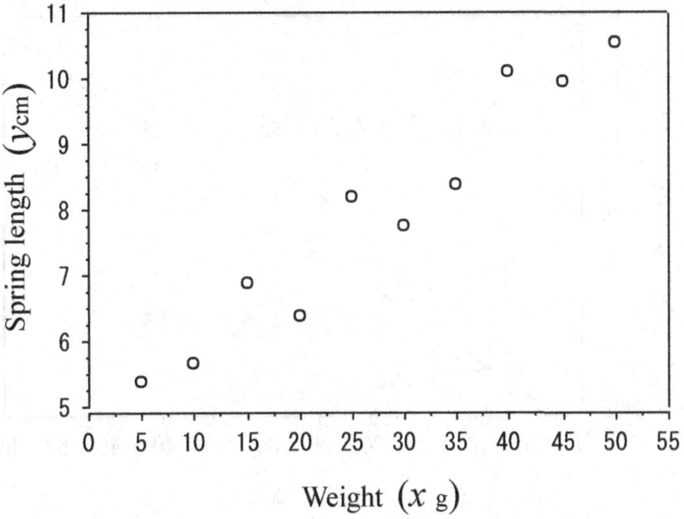

Figure 2.1 *Data obtained by measuring the length of a spring (y* cm*) under different weights (x* g*).*

Table 2.2 *The n observed data.*

No.	1	2	$\cdots$	i	$\cdots$	n
Experiment points (x)	x_1	x_2	$\cdots$	x_i	$\cdots$	x_n
Observed data (y)	y_1	y_2	$\cdots$	y_i	$\cdots$	y_n

the relation

$$\text{Spring length } = \beta_0 + \beta_1 \times \text{Weight } + \text{ Error.} \tag{2.2}$$

For the individual data points, we then have $5.4 = \beta_0+\beta_1 5+\varepsilon_1$, $\cdots$, $8.2 = \beta_0+\beta_1 25+\varepsilon_5$, $\cdots$. Figure 2.2 illustrates the relationship considering the fifth data point (25, 8.2).

In general, let us assume that measurements are performed for n experiment points, as in Table 2.2, and that a measurement at a given experiment point x_i is y_i. The general model corresponding to (2.2) is then

$$y_i = \beta_0 + \beta_1 x_i + \varepsilon_i, \qquad i = 1, 2, \cdots, n, \tag{2.3}$$

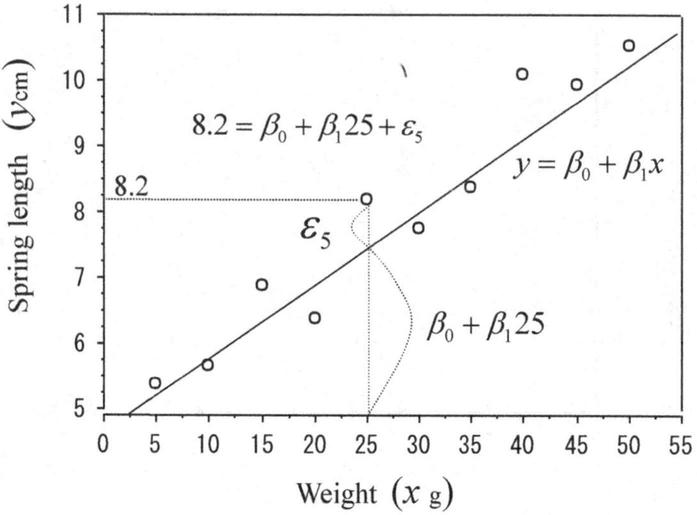

Figure 2.2 *The relationship between the spring length (y) and the weight (x).*

where β_0 and β_1 are *regression coefficients*, ε_i is the *error term*, and the equation in (2.3) is called the *linear regression model*. The variable y, which represents the length of the spring in the above experiment, is the *response variable* and the variable x, which represents the weight in that experiment, is the *predictor variable*. Variables y and x are also often referred to as the *dependent variable* and the *independent variable* or the *explanatory variable*, respectively.

This brings us to the question of how to fit a straight line to observed data in order to obtain a model that appropriately expresses the data. It is essentially a question of how to determine the regression coefficients β_0 and β_1. Various model estimation procedures can be used to determine the appropriate parameter values. One of these is the method of least squares.

2.1.2 Model Estimation by Least Squares

The underlying concept of the linear regression model (2.3) is that the true value of the response variable at the i-th point x_i is $\beta_0 + \beta_1 x_i$ and that the observed value y_i includes the error ε_i. The method of least squares consists essentially of finding the values of regression coefficients β_0 and β_1 that minimize the sum of squared errors $\varepsilon_1^2 + \varepsilon_2^2 + \cdots + \varepsilon_n^2$, which is expressed as

$$S(\beta_0, \beta_1) \equiv \sum_{i=1}^{n} \varepsilon_i^2 = \sum_{i=1}^{n} \{y_i - (\beta_0 + \beta_1 x_i)\}^2 . \tag{2.4}$$

Differentiating (2.4) with respect to the regression coefficients β_0 and β_1, and setting the resulting derivatives equal to zero, we have

$$\sum_{i=1}^{n} y_i = n\beta_0 + \beta_1 \sum_{i=1}^{n} x_i, \quad \sum_{i=1}^{n} x_i y_i = \beta_0 \sum_{i=1}^{n} x_i + \beta_1 \sum_{i=1}^{n} x_i^2. \tag{2.5}$$

The regression coefficients that minimize the sum of squared errors can be obtained by solving the above simultaneous equations. This solution is called the *least squares estimates* and is denoted by $\hat{\beta}_0$ and $\hat{\beta}_1$. The equation

$$y = \hat{\beta}_0 + \hat{\beta}_1 x, \tag{2.6}$$

having its coefficients determined by the least squares estimates, is the estimated linear regression model. We can thus find the model that best fits the data by minimizing the sum of squared errors.

The value of $\hat{y}_i = \hat{\beta}_0 + \hat{\beta}_1 x_i$ at each x_i $(i = 1, 2, \cdots, n)$ is called the *predicted value*. The difference between this value and the observed value y_i at x_i, $e_i = y_i - \hat{y}_i$, is called the *residual*, and the sum of the squares of the residuals is given by $\sum_{i=1}^{n} e_i^2$ (Figure 2.3).

Example 2.1 (Hooke's law of elasticity) For the data shown in Table 2.1, the sum of squared errors in the linear regression model is $S(\beta_0, \beta_1) = \{5.4 - (\beta_0 + \beta_1 5)\}^2 + \{5.7 - (\beta_0 + \beta_1 10)\}^2 + \cdots + \{10.5 - (\beta_0 + \beta_1 50)\}^2$, in which $S(\beta_0, \beta_1)$ is the function of the regression coefficients β_0, β_1. The least squares estimates that minimize this function are $\hat{\beta}_0 = 4.65$ and $\hat{\beta}_1 = 0.12$, and the estimated linear regression model is therefore $y = 4.65 + 0.12x$. In this way, by modeling from a set of observed data, we have derived in approximation a physical law representing the relationship between the weight and the spring length.

2.1.3 *Model Estimation by Maximum Likelihood*

In the least squares method, the regression coefficients are estimated by minimizing the sum of squared errors. *Maximum likelihood estimation*

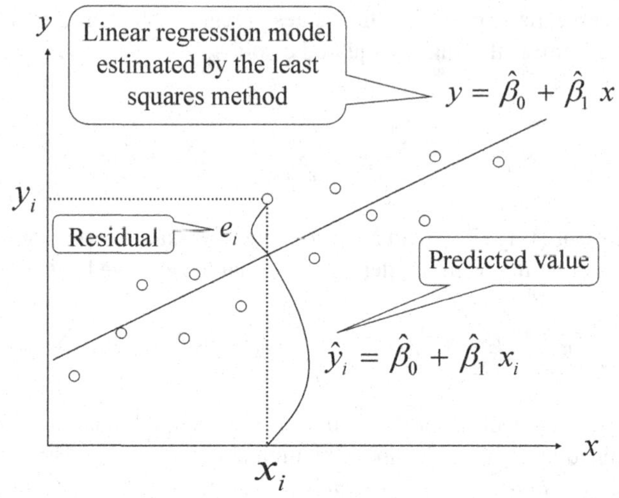

Figure 2.3 *Linear regression and the predicted values and residuals.*

is an alternative method for the same purpose in which the regression coefficients are determined so as to maximize the probability of getting the observed data, for which it is assumed that y_i observed at x_i emerges in accordance with some type of probability distribution.

Figure 2.4 (a) shows a histogram of 80 measured values obtained while repeatedly suspending a load of 25 g from one end of a spring. Figure 2.4 (b) represents the errors (i.e., noise) contained in these measurements in the form of a histogram having its origin at the mean value of the measurements. This histogram clearly shows a region containing a high proportion of the obtained measured values. A mathematical model that approximates a histogram showing the probabilistic distribution of a phenomenon is called a *probability distribution model.*

Of the various distributions that may be adopted in probability distribution models, the most representative is the *normal distribution* (*Gaussian distribution*), which is expressed in terms of mean μ and variance σ^2 and denoted by $N(\mu, \sigma^2)$. In the normal distribution model, the observed value y_i at x_i is regarded as the realization of the random variable $Y_i = y_i$, and Y_i is normally distributed with mean μ_i and variance σ^2

$$f(y_i|x_i; \mu_i, \sigma^2) = \frac{1}{\sqrt{2\pi\sigma^2}} \exp\left\{-\frac{(y_i - \mu_i)^2}{2\sigma^2}\right\}, \tag{2.7}$$

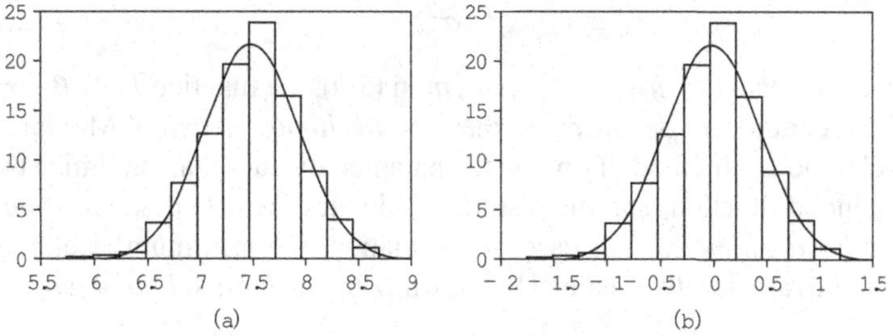

Figure 2.4 *(a) Histogram of 80 measured values obtained while repeatedly suspending a load of 25 g and its approximated probability model. (b) The errors (i.e., noise) contained in these measurements in the form of a histogram having its origin at the mean value of the measurements and its approximated error distribution.*

where μ_i for a given x_i is the conditional mean value (true value) $E[Y_i|x_i] = u(x_i) = \mu_i$ of random variable Y_i. In the normal distribution, as may be clearly seen in Figure 2.4 (a), the proportion of measured values may be expected to decline sharply with increasing distance from the true value.

In the linear regression model, it is assumed that the true values μ_1, $\mu_2, \cdots, \mu_n$ at the various data points lie on a straight line, and it follows that for any given data point, $\mu_i = \beta_0 + \beta_1 x_i$ $(i = 1, 2, \cdots, n)$. Substitution into (2.7) thus yields

$$f(y_i|x_i; \beta_0, \beta_1, \sigma^2) = \frac{1}{\sqrt{2\pi\sigma^2}} \exp\left[-\frac{\{y_i - (\beta_0 + \beta_1 x_i)\}^2}{2\sigma^2}\right]. \tag{2.8}$$

This function decreases with increasing deviation of the observed value y_i from the true value $\beta_0 + \beta_1 x_i$. Assuming that the observed data y_i around the true value $\beta_0 + \beta_1 x_i$ at x_i thus follow the probability distribution $f(y_i|x_i; \beta_0, \beta_1, \sigma^2)$, it is then an expression of the plausibility or certainty of the occurrence of a given value of y_i, called the *likelihood* of y_i.

Assuming that the observed data $y_1, y_2, \cdots, y_n$ are mutually independent and identically distributed (i.i.d.), the likelihood with n data and thus the plausibility with n specific data is given by the product of the likelihoods of all observed data

$$\prod_{i=1}^{n} f(y_i|x_i; \beta_0, \beta_1, \sigma^2) = \frac{1}{(2\pi\sigma^2)^{n/2}} \exp\left[-\frac{1}{2\sigma^2} \sum_{i=1}^{n} \{y_i - (\beta_0 + \beta_1 x_i)\}^2\right]$$

$$\equiv L(\beta_0, \beta_1, \sigma^2). \tag{2.9}$$

Given the data $\{(x_i, y_i); i = 1, 2, \cdots, n\}$ in (2.9), the function $L(\beta_0, \beta_1, \sigma^2)$ of the parameters $\beta_0, \beta_1, \sigma^2$ is then the *likelihood function.* Maximum likelihood is a method of finding the parameter values that maximize this likelihood function, and the resulting estimates are called the *maximum likelihood estimates.* For ease of calculation, the maximum likelihood estimates are usually obtained by maximizing the *log-likelihood function*

$$\begin{aligned} \ell(\beta_0, \beta_1, \sigma^2) &\equiv \log L(\beta_0, \beta_1, \sigma^2) \\ &= -\frac{n}{2}\log(2\pi\sigma^2) - \frac{1}{2\sigma^2}\sum_{i=1}^{n}\{y_i - (\beta_0 + \beta_1 x_i)\}^2 . \end{aligned} \tag{2.10}$$

The parameter values $\hat{\beta}_0, \hat{\beta}_1, \hat{\sigma}^2$ that maximize the log-likelihood function are thus obtained by solving the equations

$$\frac{\partial \ell(\beta_0, \beta_1, \sigma^2)}{\partial \beta_0} = 0, \quad \frac{\partial \ell(\beta_0, \beta_1, \sigma^2)}{\partial \beta_1} = 0, \quad \frac{\partial \ell(\beta_0, \beta_1, \sigma^2)}{\partial \sigma^2} = 0. \tag{2.11}$$

Specific solutions will be given in Section 2.2.2.

The first term of the log-likelihood function defined in (2.10) does not depend on β_0, β_1, and the sign of the second term is always negative since $\sigma^2 > 0$. Accordingly, the values of the regression coefficients β_0 and β_1 that maximize the log-likelihood function are those that minimize

$$\sum_{i=1}^{n}\{y_i - (\beta_0 + \beta_1 x_i)\}^2 . \tag{2.12}$$

With the assumption of a normal distribution model for the data, the maximum likelihood estimates of the regression coefficients are thus equivalent to the least squares estimates of the regression coefficients, that is, the minimizer of (2.4).

2.2 Relationships Involving Multiple Variables

In the case of explicating a natural or social phenomenon that may involve a number of factors, it is necessary to construct a model that links the outcome phenomena to the factors that cause it. In the case of a chemical experiment, the quantity (response variable, y) of the reaction product may be affected by temperature, pressure, pH, concentration, catalyst quantity, and various other factors. In this section, we discuss the models used to explicate the relationship between the response variable y and the various predictor variables $x_1, x_2, \cdots, x_p$ that may explain this response.

Table 2.3 *Four factors: temperature (x_1), pressure (x_2), PH (x_3), and catalyst quantity (x_4), which affect the quantity of product (y).*

No.	Product (g)	TEMP(℃)	Pressure	PH	Catalyst
1	28.7	34.1	2.3	6.4	0.1
2	32.4	37.8	2.5	6.8	0.3
⋮	⋮	⋮	⋮	⋮	⋮
i	52.9	47.6	3.8	7.6	0.7
⋮	⋮	⋮	⋮	⋮	⋮
86	65.8	52.6	4.8	7.8	1.1

Table 2.4 *The response y representing the results in n trials, each with a different combination of p predictor variables $x_1, x_2, \cdots, x_p$.*

	Response variable	Predictor variables			
No.	y	x_1	x_2	$\cdots$	x_p
1	y_1	x_{11}	x_{12}	$\cdots$	x_{1p}
2	y_2	x_{21}	x_{22}	$\cdots$	x_{2p}
⋮	⋮	⋮	⋮		⋮
i	y_i	x_{i1}	x_{i2}	$\cdots$	x_{ip}
⋮	⋮	⋮	⋮		⋮
n	y_n	x_{n1}	x_{n2}	$\cdots$	x_{np}

2.2.1 Data and Models

Table 2.3 is a partial list of the observed data in 86 experimental trials with variations in four factors that affect the quantity of product formed by a chemical reaction. Table 2.4 shows the notation corresponding to the experimental data shown in Table 2.3, with response y_i representing the results in n trials, each with a different combination of p predictor variables $x_1, x_2, \cdots, x_p$. Thus, y_i is observed as the result of the i-th experiment point (i.e., the i-th trial) $(x_{i1}, x_{i2}, \cdots, x_{ip})$.

The objective is to construct a model from the observed data that appropriately links the product quantity y to the temperature, pressure, concentration, pH, catalyst quantity, and other factors involved in the reaction. From the concept of the linear regression model for two variables in (2.3), the model for n observations in Table 2.4 is thus given by

$$y_i = \beta_0 + \beta_1 x_{i1} + \beta_2 x_{i2} + \cdots + \beta_p x_{ip} + \varepsilon_i, \quad i = 1, 2, \cdots, n. \tag{2.13}$$

Models constructed to find the relationship between a response variable and two or more predictor variables may be called *multiple linear regression models* if necessary to distinguish them from simple linear regression models concerning the relationship between just two variables. In this book, however, we simply refer to them as *linear regression models.*

The n equations in (2.13), which represent the linear regression model, may also be written as

$$\begin{pmatrix} y_1 \\ y_2 \\ \vdots \\ y_i \\ \vdots \\ y_n \end{pmatrix} = \begin{pmatrix} 1 & x_{11} & x_{12} & \cdots & x_{1p} \\ 1 & x_{21} & x_{22} & \cdots & x_{2p} \\ \vdots & \vdots & \vdots & \ddots & \vdots \\ 1 & x_{i1} & x_{i2} & \cdots & x_{ip} \\ \vdots & \vdots & \vdots & \ddots & \vdots \\ 1 & x_{n1} & x_{n2} & \cdots & x_{np} \end{pmatrix} \begin{pmatrix} \beta_0 \\ \beta_1 \\ \beta_2 \\ \vdots \\ \beta_p \end{pmatrix} + \begin{pmatrix} \varepsilon_1 \\ \varepsilon_2 \\ \vdots \\ \varepsilon_i \\ \vdots \\ \varepsilon_n \end{pmatrix}. \tag{2.14}$$

In vector and matrix notation, this can then be written as

$$\boldsymbol{y} = X\boldsymbol{\beta} + \boldsymbol{\varepsilon}, \tag{2.15}$$

where $\boldsymbol{y}$ is the n-dimensional vector of observed values for the response variable y, X is the $n \times (p+1)$ *design matrix* comprising the data obtained for the p predictor variables, with 1 added for the intercept, $\boldsymbol{\beta}$ is the $(p+1)$-dimensional vector of regression coefficients, and $\boldsymbol{\varepsilon}$ is the n-dimensional *error vector*. Let us now consider the estimation of *regression coefficient vector* $\boldsymbol{\beta}$ by the least squares and maximum likelihood methods.

2.2.2 *Model Estimation*

Model parameters can be estimated using either the least squares or maximum likelihood method for multivariable relationships, just as with relationships between two variables.

(1) Least squares method In modeling multivariable relationships using least squares, the regression model that minimizes the sum of squared errors is taken as the model best fitting the observed data. An advantage of using least squares as the method of estimation is its strong guarantee provided by the assumptions that the error terms $\varepsilon_1, \varepsilon_2, \cdots, \varepsilon_n$ in the linear regression model are mutually uncorrelated $E[\varepsilon_i\varepsilon_j] = 0$ $(i \neq j)$ and that mean $E[\varepsilon_i] = 0$ and variance $E[\varepsilon_i^2] = \sigma^2$. This shows that the following assumptions hold for the n-dimensional error vector in (2.15):

$$E[\boldsymbol{\varepsilon}] \equiv (E[\varepsilon_1],\ E[\varepsilon_2], \cdots, E[\varepsilon_n])^T = (0,\ 0, \cdots, 0)^T = \mathbf{0}, \tag{2.16}$$

$$E[\boldsymbol{\varepsilon}\boldsymbol{\varepsilon}^T] \equiv \begin{pmatrix} E[\varepsilon_1^2] & E[\varepsilon_1\varepsilon_2] & \cdots & E[\varepsilon_1\varepsilon_n] \\ E[\varepsilon_2\varepsilon_1] & E[\varepsilon_2^2] & \cdots & E[\varepsilon_2\varepsilon_n] \\ \vdots & \vdots & \ddots & \vdots \\ E[\varepsilon_n\varepsilon_1] & E[\varepsilon_n\varepsilon_2] & \cdots & E[\varepsilon_n^2] \end{pmatrix} = \sigma^2 I_n,$$

where $\boldsymbol{\varepsilon}^T$ is the transpose of vector $\boldsymbol{\varepsilon}$, $\mathbf{0}$ is an n-dimensional zero vector in which all components are 0, and I_n is an $n \times n$ identity matrix. The parameter σ^2 is called the *error variance.*

The least squares estimate of the regression coefficient vector $\boldsymbol{\beta}$ in the linear regression model (2.15) expressed in vector and matrix notation is the value $\hat{\boldsymbol{\beta}}$ that minimizes the sum of squared errors

$$S(\boldsymbol{\beta}) = \sum_{i=1}^{n} \varepsilon_i^2 = \sum_{i=1}^{n} \boldsymbol{\varepsilon}^T\boldsymbol{\varepsilon} = (\boldsymbol{y} - X\boldsymbol{\beta})^T(\boldsymbol{y} - X\boldsymbol{\beta}). \tag{2.17}$$

This equation can be written as

$$S(\boldsymbol{\beta}) = (\boldsymbol{y} - X\boldsymbol{\beta})^T(\boldsymbol{y} - X\boldsymbol{\beta}) = \boldsymbol{y}^T\boldsymbol{y} - 2\boldsymbol{y}^T X\boldsymbol{\beta} + \boldsymbol{\beta}^T X^T X\boldsymbol{\beta}. \tag{2.18}$$

We note here that this in turn yields $(X\boldsymbol{\beta})^T\boldsymbol{y} = \boldsymbol{y}^T X\boldsymbol{\beta}$. Using the results (2.42) shown in Note 2.1 of Section 2.2.3 and differentiating both sides of $S(\boldsymbol{\beta})$ with respect to $\boldsymbol{\beta}$, we have

$$\frac{\partial S(\boldsymbol{\beta})}{\partial\boldsymbol{\beta}} = \frac{\partial}{\partial\boldsymbol{\beta}}\left(\boldsymbol{y}^T\boldsymbol{y} - 2\boldsymbol{y}^T X\boldsymbol{\beta} + \boldsymbol{\beta}^T X^T X\boldsymbol{\beta}\right) = -2X^T\boldsymbol{y} + 2X^T X\boldsymbol{\beta}. \tag{2.19}$$

The least squares estimate $\hat{\boldsymbol{\beta}}$ is accordingly given by solving

$$\frac{\partial S(\boldsymbol{\beta})}{\partial\boldsymbol{\beta}} = -2X^T\boldsymbol{y} + 2X^T X\boldsymbol{\beta} = \mathbf{0}, \tag{2.20}$$

which is called the *normal equation.* If the inverse of matrix $X^T X$ exists, then the least squares estimate is given by

$$\hat{\boldsymbol{\beta}} = (X^T X)^{-1} X^T\boldsymbol{y}. \tag{2.21}$$

The equation

$$y = \hat{\beta}_0 + \hat{\beta}_1 x_1 + \cdots + \hat{\beta}_p x_p = \hat{\boldsymbol{\beta}}^T \boldsymbol{x} \tag{2.22}$$

having the least squares estimates as its coefficients is the linear regression equation, where the term $\boldsymbol{x}$ for the p predictor variables $x_1, x_2, \cdots, x_p$ is taken as $\boldsymbol{x} = (1, x_1, x_2, \cdots, x_p)^T$ includes 1 added for intercept β_0.

The predicted value ($\hat{y}_i$) is defined as the value in the linear regression equation, and the residual (e_i) is defined as the difference between the observed value (y_i) and the predicted value ($\hat{y}_i$). These are given by

$$\hat{y}_i = \hat{\beta}_0 + \hat{\beta}_1 x_{i1} + \cdots + \hat{\beta}_p x_{ip} = \hat{\boldsymbol{\beta}}^T \boldsymbol{x}_i, \quad e_i = y_i - \hat{y}_i \tag{2.23}$$

for $i = 1, 2, \cdots, n$, where $\boldsymbol{x}_i = (1, x_{i1}, \cdots, x_{ip})^T$. The n-dimensional predicted value vector $\boldsymbol{y} = (\hat{y}_1, \hat{y}_2, \cdots, \hat{y}_n)^T$ and the residual vector $\boldsymbol{e} = (e_1, e_2, \cdots, e_n)^T$ having these as the i-th components can then be expressed as

$$\hat{\boldsymbol{y}} = X\hat{\boldsymbol{\beta}} = X(X^T X)^{-1} X^T \boldsymbol{y}, \tag{2.24}$$

and

$$\boldsymbol{e} = \boldsymbol{y} - \hat{\boldsymbol{y}} = \left[I_n - X(X^T X)^{-1} X^T \right] \boldsymbol{y}. \tag{2.25}$$

The sum of squared residuals is then given by

$$\boldsymbol{e}^T \boldsymbol{e} = (\boldsymbol{y} - \hat{\boldsymbol{y}})^T (\boldsymbol{y} - \hat{\boldsymbol{y}}) = \boldsymbol{y}^T \left[I_n - X(X^T X)^{-1} X^T \right] \boldsymbol{y}, \tag{2.26}$$

in which we used $P^2 = P$ for $P = I_n - X(X^T X)^{-1} X^T$. This type of matrix is called an *idempotent matrix*.

Basic properties of the least squares estimator The least squares estimate $\hat{\boldsymbol{\beta}} = (X^T X)^{-1} X^T \boldsymbol{y}$ of the regression coefficient $\boldsymbol{\beta}$ depends on data $\boldsymbol{y}$, which varies stochastically with the error. This estimate $\hat{\boldsymbol{\beta}}$ is thus itself a random variable that varies in similar correspondence. Let us then consider its properties as a least squares estimator.

In the linear regression model $\boldsymbol{y} = X\boldsymbol{\beta} + \boldsymbol{\varepsilon}$, from the assumptions of (2.16) for the error vector $\boldsymbol{\varepsilon}$, it follows that

$$E[\boldsymbol{y}] = X\boldsymbol{\beta} + E[\boldsymbol{\varepsilon}] = X\boldsymbol{\beta}, \tag{2.27}$$

and

$$E[(\boldsymbol{y} - X\boldsymbol{\beta})(\boldsymbol{y} - X\boldsymbol{\beta})^T] = E[\boldsymbol{\varepsilon}\boldsymbol{\varepsilon}^T] = \sigma^2 I_n. \tag{2.28}$$

Accordingly, the expected value and the variance-covariance matrix of the least squares estimator $\hat{\boldsymbol{\beta}} = (X^T X)^{-1} X^T \boldsymbol{y}$ are respectively given by

$$E[\hat{\boldsymbol{\beta}}] = (X^T X)^{-1} X^T E[\boldsymbol{y}] = \boldsymbol{\beta}, \tag{2.29}$$

and

$$\begin{aligned} E[(\hat{\boldsymbol{\beta}} - \boldsymbol{\beta})(\hat{\boldsymbol{\beta}} - \boldsymbol{\beta})^T] &= E\left[\left\{(X^T X)^{-1} X^T \boldsymbol{y} - (X^T X)^{-1} X^T X\boldsymbol{\beta}\right\}\right. \\ &\qquad \left. \times \left\{(X^T X)^{-1} X^T \boldsymbol{y} - (X^T X)^{-1} X^T X\boldsymbol{\beta}\right\}^T\right] \\ &= (X^T X)^{-1} X^T E[(\boldsymbol{y} - X\boldsymbol{\beta})(\boldsymbol{y} - X\boldsymbol{\beta})^T] X (X^T X)^{-1} \\ &= \sigma^2 (X^T X)^{-1}. \end{aligned} \tag{2.30}$$

Here we used equations (2.44) and (2.45). Equation (2.29) shows that the expected value of the least squares estimator $\hat{\boldsymbol{\beta}}$ is equal to the regression coefficient vector $\boldsymbol{\beta}$, which is the parameter being estimated; in other words, it shows that $\hat{\boldsymbol{\beta}}$ is an unbiased estimator of $\boldsymbol{\beta}$.

Suppose that $\hat{\boldsymbol{\beta}} = C\boldsymbol{y}$ is in general an unbiased estimator represented as a linear combination of a random vector $\boldsymbol{y}$, where C is a $(p + 1) \times n$ constant matrix. The least squares estimator is given by taking $C = (X^T X)^{-1} X^T$. One reason for the use of a least squares estimator is that it provides minimum variance among all linear unbiased estimators. This is guaranteed by the following *Gauss-Markov theorem*.

Gauss-Markov theorem The least squares estimator $\hat{\boldsymbol{\beta}} = (X^T X)^{-1} X^T \boldsymbol{y}$ is the *best linear unbiased estimator* (BLUE), which means that the following equation holds for any linear unbiased estimator $\tilde{\boldsymbol{\beta}}$.

$$\text{cov}(\tilde{\boldsymbol{\beta}}) \geq \text{cov}(\hat{\boldsymbol{\beta}}), \tag{2.31}$$

where cov represents the variance-covariance matrix of the estimator (see Note 2.2). Inequality $A \geq B$ for $(p + 1) \times (p + 1)$ matrices A and B means that $A - B$ is a positive semidefinite matrix, that is, for any $(p + 1)$-dimensional vector $\boldsymbol{c}$

$$\boldsymbol{c}^T (A - B)\boldsymbol{c} \geq 0 \quad \Longrightarrow \quad A \geq B. \tag{2.32}$$

The proof of the Gauss-Markov theorem may be found in Rao (1973), Sen and Srivastava (1990), and so on.

(2) Maximum likelihood method In maximum likelihood modeling, the data for the response variable are assumed to have a normal distribution. The model that maximizes the likelihood function is considered as the one best fitting the observed data. For this reason, in addition to the assumptions in (2.16) for the error vector $\boldsymbol{\varepsilon} = (\varepsilon_1, \varepsilon_2, \cdots, \varepsilon_n)^T$, it is also assumed that the components of $\boldsymbol{\varepsilon}$ are mutually independent and distributed as a normal distribution $N(0, \sigma^2)$. The probability density function of the error vector $\boldsymbol{\varepsilon}$ is then given by

$$f(\boldsymbol{\varepsilon}) = \prod_{i=1}^{n} \frac{1}{\sqrt{2\pi\sigma^2}} \exp\left(-\frac{\varepsilon_i^2}{2\sigma^2}\right) = \frac{1}{(2\pi\sigma^2)^{n/2}} \exp\left(-\frac{\boldsymbol{\varepsilon}^T\boldsymbol{\varepsilon}}{2\sigma^2}\right). \tag{2.33}$$

This means that the error vector $\boldsymbol{\varepsilon}$ is distributed according to the n-dimensional normal distribution with mean vector $\mathbf{0}$ and variance-covariance matrix $\sigma^2 I_n$. Such a linear regression model

$$\boldsymbol{y} = X\boldsymbol{\beta} + \boldsymbol{\varepsilon}, \qquad \boldsymbol{\varepsilon} \sim N_n(\mathbf{0}, \sigma^2 I_n), \tag{2.34}$$

assuming a normal error distribution, is called a *Gaussian linear regression model.*

In a Gaussian linear regression model, as shown by the results in (2.27), (2.28), and (2.48) in Note 2.4, the observed data vector $\boldsymbol{y}$ follows the n-dimensional normal distribution with mean vector $X\boldsymbol{\beta}$ and variance-covariance matrix $\sigma^2 I_n$. For given data $(\boldsymbol{y}, X)$, the likelihood function as a function of the parameters $\boldsymbol{\beta}, \sigma^2$ is

$$\begin{aligned} L(\boldsymbol{\beta}, \sigma^2) &\equiv f(\boldsymbol{y}|X; \boldsymbol{\beta}, \sigma^2) \\ &= \frac{1}{(2\pi\sigma^2)^{n/2}} \exp\left\{-\frac{1}{2\sigma^2}(\boldsymbol{y} - X\boldsymbol{\beta})^T(\boldsymbol{y} - X\boldsymbol{\beta})\right\}, \end{aligned} \tag{2.35}$$

and the log-likelihood function is therefore expressed by

$$\begin{aligned} \ell(\boldsymbol{\beta}, \sigma^2) &\equiv \log L(\boldsymbol{\beta}, \sigma^2) \\ &= -\frac{n}{2}\log(2\pi\sigma^2) - \frac{1}{2\sigma^2}(\boldsymbol{y} - X\boldsymbol{\beta})^T(\boldsymbol{y} - X\boldsymbol{\beta}). \end{aligned} \tag{2.36}$$

The maximum likelihood estimates for regression coefficient vector $\boldsymbol{\beta}$ and error variance σ^2 are given by the solution that maximizes (2.36); i.e., by the solution of the *likelihood equations*

$$\begin{aligned} \frac{\partial \ell(\boldsymbol{\beta}, \sigma^2)}{\partial \boldsymbol{\beta}} &= \frac{1}{\sigma^2}\left(X^T\boldsymbol{y} - X^TX\boldsymbol{\beta}\right) = \mathbf{0}, \\ \frac{\partial \ell(\boldsymbol{\beta}, \sigma^2)}{\partial \sigma^2} &= -\frac{n}{2\sigma^2} + \frac{1}{2\sigma^4}(\boldsymbol{y} - X\boldsymbol{\beta})^T(\boldsymbol{y} - X\boldsymbol{\beta}) = 0. \end{aligned} \tag{2.37}$$

This solution yields the maximum likelihood estimates for $\boldsymbol{\beta}$ and σ^2 as follows:

$$\hat{\boldsymbol{\beta}} = (X^T X)^{-1} X^T \boldsymbol{y}, \qquad \hat{\sigma}^2 = \frac{1}{n}\left(\boldsymbol{y} - X\hat{\boldsymbol{\beta}}\right)^T \left(\boldsymbol{y} - X\hat{\boldsymbol{\beta}}\right). \tag{2.38}$$

The linear regression model estimated by the maximum likelihood method is then given by

$$y = \hat{\beta}_0 + \hat{\beta}_1 x_1 + \cdots + \hat{\beta}_p x_p = \hat{\boldsymbol{\beta}}^T \boldsymbol{x}. \tag{2.39}$$

The predicted value vector, residual vector, and the estimate of error variance are respectively given by the following equations:

$$\hat{\boldsymbol{y}} = X\hat{\boldsymbol{\beta}} = X(X^T X)^{-1} X^T \boldsymbol{y}, \quad \boldsymbol{e} = \boldsymbol{y} - \hat{\boldsymbol{y}} = [I_n - X(X^T X)^{-1} X^T]\boldsymbol{y},$$
$$\hat{\sigma}^2 = \frac{1}{n}(\boldsymbol{y} - X\hat{\boldsymbol{\beta}})^T(\boldsymbol{y} - X\hat{\boldsymbol{\beta}}) = \frac{1}{n}(\boldsymbol{y} - \hat{\boldsymbol{y}})^T(\boldsymbol{y} - \hat{\boldsymbol{y}}). \tag{2.40}$$

The Gaussian linear regression model in (2.34), based on the assumption of an n-dimensional normal distribution $N_n(\boldsymbol{0}, \sigma^2 I_n)$ for the error vector $\boldsymbol{\varepsilon}$, implies that the observed data vector $\boldsymbol{y}$ has an n-dimensional normal distribution $N_n(X\boldsymbol{\beta}, \sigma^2 I_n)$. Therefore, it follows from the results in (2.48) that the maximum likelihood estimator $\hat{\boldsymbol{\beta}} = (X^T X)^{-1} X^T \boldsymbol{y}$ for the regression coefficient vector $\boldsymbol{\beta}$ is distributed according to a $(p+1)$-dimensional normal distribution $N_{p+1}(\boldsymbol{\beta}, \sigma^2 (X^T X)^{-1})$ with mean vector $\boldsymbol{\beta}$ and variance-covariance matrix $\sigma^2 (X^T X)^{-1}$.

2.2.3 Notes

Note 2.1 (Vector differentiation) The derivative of a real-valued function $S(\boldsymbol{\beta})$ with respect to $\boldsymbol{\beta} = (\beta_0, \beta_1, \cdots, \beta_p)^T$ is defined by

$$\frac{\partial S(\boldsymbol{\beta})}{\partial \boldsymbol{\beta}} \equiv \left(\frac{\partial S(\boldsymbol{\beta})}{\partial \beta_0}, \frac{\partial S(\boldsymbol{\beta})}{\partial \beta_1}, \cdots, \frac{\partial S(\boldsymbol{\beta})}{\partial \beta_p}\right)^T. \tag{2.41}$$

Then, for a $(p+1)$ dimensional constant vector $\boldsymbol{c}$ and a $(p+1) \times (p+1)$ constant matrix A, we have the following results:

$$\text{(i)} \ \frac{\partial(\boldsymbol{c}^T \boldsymbol{\beta})}{\partial \boldsymbol{\beta}} = \boldsymbol{c},$$
$$\text{(ii)} \ \frac{\partial(\boldsymbol{\beta}^T A \boldsymbol{\beta})}{\partial \boldsymbol{\beta}} = (A + A^T)\boldsymbol{\beta}, \qquad \frac{\partial(\boldsymbol{\beta}^T A \boldsymbol{\beta})}{\partial \boldsymbol{\beta}} = 2A\boldsymbol{\beta} \quad (A; \text{symmetric}). \tag{2.42}$$

The result in (2.19) can be obtained by taking $\boldsymbol{c} = X^T\boldsymbol{y}$ and $A = X^TX$ in the above equations.

Note 2.2 (Expectation and variance-covariance matrix of a random vector) Let $\boldsymbol{Y} = (Y_1, Y_2, \cdots, Y_n)^T$ be an n-dimensional random vector of random variables. Then the expectation and variance-covariance matrix of $\boldsymbol{Y}$ are respectively defined by

$$\begin{aligned} \boldsymbol{\mu} &\equiv E[\boldsymbol{Y}] = (E[Y_1], E[Y_2], \cdots, E[Y_n])^T, \\ \text{cov}(\boldsymbol{Y}) &\equiv E\left[(\boldsymbol{Y} - E[\boldsymbol{Y}])(\boldsymbol{Y} - E[\boldsymbol{Y}])^T\right], \end{aligned} \tag{2.43}$$

where $\text{cov}(\boldsymbol{Y})$ denotes an $n \times n$ symmetric matrix with covariance $E[(Y_i - E[Y_i])(Y_j - E[Y_j])]$ as its (i, j)-th element.

Consider the linear transformation $\boldsymbol{Z} = A\boldsymbol{Y}$, where A is an $m \times n$ constant matrix. Then the expectation and variance-covariance matrix of the m-dimensional random vector $\boldsymbol{Z}$ are respectively

$$\text{(i) } E[\boldsymbol{Z}] = AE[\boldsymbol{Y}] = A\boldsymbol{\mu}, \tag{2.44}$$

$$\begin{aligned} \text{(ii) } \text{cov}(\boldsymbol{Z}) &= E[(\boldsymbol{Z} - E[\boldsymbol{Z}])(\boldsymbol{Z} - E[\boldsymbol{Z}])^T] \\ &= AE[(\boldsymbol{Y} - E[\boldsymbol{Y}])(\boldsymbol{Y} - E[\boldsymbol{Y}])^T]A^T \\ &= A\text{cov}(\boldsymbol{Y})A^T. \end{aligned} \tag{2.45}$$

Putting $A = (X^TX)^{-1}X^T$ yields the expectation (2.29) and the variance-covariance matrix (2.30) of the least squares estimator $\hat{\boldsymbol{\beta}} = (X^TX)^{-1}X^T\boldsymbol{y}$.

Note 2.3 (Multivariate normal distribution) An n-dimensional random vector $\boldsymbol{Y}$ is said to have a multivariate normal distribution with mean vector $\boldsymbol{\mu}$ and variance-covariance matrix Σ if its probability density function is

$$f(\boldsymbol{y}; \boldsymbol{\mu}, \Sigma) = \frac{1}{(2\pi)^{n/2}|\Sigma|^{1/2}} \exp\left\{-\frac{1}{2}(\boldsymbol{y} - \boldsymbol{\mu})^T\Sigma^{-1}(\boldsymbol{y} - \boldsymbol{\mu})\right\}. \tag{2.46}$$

The probability density function is denoted by $N_n(\boldsymbol{\mu}, \Sigma)$. In a particular case of $\Sigma = \sigma^2 I_n$, the probability density function of $\boldsymbol{Y}$ is

$$f(\boldsymbol{y}; \boldsymbol{\mu}, \sigma^2) = \frac{1}{(2\pi\sigma^2)^{n/2}} \exp\left\{-\frac{1}{2\sigma^2}(\boldsymbol{y} - \boldsymbol{\mu})^T(\boldsymbol{y} - \boldsymbol{\mu})\right\}. \tag{2.47}$$

Note 2.4 (Distribution of a random vector) Suppose that an n-dimensional random vector $\boldsymbol{Y}$ is distributed as $N_n(\boldsymbol{\mu}, \Sigma)$. Consider the linear transformation $\boldsymbol{Z} = \boldsymbol{c} + A\boldsymbol{Y}$, where $\boldsymbol{c}$ is an m-dimensional constant vector and A is an $m \times n$ constant matrix. Then $\boldsymbol{Z}$ is distributed as $N_m(\boldsymbol{c} + A\boldsymbol{\mu}, A\Sigma A^T)$, that is,

$$\boldsymbol{Y} \sim N_n(\boldsymbol{\mu}, \Sigma) \quad \Longrightarrow \quad \boldsymbol{Z} = \boldsymbol{c} + A\boldsymbol{\mu} \sim N_m(\boldsymbol{c} + A\boldsymbol{\mu}, A\Sigma A^T). \tag{2.48}$$

Taking $A = (X^T X)^{-1} X^T$, $\boldsymbol{c} = \boldsymbol{0}$, and $\Sigma = \sigma^2 I_n$ in the above result, it can be shown that the maximum likelihood estimator $\hat{\boldsymbol{\beta}} = (X^T X)^{-1} X^T \boldsymbol{y}$ is distributed as $N_{p+1}(\boldsymbol{\beta}, \sigma^2 (X^T X)^{-1})$.

2.2.4 *Model Selection*

Linear regression models differ according to what variables make up the p predictor variables. Selection of an appropriate model from among various models based on a criterion is called *model selection.* In cases where the purpose is to select appropriate predictor variables, the process is called *variable selection.* As will be shown in Section 5, various model evaluation criteria have been proposed for variable selection in regression models. Here we discuss the widely used Akaike's (1973) information criterion (AIC). The AIC is proposed for evaluation of many other types of modeling besides regression modeling. It is generally given by the following equation:

$$\text{AIC} = -2(\text{maximum log-likelihood}) + 2(\text{no. of free parameters}). \tag{2.49}$$

The set of variables yielding the smallest AIC value is selected as the optimum model.

Because the value of the maximum log-likelihood of the model increases with the goodness of fit of the model to the data, it might seem as though the first term alone should be sufficient as a criterion for model evaluation. The second term is included, however, because it is also necessary to consider the goodness of the model constructed from the data from the perspective of its effectiveness for future predictions. A complex model incorporating a large number of parameters is generally better for a good fit to observed data, but a model that is too complex will not be effective for predicting future phenomena. From the perspective of selecting the optimum model for prediction, as will be described in Chapter 5, it is necessary to appropriately control both the model's goodness of fit to the data and its complexity. In AIC, the maximum log-likelihood may be viewed as a measure of the goodness of fit of a model to observed data, with the number of free parameters functioning as a penalty for model complexity.

The maximum log-likelihood in the Gaussian linear regression model with p predictor variables in (2.34) is

$$\begin{aligned}\ell(\hat{\boldsymbol{\beta}}, \hat{\sigma}^2) &= -\frac{n}{2}\log(2\pi\hat{\sigma}^2) - \frac{1}{2}\frac{(\boldsymbol{y} - X\hat{\boldsymbol{\beta}})^T(\boldsymbol{y} - X\hat{\boldsymbol{\beta}})}{\hat{\sigma}^2} \\ &= -\frac{n}{2}\log(2\pi\hat{\sigma}^2) - \frac{n}{2},\end{aligned} \tag{2.50}$$

which is obtained by substituting the maximum likelihood estimates (2.38) of the parameter vector $\boldsymbol{\beta}$ and the error variance σ^2 into the log-likelihood function (2.36). The number of free parameters in the model is the number of regression coefficients $(p+1)$ plus 1 for the error variance, and thus $(p+2)$. Accordingly, AIC for the Gaussian linear regression model is given by the following equation:

$$\text{AIC} = -2\left\{\ell(\hat{\boldsymbol{\beta}}, \hat{\sigma}^2) - (p+2)\right\} = n\log(2\pi\hat{\sigma}^2) + n + 2(p+2). \tag{2.51}$$

In linear regression models, the complexity increases as their predictor variables rise in number and the sum of squared residuals $n\hat{\sigma}^2$ decreases, and the maximum log-likelihood of the model in (2.50) therefore increases. The AIC given by (2.51), however, captures the goodness of fit of the model by the maximum log-likelihood, and the number of parameters, which corresponds to the number of predictor variables, functions as a penalty for the model complexity. We give an example of model selection by AIC.

Example 2.2 (Prediction of heat evolution by cement) The data consist of 13 observations for the following four predictor variables (in wt %) related to heat evolution, with the quantity of heat evolved per gram of cement as the response variable (Draper and Smith, 1998).

x_1 : tricalcium aluminate $(3CaO \cdot Al_2O_3)$,
x_2 : tricalcium silicate $(3CaO \cdot SiO_2)$,
x_3 : tetracalcium aluminoferrite $(4CaO \cdot Al_2O_3 \cdot Fe_2O_3)$,
x_4 : dicalcium silicate $(2CaO \cdot SiO_2)$.

We consider fitting a linear regression model to these data and variable selection by AIC. A total of 15 Gaussian linear regression models may be considered for combinations of these four predictor variables. All of these models were estimated by the maximum likelihood method. The AIC defined by (2.51) was calculated for each model, with the maximum likelihood estimate of error variance σ^2 in (2.38), i.e., the sum

Table 2.5 *Comparison of the sum of squared residuals ($\hat{\sigma}^2$) divided by the number of observations, maximum log-likelihood $\ell(\hat{\boldsymbol{\beta}})$, and AIC for each combination of predictor variables.*

Predictor variables	$\hat{\sigma}^2$	$\ell(\hat{\boldsymbol{\beta}}, \hat{\sigma}^2)$	AIC
x_1, x_2, x_3, x_4	3.682	-26.918	65.837
x_1, x_2, x_3	3.701	-26.952	63.904
x_1, x_2, x_4	3.690	-26.933	**63.866**
x_1, x_3, x_4	3.910	-27.310	64.620
x_2, x_3, x_4	5.678	-29.734	69.468
x_1, x_2	4.454	-28.156	64.312
x_1, x_3	94.390	-48.005	104.009
x_1, x_4	5.751	-29.817	67.634
x_2, x_3	31.957	-40.965	89.930
x_2, x_4	66.837	-45.761	99.522
x_3, x_4	13.518	-35.373	78.745
x_1	97.361	-48.206	102.412
x_2	69.718	-46.035	98.070
x_3	149.185	-50.980	107.960
x_4	67.990	-45.872	97.744

of squared residuals divided by the number of observations and p, the number of predictor variables in the model. Table 2.5 shows the sum of squared residuals divided by the number of data, the maximum log-likelihood, and AIC for each combination of predictor variables. Among the 15 models, as shown in this table, the model yielding the smallest AIC is $y = 71.6 + 1.45x_1 + 0.41x_2 - 0.24x_4$, the model containing the three predictor variables x_1, x_2, and x_4. The model yielding the smallest sum of squared residuals and conversely the largest log-likelihood includes all of the predictor variables. This difference arises because the AIC includes consideration of how well each model will predict future phenomena, as well as the sum of squared residuals representing the model's goodness of fit. In this example, it is essential to establish a model that will be able to accurately predict the quantity of heat evolved by cement. The AIC appropriately serves this purpose. In Chapter 5, the construction of model selection and evaluation criteria as related to prediction will be discussed in some detail.

In summary, regression models are constructed in a series of modeling processes essentially comprising the assumption of data-based models that reflect the phenomenon, estimation of the parameters of the specified models, and evaluation of estimated models. The various evaluation criteria, including AIC, will be specifically described in Chapter 5.

2.2.5 *Geometric Interpretation*

The linear regression model may also be viewed geometrically. For the response variable y and the p predictor variables $x_1, x_2, \cdots, x_p$, we denote the n observations as $\{(y_i, x_{i1}, x_{i2}, \cdots, x_{ip})\ ;\ i = 1, 2, \cdots, n\}$ (see Section 2.2.1, Table 2.4). In vector and matrix notation, the linear regression model based on these observed data is $\boldsymbol{y} = X\boldsymbol{\beta} + \boldsymbol{\varepsilon}$.

For the $(p+1)$ n-dimensional vectors of the $n \times (p+1)$ design matrix X, we let $X = [\mathbf{1}, \boldsymbol{x}_1^*, \boldsymbol{x}_2^*, \cdots, \boldsymbol{x}_p^*]$, where $\mathbf{1}$ is an n-dimensional vector having all of its components equal to 1 and $\boldsymbol{x}_i^*$ are n-dimensional column vectors. The linear regression model can then be expressed as

$$\begin{aligned} \boldsymbol{y} &= [\mathbf{1}, \boldsymbol{x}_1^*, \boldsymbol{x}_2^*, \cdots, \boldsymbol{x}_p^*]\boldsymbol{\beta} + \boldsymbol{\varepsilon} \\ &= \beta_0 \mathbf{1} + \beta_1 \boldsymbol{x}_1^* + \beta_2 \boldsymbol{x}_2^* + \cdots + \beta_p \boldsymbol{x}_p^* + \boldsymbol{\varepsilon}. \end{aligned} \tag{2.52}$$

The equation on the right constitutes the $(p+1)$-dimensional linear subspace (vector space)

$$M(X) = \{\boldsymbol{x}^* : \boldsymbol{x}^* = \beta_0 \mathbf{1} + \beta_1 \boldsymbol{x}_1^* + \beta_2 \boldsymbol{x}_2^* + \cdots + \beta_p \boldsymbol{x}_p^*\}, \tag{2.53}$$

spanned by the $(p+1)$ n-dimensional column vectors of the design matrix X (Figure 2.5). It may be seen that changing the regression coefficients will move $X\boldsymbol{\beta}$ in the linear subspace $M(X)$.

In the least squares method, the estimate $\hat{\boldsymbol{\beta}}$ of the regression coefficient vector is the value that minimizes the sum of squared errors $\boldsymbol{\varepsilon}^T \boldsymbol{\varepsilon} = (\boldsymbol{y} - X\boldsymbol{\beta})^T(\boldsymbol{y} - X\boldsymbol{\beta})$. For $\boldsymbol{y} = X\boldsymbol{\beta} + \boldsymbol{\varepsilon}$, accordingly, the sum of squared errors is minimal when $X\boldsymbol{\beta}$, within the linear subspace $M(X)$, matches the projection of the n-dimensional observed data vector $\boldsymbol{y}$ onto the linear subspace $M(X)$, thus yielding $\hat{\boldsymbol{y}} = X\hat{\boldsymbol{\beta}}$. The residual vector $\boldsymbol{e}$, representing the error, then extends perpendicularly from vector $\boldsymbol{y}$ to $M(X)$. The relation $\boldsymbol{y}^T\boldsymbol{y} = \hat{\boldsymbol{y}}^T\hat{\boldsymbol{y}} + \boldsymbol{e}^T\boldsymbol{e}$ is thus established (Figure 2.5). This may be summarized as follows, using the concept of projection matrix described below.

The n-dimensional observed data vector $\boldsymbol{y}$ can be uniquely decomposed to the projection $\hat{\boldsymbol{y}}$ and the perpendicular $\boldsymbol{e}$ as $\boldsymbol{y} = \hat{\boldsymbol{y}} + \boldsymbol{e}$. Since projection $\hat{\boldsymbol{y}}$ is a vector in the $(p+1)$-dimensional linear subspace $M(X)$ spanned by the column vectors of the $n \times (p+1)$ design matrix X, it can be expressed in terms of the projection matrix $X(X^TX)^{-1}X^T$ onto $M(X)$

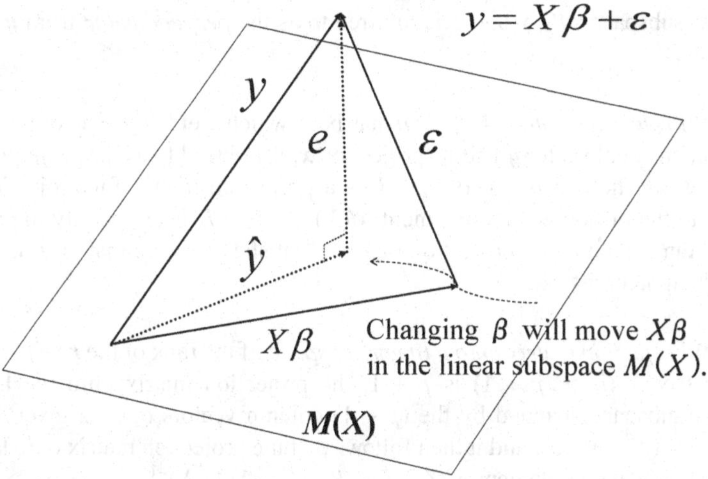

Figure 2.5 *Geometrical interpretation of the linear regression model* $\boldsymbol{y} = X\boldsymbol{\beta}+\boldsymbol{\varepsilon}$. *$M(X)$ denotes the $(p + 1)$-dimensional linear subspace spanned by the $(p + 1)$ n-dimensional column vectors of the design matrix X.*

as $\hat{\boldsymbol{y}} = X(X^TX)^{-1}X^T\boldsymbol{y}$, and we then have

$$\hat{\boldsymbol{y}} = X(X^TX)^{-1}X^T\boldsymbol{y} = X\hat{\boldsymbol{\beta}}. \tag{2.54}$$

Thus it can be shown geometrically that the regression coefficient vector yielding the least squares estimate is given by $\hat{\boldsymbol{\beta}} = (X^TX)^{-1}X^T\boldsymbol{y}$, comparing with the predicted value vector $\hat{\boldsymbol{y}}$ defined in (2.24). It may also be seen that the perpendicular $\boldsymbol{e}$ that corresponds to the residual vector is given by $\boldsymbol{e} = [I_n - X(X^TX)^{-1}X]\boldsymbol{y}$, using the projection matrix $I_n - X(X^TX)^{-1}X^T$ onto the orthogonal complement $M(X)^{\perp}$, since $\boldsymbol{e}$ is a projection onto $M(X)^{\perp}$.

Projection, perpendicular, and projection matrix

(a) *Projection and perpendicular* Let the $(p+1)$-dimensional linear subspace spanned by the column vectors of $n \times (p + 1)$ matrix X be $M(X)$ and its orthogonal complement be $M(X)^{\perp}$. Then any given n-dimensional vector $\boldsymbol{y}$ can be uniquely decomposed as

$$\boldsymbol{y} = \boldsymbol{x} + \boldsymbol{z}, \qquad \boldsymbol{x} \in M(X), \quad \boldsymbol{z} \in M(X)^{\perp}. \tag{2.55}$$

The n-dimensional vector $\boldsymbol{x}$ is referred to as the *projection* onto the linear subspace $M(X)$, and $\boldsymbol{z}$ is referred to as the *perpendicular* from $\boldsymbol{y}$ to $M(X)^{\perp}$.

(b) *Projection matrix* The $n \times n$ matrix P, which yields $P\boldsymbol{y} = \boldsymbol{x}$ for the n-dimensional vector $\boldsymbol{y}$ and its projection $\boldsymbol{x}$, is referred to as the *projection matrix*. The $n \times n$ matrix $I_n - P$ is a projection matrix for projection onto the orthogonal complement $M(X)^{\perp}$, where I_n is an identity matrix of dimension n and projection matrix P satisfies $P^2 = P$, that is, P is an idempotent matrix.

(c) *Projection matrix onto a linear subspace* If the rank of the $n \times (p+1)$ matrix X ($n > p + 1$) is $p + 1$, the projection matrix onto the linear subspace spanned by the $(p + 1)$ column vectors of X is given by $P = X(X^T X)^{-1} X^T$, and it then follows that the projection matrix onto the orthogonal complement of X is $I_n - P = I_n - X(X^T X)^{-1} X^T$.

Proof Suppose that the n-dimensional vector $\boldsymbol{y}$ can be uniquely decomposed as $\boldsymbol{y} = \boldsymbol{x} + \boldsymbol{z}$, where $\boldsymbol{x} \in M(X)$ and $\boldsymbol{z} \in M(X)^{\perp}$. Then there exists the $(p + 1)$-dimensional vector $\hat{\boldsymbol{\beta}}$, and the projection $\boldsymbol{x}$ ($\in M(X)$) can be written as $\boldsymbol{x} = X\hat{\boldsymbol{\beta}}$, yielding $\boldsymbol{y} = X\hat{\boldsymbol{\beta}} + \boldsymbol{z}$. If we then left multiply both sides by X^T, we have $X^T\boldsymbol{y} = X^T X\hat{\boldsymbol{\beta}} + X\boldsymbol{z} = X^T X\hat{\boldsymbol{\beta}}$ and accordingly obtain $X(X^T X)^{-1} X^T \boldsymbol{y} = X\hat{\boldsymbol{\beta}} = \boldsymbol{x}$. This implies that $P = X(X^T X)^{-1} X^T$ is a projection matrix onto linear subspace $M(X)$.

2.3 Regularization

In estimating and evaluating a linear regression model having a large number of predictor variables in comparison with the number of data, the usual methods of separating model estimation and evaluation are ineffectual for the selection of factors affecting the outcome of the phenomena. Various regularization methods with an L_1 norm penalty term (the sum of absolute values of regression coefficients) have been proposed for effective regression modeling based on high-dimensional data. A distinctive feature of the L_1 norm regularizations is their capability for simultaneous model estimation and variable selection. Their theoretical and numerical aspects have come under intense study, and vigorous investigation has begun on their application to many fields (see, e.g., Hastie et al., 2009).

In this section, we first discuss ridge regression within the framework of linear regression models. Ridge regression was initially proposed by Hoerl and Kennard (1970) as a method for cases involving multicollinearity among predictor variables. Although ridge regression provides no means of L_1 regularization, since the sum of squares of the regression coefficients is taken as the penalty term, in its basic concept it has strongly influenced subsequent studies. We next discuss lasso, which was proposed by Tibshirani (1996) and holds the most fundamental position among L_1 regularization methods. We then turn to several other regularized estimation methods based on the fundamental lasso concept, and discuss their distinctive features.

Because of the nondifferentiability of the L_1 norm constraint on the regression coefficients that is used as a penalty term in lasso, analytical derivation of the estimates is difficult. Various algorithms have therefore been proposed for their approximation, such as the shooting algorithm of Fu (1998), the least angle regression (LARS) algorithm of Efron et al. (2004), and the coordinate descent algorithm of Friedman et al. (2007), Wu and Lange (2008), and Friedman et al. (2010). In addition, software packages in *R* have been introduced as programs for the implementation of various L_1 regularized estimation methods (see, e.g., Friedman et al., 2013; Hastie and Efron, 2013; James et al., 2013; Park and Hastie, 2013; Zou and Hastie, 2013).

2.3.1 Ridge Regression

Consider the linear regression model

$$y_i = \beta_0 + \beta_1 x_{i1} + \beta_2 x_{i2} + \cdots + \beta_p x_{ip} + \varepsilon_i, \quad i = 1, 2, \cdots, n, \tag{2.56}$$

based on the observed data $\{(y_i, x_{i1}, x_{i2}, \cdots, x_{ip});\ i = 1, 2, \cdots, n\}$ for response variable y and p predictor variables $\boldsymbol{x} = (x_1, x_2, \cdots, x_p)$. The ridge regression proposed by Hoerl and Kennard (1970) is known to be a method of avoiding instability of estimates in linear regression models caused by multicollinearity among predictor variables. This method is a regularization in which the sum of squares of the regression coefficients, excluding the intercept, is the penalty term, and the estimates of regression coefficients are obtained as follows.

First, we obtain the mean $\overline{x}_j = n^{-1} \sum_{i=1}^n x_{ij}$ and the variance $s_j^2 = n^{-1} \sum_{i=1}^n (x_{ij} - \overline{x}_j)^2$ $(j = 1, 2, \cdots, p)$ of the data for the predictor variables and standardize the data as follows:

$$z_{ij} = \frac{x_{ij} - \overline{x}_j}{s_j}, \qquad i = 1, 2, \cdots, n, \quad j = 1, 2, \cdots, p. \tag{2.57}$$

The linear regression model based on the standardized data can then be expressed as

$$\begin{aligned} y_i &= \beta_0 + \beta_1\overline{x}_1 + \beta_2\overline{x}_2 + \cdots + \beta_p\overline{x}_p + \beta_1^* z_{i1} + \beta_2^* z_{i2} + \cdots + \beta_p^* z_{ip} + \varepsilon_i \\ &= \beta_0^* + \beta_1^* z_{i1} + \beta_2^* z_{i2} + \cdots + \beta_p^* z_{ip} + \varepsilon_i, \qquad i = 1, 2, \cdots, n, \end{aligned} \tag{2.58}$$

where $\beta_0^* = \beta_0 + \beta_1\overline{x}_1 + \beta_2\overline{x}_2 + \cdots + \beta_p\overline{x}_p$ and $\beta_j^* = s_j\beta_j$. We can therefore express the linear regression model based on the standardized data for predictor variables as

$$\boldsymbol{y} = \beta_0^*\mathbf{1} + Z\boldsymbol{\beta}_s + \boldsymbol{\varepsilon}, \tag{2.59}$$

where $\mathbf{1}$ is an n-dimensional vector of 1s, $\boldsymbol{\beta}_s = (s_1\beta_1, s_2\beta_2, \cdots, s_p\beta_p)^T$, and Z is an $n \times p$ matrix having the standardized data $z_{ij} = (x_{ij} - \overline{x}_j)/s_j$ $(i = 1, \cdots, n;\ j = 1, \cdots, p)$ as its (i, j)-th element.

The ridge estimator for the coefficient vector is then given by minimization of

$$S_\lambda(\beta_0^*, \boldsymbol{\beta}_s) = (\boldsymbol{y} - \beta_0^*\mathbf{1} - Z\boldsymbol{\beta}_s)^T(\boldsymbol{y} - \beta_0^*\mathbf{1} - Z\boldsymbol{\beta}_s) + \lambda\boldsymbol{\beta}_s^T\boldsymbol{\beta}_s, \tag{2.60}$$

in which the L_2 regularization term with regularization parameter λ has been added to the regression coefficient vector, excluding the intercept. In machine learning, this is referred to as *weight decay*. By differentiation of both sides with respect to β_0^* and $\boldsymbol{\beta}_s$, and then setting the derivatives to 0, we obtain

$$\begin{aligned} \frac{\partial S_\lambda(\beta_0^*, \boldsymbol{\beta}_s)}{\partial \beta_0^*} &= -2n\overline{y} + 2n\beta_0^* = 0, \\ \frac{\partial S_\lambda(\beta_0^*, \boldsymbol{\beta}_s)}{\partial \boldsymbol{\beta}_s} &= -2Z^T\boldsymbol{y} + 2Z^TZ\boldsymbol{\beta}_s + 2\lambda\boldsymbol{\beta}_s = \mathbf{0}. \end{aligned} \tag{2.61}$$

We note here that $Z^T\mathbf{1} = \mathbf{0}$ and $\mathbf{1}^TZ = \mathbf{0}^T$. The estimator for the intercept and the ridge estimator for the regression coefficient vector in the linear regression model (2.59) based on the standardized data are therefore given by

$$\hat{\beta}_0^* = \overline{y}, \qquad \hat{\boldsymbol{\beta}}_s = (Z^TZ + \lambda I_p)^{-1}Z^T\boldsymbol{y}. \tag{2.62}$$

As shown by these results, the intercept of the linear regression model in (2.56) is estimated as $\hat{\beta}_0 = \overline{y} - \hat{\beta}_1\overline{x}_1 - \cdots - \hat{\beta}_p\overline{x}_p$, and the ridge estimator of the regression coefficient vector is separately given by minimization of the function

$$S_\lambda(\boldsymbol{\beta}_s) = (\boldsymbol{y} - Z\boldsymbol{\beta}_s)^T(\boldsymbol{y} - Z\boldsymbol{\beta}_s) + \lambda\boldsymbol{\beta}_s^T\boldsymbol{\beta}_s, \tag{2.63}$$

based on the design matrix Z composed of the standardized data in (2.57), excluding the intercept estimation. By replacing β_0^* with its estimate $\hat{\beta}_0^* = \bar{y}$ in (2.59), we also note that $\boldsymbol{y} - \bar{y}\mathbf{1}$ is an observed value vector that centers the data relative to the response variable. Standardizing the response and predictor variables, y_i and x_{ij} are obtained from $y_i - \bar{y}$ and $(x_{ij} - \bar{x}_j)/s_j$, respectively. Then the standardized data satisfy

$$\sum_{i=1}^{n} y_i = 0, \quad \sum_{i=1}^{n} x_{ij} = 0, \quad \sum_{i=1}^{n} x_{ij}^2 = n, \quad j = 1, 2, \cdots, p. \tag{2.64}$$

This now allows us, without loss of generality, to consider the following linear regression model exclusive of the intercept:

$$\boldsymbol{y} = X\boldsymbol{\beta} + \boldsymbol{\varepsilon}, \tag{2.65}$$

where X is the $n \times p$ design matrix, $\boldsymbol{\beta}$ is the p-dimensional regression coefficient vector, and $E[\boldsymbol{\varepsilon}] = \mathbf{0}$, $\mathrm{cov}(\boldsymbol{\varepsilon}) = \sigma^2 I_n$. The ridge estimator is then given by

$$\hat{\boldsymbol{\beta}}_R = (X^T X + \lambda I_p)^{-1} X^T \boldsymbol{y}. \tag{2.66}$$

The expectation and the variance-covariance matrix of the ridge estimator are

$$E[\hat{\boldsymbol{\beta}}_R] = (X^T X + \lambda I_p)^{-1} X^T X \boldsymbol{\beta}, \tag{2.67}$$

and

$$\begin{aligned} \mathrm{cov}(\hat{\boldsymbol{\beta}}_R) &= E\left[(\hat{\boldsymbol{\beta}}_R - E[\hat{\boldsymbol{\beta}}_R])(\hat{\boldsymbol{\beta}}_R - E[\hat{\boldsymbol{\beta}}_R])^T\right] \\ &= \sigma^2 (X^T X + \lambda I_p)^{-1} X^T X (X^T X + \lambda I_p)^{-1}. \end{aligned} \tag{2.68}$$

Using the equation $(X^T X + \lambda I_p)^{-1} X^T X = I_p - \lambda (X^T X + \lambda I_p)^{-1}$, we have the bias

$$\begin{aligned} E[\hat{\boldsymbol{\beta}}_R - \boldsymbol{\beta}] &= -\{I_p - (X^T X + \lambda I_p)^{-1} X^T X\}\boldsymbol{\beta} \\ &= -\lambda (X^T X + \lambda I_p)^{-1} \boldsymbol{\beta}. \end{aligned} \tag{2.69}$$

As shown by this equation, the bias diminishes with decreasing values of λ, and when $\lambda = 0$ the ridge estimator becomes unbiased and equivalent to the ordinary least squares estimator.

The mean squared error for the regression coefficient vector in the ridge estimator is given by

$$\begin{aligned}
&E[(\hat{\beta}_R - \beta)^T(\hat{\beta}_R - \beta)] \\
&\quad = E[(\hat{\beta}_R - E[\hat{\beta}_R])^T(\hat{\beta}_R - E[\hat{\beta}_R])] + (E[\hat{\beta}_R] - \beta)^T(E[\hat{\beta}_R] - \beta) \\
&\quad = \sigma^2 \mathrm{tr}\left\{(X^TX + \lambda I_p)^{-1}X^TX(X^TX + \lambda I_p)^{-1}\right\} + \lambda^2\beta^T(X^TX + \lambda I_p)^{-2}\beta \\
&\quad = \sigma^2 \sum_{j=1}^{p} \frac{\ell_j}{(\ell_j + \lambda)^2} + \lambda^2\beta^T(X^TX + \lambda I_p)^{-2}\beta, \qquad (2.70)
\end{aligned}$$

where $\ell_1 \geq \ell_2 \geq \cdots \geq \ell_p$ are the ordered eigenvalues of X^TX. The first term on the right-hand side of the final equation represents the sum of variances of the ridge estimator components, and the second term is the square of the bias. As shown in this equation, the bias increases and the sum of variance of the components decreases as the regularization parameter λ increases. Hoerl and Kennard (1970) showed that the mean squared error of the ridge estimator can be made smaller than that of the least squares estimator by choosing an appropriate value of the regularization parameter. Note also that the expression of X^TX in eigenvalues in the first term on the right-hand side of the final equation of (2.70) can be obtained by singular value decomposition of the design matrix X. For information on singular value decomposition, see Section 9.3.

2.3.2 Lasso

Lasso is a method of estimating model parameters by minimization of the following objective function, which imposes the sum of absolute values (L_1 norms) of the regression coefficients as a constraint on the sum of squared errors:

$$S_\lambda(\beta) = (y - X\beta)^T(y - X\beta) + \lambda \sum_{j=1}^{p} |\beta_j|, \qquad (2.71)$$

where the observed data are standardized as in (2.64). In contrast to the shrinkage of regression coefficients toward 0 that occurs in ridge regression, lasso results in an estimation of exactly 0 for some of the coefficients. An advantage of ridge regression is that if $p < n$, then with appropriate selection of the regularization parameter λ, it is possible to obtain stable estimates of regression coefficients even in cases involving multicollinearity among predictor variables or nearly singular X^TX

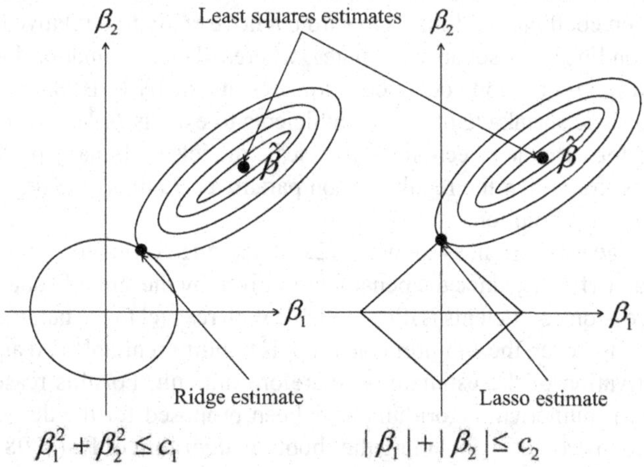

Figure 2.6 *Ridge estimate (left panel) and lasso estimate (right panel): Ridge estimation shrinks the regression coefficients β_1, β_2 toward but not exactly to 0 relative to the corresponding least squares estimates $\hat{\beta}$, whereas lasso estimates the regression coefficient β_1 at exactly 0.*

for the design matrix X. However, because, unlike lasso, it cannot yield estimates of exactly 0, ridge regression cannot be used as a method of variable selection.

The difference between the lasso estimate and the ridge estimate can be demonstrated, for simplicity, for the case of just two predictor variables x_1 and x_2. In ridge estimation, the solution is based on the constraint $\beta_1^2 + \beta_2^2 \le c_1$ of minimizing

$$S(\beta_1, \beta_2) = \sum_{i=1}^{n} (y_i - \sum_{j=1}^{2} \beta_j x_{ij})^2 \qquad (2.72)$$

for centered data, whereas in lasso estimation it is based on the constraint $|\beta_1| + |\beta_2| \le c_2$. Since the least squares estimate is the solution that minimizes $S(\beta_1, \beta_2)$, it occurs at the center of an ellipse. As shown in Figure 2.6, however, solutions that satisfy the constraints in ridge estimates are found in different regions from those that satisfy the constraints in lasso estimates.

The essential difference between ridge estimation and lasso estimation, as shown in Figure 2.6, is thus that ridge estimation shrinks all of the regression coefficient estimates toward but not exactly to 0 relative to the corresponding least squares estimates, whereas lasso estimation locates some of the regression coefficient estimates at exactly 0. Because of its characteristic shrinkage of some coefficients to exactly 0, lasso can also be used for variable selection in large-scale models with many predictor variables, for which the regularization parameter λ affects the degree of sparsity of the solution.

In ridge estimation, as shown in (2.66), the ridge estimator can be obtained analytically, since the penalty term given by the sum of squares of the regression coefficients is differentiable with respect to the parameters. In lasso, however, the L_1 norm constraint is nondifferentiable and analytical derivation of the estimator is therefore difficult. For this reason, a number of numerical algorithms have been proposed for the derivation of the lasso estimator, including the shooting algorithm of Fu (1998), the LARS (least angle regression) algorithm of Efron et al. (2004), and the coordinate descent algorithm of Friedman et al. (2007).

LARS is an algorithm for linear regression model estimation and variable selection, and characterized by extremely high-speed execution. Its estimates are generally fairly close to those of lasso and with small adjustments in the algorithm become equal. Unlike the shooting algorithm, which presents several candidates for regularization parameter λ and requires iterative computation for each candidate to obtain its estimate, LARS automatically selects the λ candidate. In LARS, moreover, variables can be added or deleted and an estimate obtained in one iteration, and the estimation can therefore be completed in approximately p iterations. The coordinate descent algorithm has been proposed as an algorithm no less powerful than LARS and has become the subject of studies for its application to various forms of L_1 regularization.

Lasso results in an estimation of exactly 0 for some of the regression coefficients, and the regularization parameter λ affects the degree of sparsity of the solution; large values produce sparser results. This property can be seen by using a method of drawing the entire regularization path for the estimated regression coefficients with λ varying from large values to 0. Efron et al. (2004) provided an algorithm for computing the entire regularization path for the lasso. Hastie et al. (2004) presented a method of drawing the entire regularization path for the support vector machine described in Chapter 8. Regularization path algorithms for generalized linear models were introduced by Park and Hastie (2007) and Friedman et al. (2010).

The lasso and ridge regression differ only in the penalty function imposed on the regression coefficients $\boldsymbol{\beta}$. The objective function with L_1 or L_2 penalty can be extended to the regularization with L_q penalty as

$$(\boldsymbol{y} - X\boldsymbol{\beta})^T(\boldsymbol{y} - X\boldsymbol{\beta}) + \lambda \sum_{j=1}^{p} |\beta_j|^q, \tag{2.73}$$

called *bridge regularization* (Frank and Friedman, 1993). It is known that for the values $q \in (0, 1]$, the bridge penalties yield estimates that are 0 for some of the regression coefficients, and that the values $q \in (0, 1)$ will lead to non-concave minimization problems.

Example 2.3 (Prediction of housing prices) In this example, we perform modeling of the relationship between median housing price, as the response value, and 13 predictor variables that might affect it, in 506 districts of Boston, the capital of Massachusetts, U.S.A. (Harrison and Rubenfeld, 1978). The predictor variables are as follows: x_1, crime rate; x_2, residential area proportion; x_3, nonretail business area proportion; x_4, boundary at or away from Charles River; x_5, NOx concentration; x_6, number of rooms; x_7, year constructed; x_8, distance from business district; x_9, highway access index; x_{10}, property tax; x_{11}, pupil-to-teacher ratio; x_{12}, non-white proportion of population; and x_{13}, low-income proportion of population.

For the models, we assumed a linear relationship between the response variable y and the 13 predictor variables. We performed linear regression by least squares and by lasso and then compared the regression coefficient estimates in Table 2.6. As shown in this table, lasso estimated coefficients of 0 for both nonretail business area proportion (x_3) and year constructed (x_7), indicating that these variables are unnecessary in the modeling. The lasso estimation here thus clearly demonstrates its characteristic feature of eliminating some coefficients by taking them to exactly 0, as an effect of its imposed L_1 norm constraint on the regression coefficients, i.e., a constraint based on the sum of their absolute values. With this feature, it can be used to perform variable selection and model stabilization, simultaneously.

Figure 2.7 shows the profiles of estimated regression coefficients for different values of the L_1 norm $= \sum_{i=1}^{13} |\beta_i(\lambda)|$ with λ varying from 6.78 to 0, in which each path corresponds to a predictor variable. The axis above indicates the number of nonzero coefficients. We see that the regularization parameter λ affects the degree of sparsity of the solution, and making λ sufficiently large will cause some of the estimated coefficients to be ex-

Table 2.6 *Comparison of the estimates of regression coefficients by least squares (LS) and lasso L_1.*

	$\hat{\beta}_0$	x_1	x_2	x_3	x_4	x_5	x_6
LS	36.46	−0.108	0.046	0.021	2.687	−17.767	3.810
L_1	35.34	−0.103	0.044	0.0	2.670	−16.808	3.835

	x_7	x_8	x_9	x_{10}	x_{11}	x_{12}	x_{13}
LS	0.001	−1.476	0.306	−0.012	−0.953	0.009	−0.525
L_1	0.0	−1.441	0.275	−0.011	−0.938	0.009	−0.523

actly zero. The optimal λ may be determined via model evaluation and selection criteria described in Chapter 5.

2.3.3 L_1 Norm Regularization

For selection of effective variables from among many that might conceivably affect the outcome of a phenomenon and thus the response variable, a round-robin algorithm poses major problems in terms of computational time. Sequential selection, on the other hand, may result in selection of non-optimum variables and instability in the estimated model, as noted by Breiman (1996). This is a major reason for the widespread interest in lasso as an effective modeling method based on high-dimensional data, as well as its capability for simultaneous model estimation and variable selection.

Lasso itself, however, is subject to certain limitations. It is known that lasso can select at most n variables in analysis of high-dimensional data with small sample sizes where $p > n$, that is, the number of predictor variables is larger than the number of data (Efron et al., 2004; Rosset et al., 2004). In variable selection for genomic data and other cases involving highly correlated predictor variables, moreover, lasso tends to overlook the correlation.

Elastic net To overcome these problems, Zou and Hastie (2005) proposed the following regularization method for minimizing the objective

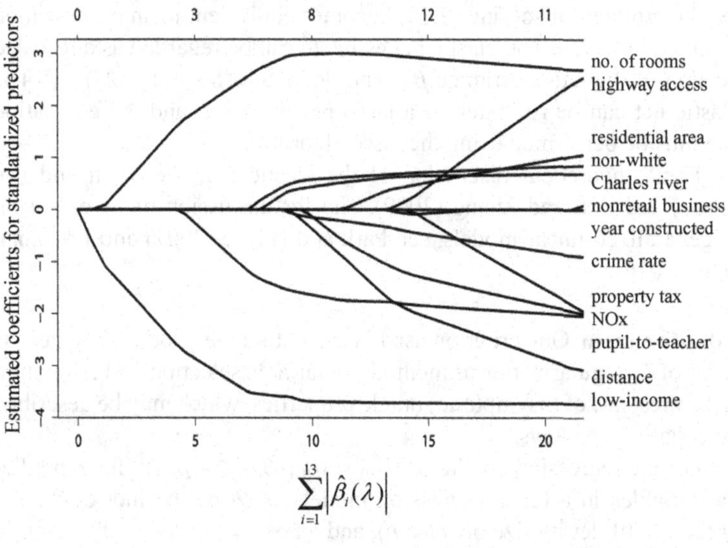

Figure 2.7 *The profiles of estimated regression coefficients for different values of the* L_1 *norm* $= \sum_{i=1}^{13} |\beta_i(\lambda)|$ *with* λ *varying from 6.78 to 0. The axis above indicates the number of nonzero coefficients.*

function, which they referred to as the *elastic net*:

$$(\boldsymbol{y} - X\boldsymbol{\beta})^T(\boldsymbol{y} - X\boldsymbol{\beta}) + \lambda \sum_{j=1}^{p} \{\alpha\beta_j^2 + (1-\alpha)|\beta_j|\}, \qquad (2.74)$$

where λ (> 0) and α ($0 \leq \alpha \leq 1$) are the regularization parameters. The elastic net is a sparse regularization that incorporates the penalty term $\alpha\beta_j^2 + (1-\alpha)|\beta_j|$, which simultaneously includes both L_2 and L_1 penalty terms. It thus combines the properties of ridge regression with regard to model estimation containing highly correlated variables and the properties of lasso with regard to variable selection.

To avoid the problem of overshrinkage, Zou and Hastie (2005) suggested a two-stage estimation method in which the regression coefficient estimate $\hat{\boldsymbol{\beta}}_N$, known as the naïve estimate, is first obtained by the following regularization formula, which includes the penalty term $\lambda_1|\beta_j| + \lambda_2\beta_j^2$,

$$(\boldsymbol{y} - X\boldsymbol{\beta})^T(\boldsymbol{y} - X\boldsymbol{\beta}) + \sum_{j=1}^{p} \left\{\lambda_1|\beta_j| + \lambda_2\beta_j^2\right\}. \qquad (2.75)$$

From this it may be seen that, by setting $\alpha = \lambda_2/(\lambda_1 + \lambda_2)$ and $\lambda = \lambda_1 + \lambda_2$, the problem of solving (2.74) becomes equivalent to an optimization problem in (2.75). The elastic net estimate can be regarded as a rescaled version of the naïve estimate $\hat{\boldsymbol{\beta}}_N$, and defined as $\hat{\boldsymbol{\beta}}_E = (1 + \lambda_2)\, \hat{\boldsymbol{\beta}}_N$. The elastic net can be rewritten as a lasso penalty term, and the estimation can thus be performed using the lasso algorithm.

For a theoretical discussion of the elastic net, see Yuan and Lin (2007) and Zou and Zhang (2009), and for discussion of its extension to generalized linear models, see Park and Hastie (2007) and Friedman et al. (2010).

Adaptive lasso One criterion used in evaluating the goodness of various types of L_1 regularization as methods of variable selection and estimation is the presence of (asymptotic) oracle properties, which may be described essentially as follows.

For the regression coefficients $\boldsymbol{\beta}_0 = (\beta_{01}, \beta_{02}, \cdots, \beta_{0p})^T$ for p predictor variables in a linear regression model, let $\mathcal{A}$ be the index set $\mathcal{A} = \{j : \beta_{0j} \neq 0\}$, let its size be q $(< p)$, and let $\mathcal{A}^c = \{j : \beta_{0j} = 0\}$. The true model thus comprises a partial set of the p predictor variables. The regularization procedure for linear regression models is oracle if the estimator $\hat{\boldsymbol{\beta}}$ of the regression coefficients satisfies the following properties:

(1) Consistency in variable selection: The procedure identifies the true model asymptotically;

$$\{j : \hat{\beta}_j \neq 0\} = \mathcal{A} \quad \text{and} \quad \{j : \hat{\beta}_j = 0\} = \mathcal{A}^c.$$

(2) Asymptotic normality: The estimator $\hat{\boldsymbol{\beta}}_{\mathcal{A}}$ has asymptotic normality;

$$\sqrt{n}(\hat{\boldsymbol{\beta}}_{\mathcal{A}} - \boldsymbol{\beta}_{\mathcal{A}}) \to_d N(\mathbf{0}, \boldsymbol{\Sigma}_{\mathcal{A}}) \quad \text{as } n \to +\infty,$$

where $\Sigma_{\mathcal{A}}$ is the covariance matrix corresponding to the true subset model $\mathcal{A}$.

It should be noted that it is necessary to choose an appropriate regularization parameter to satisfy asymptotic oracle properties, and that the rate of convergence depends on the regularization parameter.

Lasso is useful as a method enabling simultaneous performance of parameter estimation and variable selection. As noted by Meinshausen and Bühlmann (2006) and Zou (2006), however, it generally lacks consistency in variable selection. To impart the desired consistency, Zou (2006) proposed a generalization of lasso, referred to as adaptive lasso, by minimization of the objective function

$$(\boldsymbol{y} - X\boldsymbol{\beta})^T(\boldsymbol{y} - X\boldsymbol{\beta}) + \lambda \sum_{j=1}^{p} w_j|\beta_j|, \tag{2.76}$$

where λ (> 0) is the regularization parameter and $w_j > 0$ are weightings applied to the coefficient parameters in the penalty term. The concept is one of applying different weights to different coefficients and thereby imposing small penalties on important variables and larger penalties on unimportant variables. As a means of determining the weightings, Zou (2006) suggested the use of $w_j = 1/|\hat{\beta}_j|^\gamma$ for $\sqrt{n}$-consistent estimator $\hat{\boldsymbol{\beta}}$ with $\gamma > 0$ as a fixed constant, and if a least squares estimator is obtained, then taking it as $\hat{\boldsymbol{\beta}}$. The lasso algorithm can be used as the adaptive lasso estimation algorithm, and Zou (2006) recommended the use of the LARS algorithm because of its relatively low computational complexity.

SCAD Let us now turn to the question of what form of penalty term to apply in estimating the regression coefficients of a linear regression model based on high-dimensional data in a small sample in order to obtain an estimator characterized by both sparsity and oracle properties. In this regard, Fan and Li (2001) proposed a penalty function, referred to as SCAD (smoothly clipped absolute deviation), that yields a sparse and nearly unbiased estimator that is continuous in data. The SCAD penalty is given by

$$p_\lambda(|\beta|) = \begin{cases} \lambda|\beta| & (|\beta| \le \lambda), \\ -\dfrac{|\beta|^2 - 2a\lambda|\beta| + \lambda^2}{2(a-1)} & (\lambda < |\beta| \le a\lambda), \\ \dfrac{(a+1)\lambda^2}{2} & (a\lambda < |\beta|), \end{cases} \tag{2.77}$$

where λ (> 0) and a (> 2) are tuning parameters that must be assigned appropriate values. Fan and Li (2001) used $a = 3.7$ for Bayes risk minimization. As shown in Figure 2.8, this penalty function is a symmetric quadratic spline with knots at λ and $a\lambda$ (see Section 3.2.2). The regression coefficient estimator in regularization with the penalty function in (2.77) is given by minimization of

$$(\boldsymbol{y} - X\boldsymbol{\beta})^T(\boldsymbol{y} - X\boldsymbol{\beta}) + \sum_{j=1}^{p} p_\lambda(|\beta_j|). \tag{2.78}$$

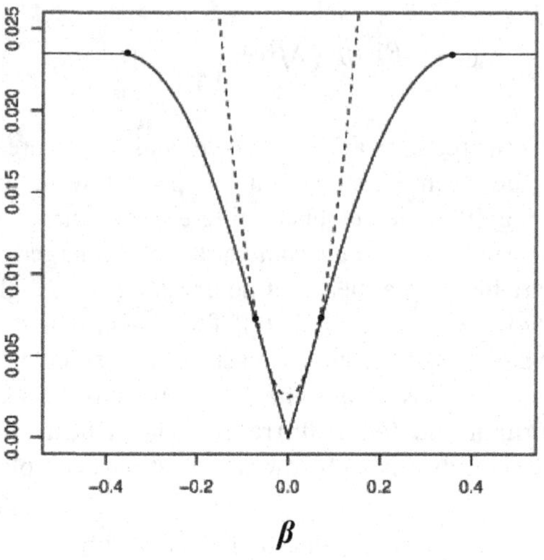

Figure 2.8 *The function* $p_\lambda(|\beta_j|)$ *(solid line) and its quadratic approximation (dotted line) with the values of* β_j *along the x axis, together with the quadratic approximation for a* β_{j0} *value of* 0.15.

The objective function imposing the SCAD penalty is a nonconvex optimization problem and, because it includes the constraint L_1, is nondifferentiable at 0. This has led to the proposal of several algorithms for finding SCAD estimators. One of these is *local quadratic approximation* (LQA), which is performed essentially as follows. Let $\boldsymbol{\beta}_0$ be the initial value of $\boldsymbol{\beta}$. In any case where the j-th component β_{j0} of $\boldsymbol{\beta}_0$ is very close to 0, set $\hat{\beta}_j = 0$ and for the other components perform LQA of the derivative of the penalty function $p_\lambda(|\beta_j|)$ expressed as

$$\begin{aligned}[p_\lambda(|\beta_j|)]' &= p'_\lambda(|\beta_j|)\text{sign}(\beta_j) \\ &= \left\{p'_\lambda(|\beta_j|)/|\beta_j|\right\}\beta_j \approx \left\{p'_\lambda(|\beta_{j0}|)/|\beta_{j0}|\right\}\beta_j, \qquad (2.79)\end{aligned}$$

where $\beta_j \neq 0$. By integrating both sides of (2.79), we obtain the approximation

$$p_\lambda(|\beta_j|) \approx p_\lambda(|\beta_{j0}|) + \frac{1}{2}\left\{p'_\lambda(|\beta_{j0}|)/|\beta_{j0}|\right\}(\beta_j^2 - \beta_{j0}^2). \qquad (2.80)$$

This approximation coincides at $\beta_j = |\beta_{j0}|$, and the function $p_\lambda(|\beta_j|)$ is thus approximated by the quadratic function contacting it at β_{j0}. Figure 2.8 shows the function $p_\lambda(|\beta_j|)$ (solid line) and its quadratic approximation (dotted line) with the values of β_j along the x axis, together with the quadratic approximation for a β_{j0} value of 0.15. As shown here, the approximation performs well in regions where β_j is near β_{j0} so long as β_{j0} is not 0.

The right-hand side of (2.80), as a quadratic function for β_j, is differentiable. The problem therefore becomes one of optimizing the quadratic function

$$(\boldsymbol{y} - X\boldsymbol{\beta})^T(\boldsymbol{y} - X\boldsymbol{\beta}) + \frac{1}{2}\boldsymbol{\beta}^T\Sigma_\lambda(\boldsymbol{\beta}_0)\boldsymbol{\beta}, \tag{2.81}$$

where $\Sigma_\lambda(\boldsymbol{\beta}_0) = \text{diag}[p'_\lambda(|\beta_{10}|)/|\beta_{10}|, p'_\lambda(|\beta_{20}|)/|\beta_{20}|, \cdots, p'_\lambda(|\beta_{p0}|)/|\beta_{p0}|]$ is a p-dimensional diagonal matrix that is dependent on initial values. The SCAD estimator can therefore be obtained by solving iteratively. For details, see Fan and Li (2001).

Solution of (2.81) generally does not yield an estimate of exactly 0 for the regression coefficient. It is therefore necessary to apply a thresholding rule, by which any estimate of $|\hat{\beta}_j|$ obtained during the iteration that is smaller than the threshold is taken as 0. It must be noted, however, that the drawback to this approach is that variables do not reenter the model following their elimination in the iteration.

Figure 2.9 shows the relationship between the least squares estimator (dotted line) and three shrinkage estimators (solid lines) for a linear regression model with a design matrix X of presumed orthonormality: (a) hard thresholding, (b) lasso, and (c) SCAD. As illustrated in this figure, problems that tend to occur in shrinkage with hard thresholding and lasso may be mitigated with SCAD. Hard thresholding (Donoho and Johnstone, 1994) is consistent with least squares estimate for nonzero estimates but is discontinuous in the boundary regions and therefore unstable. With lasso, the estimate varies continuously but the estimator is biased because of the constraint imposed on nonzero estimates. SCAD deals with such problems through its consideration of continuity and unbiasedness as related to the shrinkage level.

SCAD is also used for semiparametric, proportional hazard, and other models (e.g., Fan and Li, 2002, 2004; Cai et al., 2005). Available studies on the theoretical aspects of SCAD include Fan and Li (2001), Fan and Peng (2004), and Kim et al. (2008), among others.

Fused lasso Tibshirani et al. (2005) have proposed what they refer to as *fused lasso* for analysis of observed data relating to predictor vari-

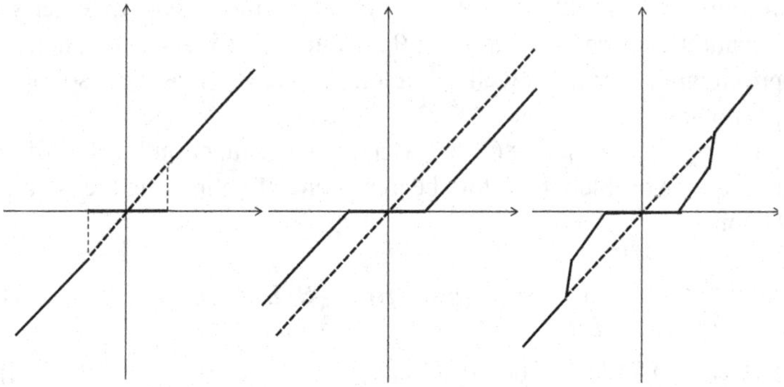

(a) Hard thresholding (b) Lasso (c) SCAD

Figure 2.9 *The relationship between the least squares estimator (dotted line) and three shrinkage estimators (solid lines): (a) hard thresholding, (b) lasso, and (c) SCAD.*

ables that are in some sense ordered, as a regularization procedure that minimizes the objective function

$$(\boldsymbol{y} - X\boldsymbol{\beta})^T(\boldsymbol{y} - X\boldsymbol{\beta}) + \lambda_1 \sum_{j=1}^{p} |\beta_j| + \lambda_2 \sum_{j=2}^{p} |\beta_j - \beta_{j-1}|, \qquad (2.82)$$

where λ_1 ($>$ 0) and λ_2 ($>$ 0) are the regularization parameters. This may be regarded as an L_1 regularization in which the first penalty term assesses the sparsity of the variables and the second penalty term assesses the sparsity between neighboring variables. It corresponds to L_1 with the penalty term expressed in quadratic form using the difference matrix discussed in Section 3.4.1. Tibshirani et al. (2005) proposed fused lasso modeling for analysis of mass spectrometric data on proteins. The data comprised intensities observed in blood serum samples at a large number of charge transfer rates.

In recent years, fused lasso has become the focus of increasing interest as a useful technique in life sciences, image processing, and many other fields (see, e.g., Friedman et al., 2007; Tibshirani and Wang, 2008). It is a particularly useful method of regression modeling for analysis of high-dimensional, small-sample data in cases where the dimensionality is extremely high in comparison with the number of data. A number of useful algorithms have been proposed for fused lasso estimation. For details, see Tibshirani et al. (2005) and Friedman et al. (2007).

Group lasso Certain situations may be conceptually regarded as possibly involving the presence of a number of groups each having a number of mutually correlated predictor variables that affect the response variable. Such group variables are in fact involved in analyses of genetic data from the same biological pathways or the data of dummy variables representing categorical data levels. In these situations, it is desirable to perform variable selection for each group. *Group lasso*, proposed by Yuan and Lin (2006), is a sparse regularization that enables such group-by-group variable selection.

Suppose that p explanatory variables are partitioned into J groups, with p_j denoting the number of predictor variables in each group. With X_j the $n \times p_j$ design matrix of the j-th group, the regression model may be expressed as

$$\boldsymbol{y} = \sum_{j=1}^{J} X_j \boldsymbol{\beta}_j + \boldsymbol{\varepsilon}, \tag{2.83}$$

where $\boldsymbol{\beta}_j$ is a p_j-dimensional regression coefficient vector. The grouped lasso may then be formalized by minimization of the objective function (Yuan and Lin, 2006)

$$\left(\boldsymbol{y} - \sum_{j=1}^{J} X_j \boldsymbol{\beta}_j\right)^T \left(\boldsymbol{y} - \sum_{j=1}^{J} X_j \boldsymbol{\beta}_j\right) + \lambda \sum_{j=1}^{J} \sqrt{p_j} \|\boldsymbol{\beta}_j\|_2, \tag{2.84}$$

where λ (> 0) is the regularization parameter and $\|\cdot\|_2$ represents the Euclidean norm ($\|\boldsymbol{\beta}\|_2 = (\boldsymbol{\beta}^T \boldsymbol{\beta})^{1/2}$). If we now set $\eta_j^2 = \boldsymbol{\beta}_j^T \boldsymbol{\beta}_j$ as the sum of squares of the group regression coefficients, the penalty term in (2.84) can then be transformed to

$$\sum_{j=1}^{J} \sqrt{p_j} \|\boldsymbol{\beta}_j\|_2 = \sum_{j=1}^{J} \sqrt{p_j} |\eta_j|. \tag{2.85}$$

As this shows, the penalty term thus imposes an L_1 norm constraint on the sum of squares of the group coefficients. Since the Euclidean norm of the vector $\boldsymbol{\beta}_j$ is 0 only if all of the vector components are 0, variable selection can then be performed for each group, using the penalty term in (2.84). The algorithm used in ordinary lasso estimation can be used without change in group lasso. Yuan and Lin (2006) in particular use the LARS algorithm.

Exercises

2.1 Show that the regression coefficients that minimize the sum of squared errors in (2.4) are given by solving the following simultaneous equations:

$$\sum_{i=1}^{n} y_i = n\beta_0 + \beta_1 \sum_{i=1}^{n} x_i, \qquad \sum_{i=1}^{n} x_i y_i = \beta_0 \sum_{i=1}^{n} x_i + \beta_1 \sum_{i=1}^{n} x_i^2.$$

2.2 Show that the least squares estimates $\hat{\beta}_0$ and $\hat{\beta}_1$ are given by

$$\hat{\beta}_0 = \overline{y} - \hat{\beta}_1 \overline{x}, \qquad \hat{\beta}_1 = \frac{\sum_{i=1}^{n}(x_i - \overline{x})(y_i - \overline{y})}{\sum_{i=1}^{n}(x_i - \overline{x})^2},$$

where $\overline{x} = n^{-1} \sum_{i=1}^{n} x_i$ and $\overline{y} = n^{-1} \sum_{i=1}^{n} y_i$.

2.3 By differentiating the sum of squared errors $S(\boldsymbol{\beta}) = (\boldsymbol{y} - X\boldsymbol{\beta})^T(\boldsymbol{y} - X\boldsymbol{\beta})$ with respect to $\boldsymbol{\beta}$, derive the normal equation given by (2.20).

2.4 Derive the maximum likelihood estimates for regression coefficient vector $\boldsymbol{\beta}$ and error variance σ^2 of the Gaussian linear regression model defined by (2.34).

2.5 Show that the maximum likelihood estimator of the regression coefficient vector $\boldsymbol{\beta}$ of the Gaussian linear regression model (2.34) equals the least squares estimator.

2.6 Let $\boldsymbol{c}$ be a $(p+1)$ dimensional constant vector and A a $(p+1)\times(p+1)$ constant matrix. Show that

$$\text{(a)} \quad \frac{\partial(\boldsymbol{c}^T\boldsymbol{\beta})}{\partial\boldsymbol{\beta}} = \boldsymbol{c}, \qquad \text{(b)} \quad \frac{\partial(\boldsymbol{\beta}^T A\boldsymbol{\beta})}{\partial\boldsymbol{\beta}} = (A + A^T)\boldsymbol{\beta},$$

$$\text{(c)} \quad \frac{\partial(\boldsymbol{\beta}^T A\boldsymbol{\beta})}{\partial\boldsymbol{\beta}} = 2A\boldsymbol{\beta} \quad \text{if } A \text{ is symmetric.}$$

2.7 Let $\boldsymbol{Y}$ be an n-dimensional random vector. Consider the linear transformation $\boldsymbol{Z} = \boldsymbol{c} + A\boldsymbol{Y}$, where $\boldsymbol{c}$ is an m-dimensional constant vector and A is an $m \times n$ constant matrix. Show that the expectation and variance-covariance matrix of the m-dimensional random vector $\boldsymbol{Z}$ are respectively given by

$$E[\boldsymbol{Z}] = \boldsymbol{c} + AE[\boldsymbol{Y}], \quad E[(\boldsymbol{Z} - E[\boldsymbol{Z}])(\boldsymbol{Z} - E[\boldsymbol{Z}])^T] = A\mathrm{cov}(\boldsymbol{Y})A^T.$$

2.8 A square matrix P is said to be idempotent if $P^2 = P$. Show that the $n \times n$ symmetric matrix P is idempotent of rank r if and only if all its eigenvalues are either 0 or 1, and the number of eigenvalues equal to 1 is then tr $P = r$.

2.9 Let $P = X(X^TX)^{-1}X^T$, where X is an $n\times(p+1)$ design matrix defined by (2.15). Show that

$$\text{(a)}\quad P^2 = P, \qquad\qquad \text{(b)}\quad (I_n - P)^2 = I_n - P,$$

$$\text{(c)}\quad \mathrm{tr}\,P = p + 1, \qquad\qquad \text{(d)}\quad \mathrm{tr}\,(I_n - P) = n - p - 1.$$

2.10 Show that AIC for the Gaussian linear regression model (2.34) is given by the following equation:

$$\mathrm{AIC} = n\log(2\pi\hat{\sigma}^2) + n + 2(p + 2),$$

where $\hat{\sigma}^2 = (\boldsymbol{y} - X\hat{\boldsymbol{\beta}})^T(\boldsymbol{y} - X\hat{\boldsymbol{\beta}})/n = (\boldsymbol{y} - \hat{\boldsymbol{y}})^T(\boldsymbol{y} - \hat{\boldsymbol{y}})/n$.

2.11 Show that the ridge estimator of the regression coefficient vector is separately given by minimization of the function given by (2.63).

2.12 Show that solving (2.73) is equivalent to minimization of the following equation with the L_q norm constraint imposed:

$$(\boldsymbol{y} - X\boldsymbol{\beta})^T(\boldsymbol{y} - X\boldsymbol{\beta}) \quad \text{subject to} \quad \sum_{j=1}^{m} |\beta_j|^q \leq \eta.$$

Find the relationship between the regularization parameter λ in (2.73) and η.

Chapter 3

Nonlinear Regression Models

In Chapter 2 we discussed the basic concepts of modeling and constructed linear regression models for phenomena with linear structures. We considered, in particular, the series of processes in regression modeling; specification of linear models to ascertain relations between variables, estimation of the parameters of the specified models by least squares or maximum likelihood, and evaluation of the estimated models to select the most appropriate model. For linear regression models having a large number of predictor variables, the usual methods of separating model estimation and evaluation are inefficient for the construction of models with high reliability and prediction. We introduced various regularization methods with an L_1 penalty term in addition to the sum of squared errors and log-likelihood functions.

We now consider the types of models to assume for analysis of phenomena containing complex nonlinear structures. In assessing the degree of impact on the human body in a car collision, as a basic example, it is necessary to accumulate and analyze experimental data and from them develop models. The measured and observed data are quite complex and therefore difficult to comprehend, however, in contrast to the types of data amenable to the specific functional forms such as linear and polynomial models. In this chapter, we consider the resolution of this problem by organizing the basic concepts of regression modeling into a more general framework and extending linear models to nonlinear models for the extraction of information from data with complex structures.

3.1 Modeling Phenomena

We first consider the general modeling process for a certain phenomenon. Suppose that the observed n-set of data relating to the predictor variable x and the response variable y are given by $\{(x_i, y_i);\ i = 1, 2, \cdots, n\}$. It is assumed that the value y_i at each data point x_i is observed as

$$y_i = u(x_i) + \varepsilon_i, \qquad i = 1, 2, \cdots, n \tag{3.1}$$

with the error or noise ε_i.

In the relation between weight and spring elongation, for example, the true value $u(x_i)$ with no error can be taken as the value derived from Hooke's law for the predictor data point x_i, and thus as $\mu_i = E[Y_i|x_i]$, where $E[Y_i|x_i]$ is the conditional expectation given x_i. If the true values $u(x_1), u(x_2), \cdots, u(x_n)$ at the given data points lie on a straight line, then the relationship $u(x_i) = \beta_0 + \beta_1 x_i$ $(i = 1, 2, \cdots, n)$ can be established. If the n true values actually lie on a quadratic curve, then the relationship is $u(x_i) = \beta_0 + \beta_1 x_i + \beta_2 x_i^2$ $(i = 1, 2, \cdots, n)$. In general, a p-th degree *polynomial regression model* may be given as

$$y_i = \beta_0 + \beta_1 x_i + \beta_2 x_i^2 + \cdots + \beta_p x_i^p + \varepsilon_i, \qquad i = 1, 2, \cdots, n. \quad (3.2)$$

In a similar manner, a model in which the true values lie approximately on an exponential curve is given as

$$y_i = \beta_0 e^{\beta_1 x_i} + \varepsilon_i, \qquad i = 1, 2, \cdots, n, \quad (3.3)$$

which is referred to as a *growth curve model.*

In the case of Hooke's law the true model is linear. In practice, however, phenomena are usually more complex, and it is difficult to specify a true model. It is then necessary to perform data-based construction of a model that adequately approximates the structure of the phenomenon. Given the value of its predictor variable x, the true structure of a phenomenon is the conditional expectation $E[Y|x] = u(x)$, and it is approximated by the model $u(x; \boldsymbol{\beta})$ characterized by parameters $\boldsymbol{\beta} = (\beta_0, \beta_1, \cdots, \beta_p)^T$ as $E[Y|x] = u(x) \approx u(x; \boldsymbol{\beta})$.

In general, a *regression model* may be given as

$$y_i = u(x_i; \boldsymbol{\beta}) + \varepsilon_i, \qquad i = 1, 2, \cdots, n. \quad (3.4)$$

To capture the true structure $u(x)$ of the phenomenon on the basis of observed data, we approximate it with the function $u(x; \boldsymbol{\beta})$, referred to as a model characterized by several parameters, and thus convert the problem to the parameter estimation of this model.

The modeling problem can be considered similarly for estimating the relation between multiple (p) predictor variables $\boldsymbol{x} = (x_1, x_2, \cdots, x_p)^T$ and a response variable y on the basis of n set of observed data $\{(\boldsymbol{x}_i, y_i);\ i = 1, 2, \cdots, n\}$ (see data in Table 2.4). We assume that at each data point $\boldsymbol{x}_i$, y_i is observed as

$$y_i = u(\boldsymbol{x}_i; \boldsymbol{\beta}) + \varepsilon_i, \qquad i = 1, 2, \cdots, n \quad (3.5)$$

with the error (noise) ε_i. Here, the true structure is approximated by the model $u(x;\beta)$, in which the higher-dimensional structure of a phenomenon is characterized by multiple parameters, for example, $u(x;\beta) = \beta_0 + \beta_1 x_1 + \cdots + \beta_p x_p$, the model approximated in a hyperplane as a linear regression model.

3.1.1 *Real Data Examples*

Figure 3.1(left) shows a plot of 104 tree data obtained by measurement of tree trunk girth (inch) and tree weight above ground (kg) (Andrews and Herzberg, 1985, p. 357). We model the relation between the girth and the weight so that weight can be predicted from girth, which avoids the necessity of felling any trees. We take the girth as the predictor variable x and the weight as the response variable y.

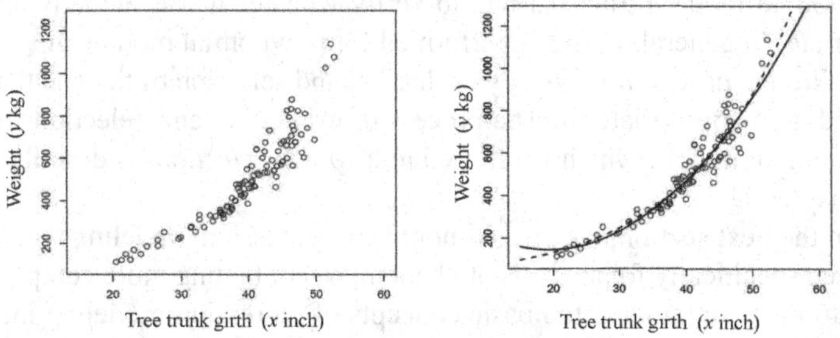

Figure 3.1 *Left panel: The plot of 104 tree data obtained by measurement of tree trunk girth (inch) and tree weight above ground (kg). Right panel: Fitting a polynomial of degree 2 (solid curve) and a growth curve model (dashed curve).*

Let us take the representation of the relation between the 104 set of observed data for weight (y_i) and girth (x_i) as

$$y_i = u(x_i) + \varepsilon_i, \qquad i = 1, 2, \cdots, 104. \tag{3.6}$$

From Figure 3.1 (left), it appears that if the measured girth is approximately 40 inches or less, then a good approximation may be obtained using the linear model $u(x_i;\beta_0,\beta_1) = \beta_0 + \beta_1 x_i$ for $u(x_i)$. For all of the data, however, a linear model is apparently inappropriate. We therefore first consider the fit of a quadratic polynomial regression model

$$y_i = \beta_0 + \beta_1 x_i + \beta_2 x_i^2 + \varepsilon_i, \qquad i = 1, 2, \cdots, 104, \tag{3.7}$$

assuming that $u(x_i) \approx \beta_0 + \beta_1 x_i + \beta_2 x_i^2$. Estimation of the regression coefficients by least squares then yields $y = 946.87 - 6.49x + 0.02x^2$ (represented by the solid curve in Figure 3.1 (right)). Prediction with this model is apparently not effective, however, particularly in view of its overestimation of tree weight in the boundary region of small measured girth.

We therefore next consider fitting a regression model

$$y_i = \beta_0 e^{\beta_1 x_i} + \varepsilon_i, \qquad i = 1, 2, \cdots, 104, \tag{3.8}$$

using a growth curve model $u(x_i) \approx \beta_0 e^{\beta_1 x_i}$ in (3.3). Estimation of the parameters $\beta_0, \beta_1, \beta_2$ by least squares yields $y = 75e^{0.0006x}$ (represented by the dashed curve in Figure 3.1 (right)). This growth curve model, as compared with the quadratic polynomial model, clearly yields a better fit in the boundary regions and is more appropriate for tree weight prediction.

In the analysis of tree data, in short, we first attempted quadratic polynomial model fitting but had to verify whether it was actually appropriate. In general, fitting is performed for polynomial models of several different orders, followed by evaluation and selection of the one that provides an appropriate fit. The process of evaluation and selection for polynomial models, which is referred to as *order selection*, is described in Chapter 5.

In the next section, we discuss nonlinear regression modeling that is utilized specifically for analysis of phenomena containing more complex structures by organizing the basic concepts of regression modeling into a more general framework.

3.2 Modeling by Basis Functions

Figure 3.2 shows a plot of the measured acceleration y (in terms of g, the acceleration due to gravity) of a crash-dummy's head at time x (in milliseconds, ms) from the moment of impact in repeated motorcycle crash trials (Härdle, 1990). With such data, which involve complex nonlinear structures, it is difficult to effectively capture the structure of the phenomena by modeling based on a polynomial model or a specific nonlinear function.

In this section, we describe more flexible models for explication of such nonlinear relationships. The models are based on spline, B-spline, radial, and other basis functions. In Section 3.3, we show that they can be uniformly organized by the method known as basis expansions. We are chiefly concerned here with models that explicate the relationship between a single predictor variable x and a response variable y, and consider these nonlinear methods in the following framework.

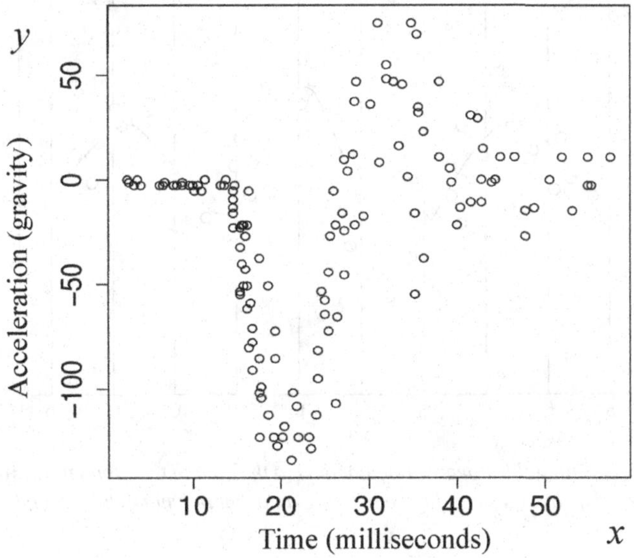

Figure 3.2 *Motorcycle crash trial data* $(n = 133)$.

Suppose that we have n set of observations $\{(x_i, y_i);\ i = 1, 2, \cdots, n\}$ for predictor variable x and response variable y, and the n values x_i of the predictor variable are given in order of increasing size in the interval $[a, b]$ such as $a < x_1 < x_2 < \cdots < x_n < b$. We assume that y_i at each point x_i is observed as

$$y_i = u(x_i) + \varepsilon_i, \qquad i = 1, 2, \cdots, n \tag{3.9}$$

with the noise ε_i. Several types of flexible models that reflect the structures ($u(x)$) of phenomena have been proposed for the separation of noise from data and explication of these structures.

3.2.1 Splines

In basic concept, a spline is a special function that smoothly connects several low-degree polynomials to fit a model to the observed data, rather than fitting a single polynomial model to the data. The interval containing all n observed values $\{x_1, x_2, \cdots, x_n\}$ is divided into subintervals, or segments, and piecewise fitting of polynomial models to the segments is

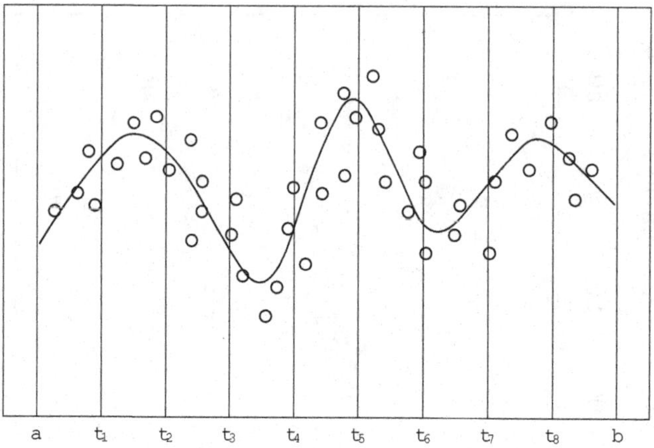

Figure 3.3 *Fitting third-degree polynomials to the data in the subintervals* $[a, t_1]$, $[t_1, t_2]$, $\cdots$, $[t_m, b]$ *and smoothly connecting adjacent polynomials at each knot.*

performed. Explication of a complex structure with a single polynomial model would invariably require the fitting of a high-order polynomial to all observed data, which would then lead to difficulty in obtaining a stable model suitable for making predictions. In contrast, a spline, as illustrated in Figure 3.3, performs piecewise fitting of low-order polynomial models to data in subintervals and forms smooth connections between the models of adjacent intervals, in the manner next described.

Let us divide the interval containing all values $\{x_1, x_2, \cdots, x_n\}$ of the predictor variable into subintervals at $t_1 < t_2 < \cdots < t_m$. In the spline, these m $(\leq n)$ points are known as *knots*. Suppose the spline is used to fit third-degree polynomials to the data in the subintervals $[a, t_1], [t_1, t_2], \cdots, [t_m, b]$ and smoothly connect adjacent polynomials at each knot (Figure 3.3). This means that the model fitting is performed under the constraint that the first and second derivatives of both third-degree polynomials at any given knot are continuous.

In practice, *cubic splines* are the most commonly used, for which the spline function having knots $t_1 < t_2 < \cdots < t_m$ is given by

$$u(x; \boldsymbol{\theta}) = \beta_0 + \beta_1 x + \beta_2 x^2 + \beta_3 x^3 + \sum_{i=1}^{m} \theta_i (x - t_i)_+^3, \qquad (3.10)$$

where $\boldsymbol{\theta} = (\beta_0, \beta_1, \beta_2, \beta_3, \theta_1, \theta_2, \cdots, \theta_m)^T$ and $(x - t_i)_+ = \max\{0, x - t_i\}$

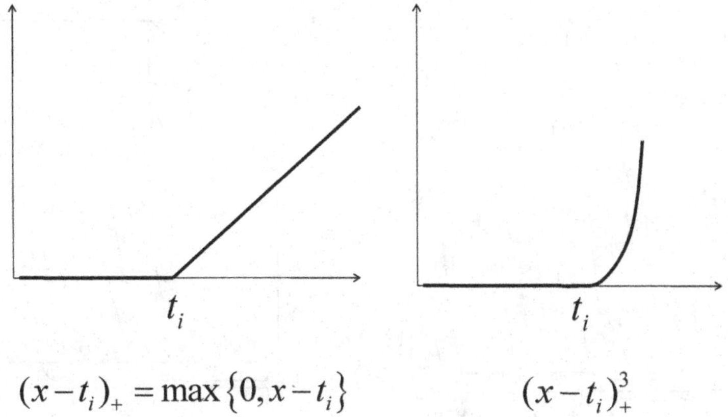

Figure 3.4 *Functions* $(x - t_i)_+ = \max\{0, x - t_i\}$ *and* $(x - t_i)_+^3$ *included in the cubic spline given by (3.10).*

(Figure 3.4). One further condition is applied at the two ends of the overall interval. It is known that third-degree polynomial fitting is unsuitable near such boundaries, as it generally tends to induce large variations in the estimated curve. For cubic splines, the condition is therefore added that a linear function be used in the subintervals $[a, t_1]$ and $[t_m, b]$, the two end intervals. Such a cubic spline is known as a *natural cubic spline* and is given by the equation

$$u(x; \boldsymbol{\theta}) = \beta_0 + \beta_1 x + \sum_{i=1}^{m-2} \theta_i \{d_i(x) - d_{m-1}(x)\}, \tag{3.11}$$

where $\boldsymbol{\theta} = (\beta_0, \beta_1, \theta_1, \theta_2, \cdots, \theta_{m-2})^T$ and

$$d_i(x) = \frac{(x - t_i)_+^3 - (x - t_m)_+^3}{t_m - t_i}. \tag{3.12}$$

A key characteristic of (3.10) and (3.11) is that, although the spline function itself is a nonlinear model, it is a linear model in terms of its parameters $\beta_0, \beta_1, \cdots, \theta_1, \theta_2, \cdots$. Cubic splines are thus models represented

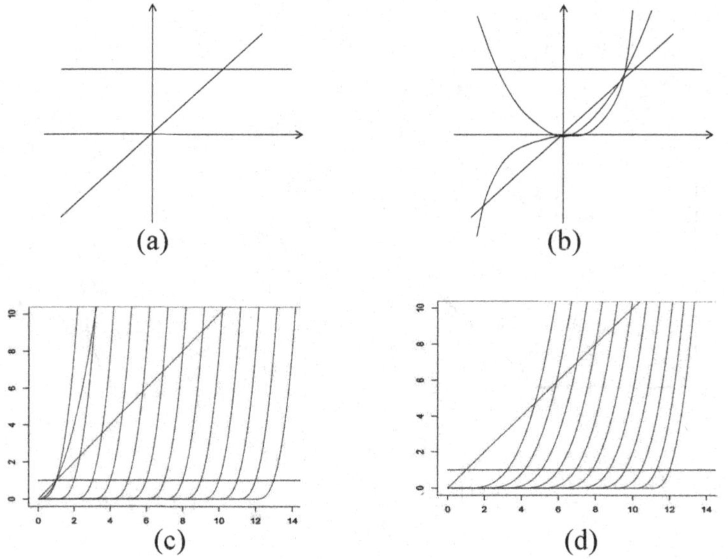

Figure 3.5 *Basis functions: (a)* $\{1, x\}$*; linear regression, (b) polynomial regression;* $\{1, x, x^2, x^3\}$*, (c) cubic splines, (d) natural cubic splines.*

by a linear combination of

$$1,\ x,\ x^2,\ x^3,\ (x-t_1)_+^3,\ (x-t_2)_+^3, \cdots, (x-t_m)_+^3, \tag{3.13}$$

and natural cubic splines are models represented by a linear combination of

$$1,\ x,\ d_1(x)-d_{m-1}(x),\ d_2(x)-d_{m-1}(x), \cdots, d_{m-2}(x)-d_{m-1}(x). \tag{3.14}$$

The functions given in these equations are referred to as *basis functions*. Thus a simple linear regression model is a linear combination of the basis functions $\{1, x\}$, and a polynomial regression model can be represented by a linear combination of the basis functions $\{1, x, x^2, \cdots, x^p\}$.

Figure 3.5 (a) – (d) shows the basis functions of a linear regression model, a polynomial regression model, a cubic spline given in (3.13), and a natural cubic spline given in (3.14). In contrast to the polynomial model, the spline models are characterized by the allocation of basis functions to specific regions of the observed data. The influence of any given weight estimate on the overall model is therefore small, which may be seen to facilitate estimation by a model that can capture the structure

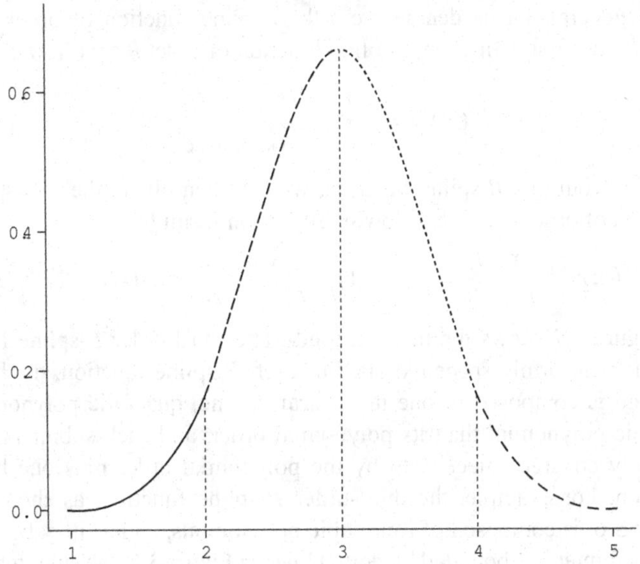

Figure 3.6 *A cubic B-spline basis function connected four different third-order polynomials smoothly at the knots 2, 3, and 4.*

of the phenomenon throughout the entire region. Model estimation based on splines is described in Section 3.3.

3.2.2 B-splines

A spline, as described above, is constructed from piecewise fitting of polynomials and smooth connection at knots between the polynomials in adjacent subintervals. A *B*-spline basis function, in contrast, consists of multiple polynomials connected smoothly. Figure 3.6 shows a cubic *B*-spline basis connected different third-order polynomials smoothly at the knots 2, 3, and 4.

In order to construct m basis functions $\{b_1(x), b_2(x), \cdots, b_m(x)\}$, we set the knots t_i as follows:

$$t_1 < t_2 < t_3 < t_4 = x_1 < \cdots < t_{m+1} = x_n < \cdots < t_{m+4}. \quad (3.15)$$

Then, n data points are partitioned in the $(m-3)$ subintervals $[t_4, t_5]$, $[t_5, t_6]$, $\cdots$, $[t_m, t_{m+1}]$. Given these knots, we use the following de Boor's (2001) algorithm to construct the *B*-spline basis functions.

In general, let us denote the j-th B-spline function of order r as $b_j(x;r)$. We first define the B-spline function of order 0 as follows:

$$b_j(x;0) = \begin{cases} 1, & t_j \leq x < t_{j+1} \\ 0, & \text{otherwise} \end{cases} \tag{3.16}$$

Starting from this B-spline function, we can then obtain the j-th spline function of order r by the following recursion formula.

$$b_j(x;r) = \frac{x - t_j}{t_{j+r} - t_j} b_j(x;r-1) + \frac{t_{j+r+1} - x}{t_{j+r+1} - t_{j+1}} b_{j+1}(x;r-1). \tag{3.17}$$

Figure 3.7 shows the first-, second-, and third-order B-spline functions for uniformly spaced knots. For each B-spline function, the basis function is composed of one more straight line, quadratic polynomial, or cubic polynomial than its polynomial order, and each subinterval is similarly covered (piecewise) by the polynomial order plus one basis function. For example, the third-order B-spline function, as shown in Figure 3.6, is composed of four cubic polynomials, and as may be seen in the subintervals bounded by dotted lines in Figure 3.7, the subintervals $[t_i, t_{i+1}]$ $(i = 4, \cdots, m)$ are respectively covered by the four cubic B-spline basis functions $b_{i-2}(x;3)$, $b_{i-1}(x;3)$, $b_i(x;3)$, $b_{i+1}(x;3)$.

A third-order B-spline regression model approximates the structure of a phenomenon by a linear combination of cubic B-spline basis functions and is given by

$$y_i = \sum_{j=1}^{m} w_j b_j(x_i;3) + \varepsilon_i, \qquad i = 1, 2, \cdots, n. \tag{3.18}$$

Figure 3.8 shows a curve fitting, in which a third-order B-spline regression model is fitted to a set of simulated data. With $u(x) = \exp\{-x\sin(2\pi x)\} + 0.5$ as the true structure, which is represented in the figure by the dotted line, we generated two-dimensional data using $y = u(x) + \varepsilon$ with Gaussian noise. The solid line represents the fitted curve. With good estimates of the model, it is then possible to capture the nonlinear structure of the data.

When applying splines in practical situations, we still need to determine the appropriate number and position of the knots. Moving the knots to various positions and then estimating them as parameters can result in extremely computational difficulties. One approach to this problem is to position the knots at equal intervals in the observed range of the data and refine the smoothness of the fitted curve by changing the number of knots. In the following section, we consider this problem within the framework of model selection.

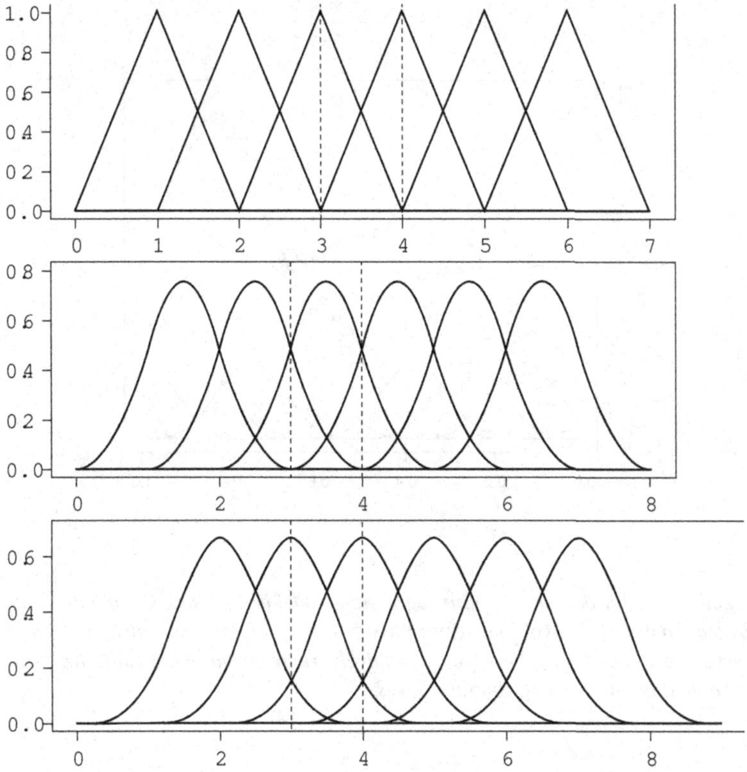

Figure 3.7 *Plots of the first-, second-, and third-order B-spline functions. As may be seen in the subintervals bounded by dotted lines, each subinterval is covered (piecewise) by the polynomial order plus one basis function.*

3.2.3 Radial Basis Functions

Here let us consider a model based on a set of n observations $\{(\boldsymbol{x}_i, y_i); i = 1, 2, \cdots, n\}$ for a p-dimensional vector of predictor variables $\boldsymbol{x} = (x_1, x_2, \cdots, x_p)^T$ and a response variable Y, where $\boldsymbol{x}_i = (x_{i1}, x_{i2}, \cdots, x_{ip})^T$. In general, a nonlinear function $\phi(z)$ depending on Euclidean distance $z = \|\boldsymbol{x} - \boldsymbol{\mu}\|$ between p-dimensional vector $\boldsymbol{x}$ and $\boldsymbol{\mu}$ is known as *a radial basis function*, and a regression model based on radial basis functions is given by

$$y_i = w_0 + \sum_{j=1}^{m} w_j \phi_j\left(\|\boldsymbol{x}_i - \boldsymbol{\mu}_j\|\right) + \varepsilon_i, \qquad i = 1, 2, \cdots, n, \quad (3.19)$$

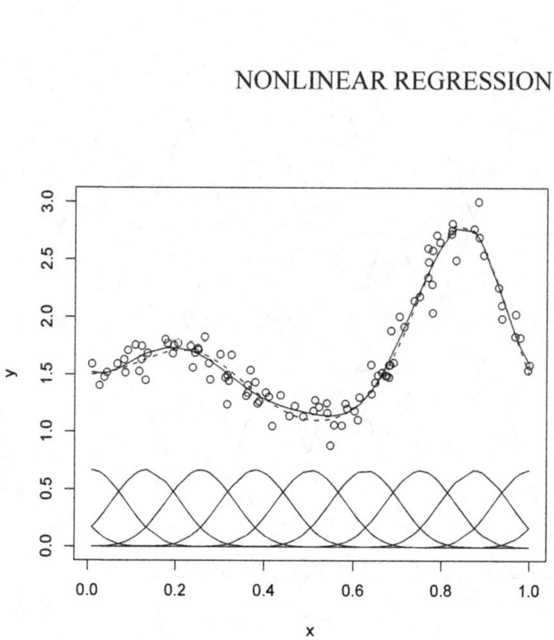

Figure 3.8 *A third-order B-spline regression model is fitted to a set of data, generated from* $u(x) = \exp\{-x\sin(2\pi x)\} + 0.5 + \varepsilon$ *with Gaussian noise. The fitted curve and the true structure are, respectively, represented by the solid line and the dotted line with cubic B-spline bases.*

(Bishop, 1995, Chapter 5; Ripley, 1996, Section 4.2), where μ_j is a p-dimensional vector of center that determines the position of the basis function. The function often employed in practice is the *Gaussian basis function* given by

$$\phi_j(\boldsymbol{x}) \equiv \exp\left(-\frac{||\boldsymbol{x}-\boldsymbol{\mu}_j||^2}{2h_j^2}\right), \qquad j = 1, 2, ..., m, \tag{3.20}$$

where the quantity h_j^2 represents the spread of the function and, together with the number of basis functions, plays the role of a parameter that controls the smoothness of the fitted model. Other commonly used nonlinear functions include the thin plate spline function $\phi(z) = z^2 \log z$ and the inverse polynomial function $\phi(z) = (z^2 + h^2)^{-\gamma}$ $(\gamma > 0)$.

The unknown parameters included in the nonlinear regression model with Gaussian basis functions are $\{\boldsymbol{\mu}_1, \cdots, \boldsymbol{\mu}_m, h_1^2, \cdots, h_m^2\}$ in addition to the coefficients $\{w_0, w_1, \cdots, w_m\}$. A method of estimating all of these parameters simultaneously might be considered, but this would lead to questions concerning the uniqueness of the estimates and localization of solutions in numerical optimization, which, along with the selection of the number of basis functions, would require an extremely large computation time. In practice, a two-step estimation method is used as an effective tool of avoiding these problems: first the basis functions are determined from the data on the predictor variables and then a model having the known basis functions is fit to the data (Moody and Darken, 1989; Kawano and Konishi, 2007; Ando et al., 2008).

One such method employs clustering as a tool of determining the basis functions, for which a technique such as k-means clustering or self-organizing mapping described in Chapter 10 may be used to partition the data $\{\boldsymbol{x}_1, \boldsymbol{x}_2, \cdots, \boldsymbol{x}_n\}$ on p predictor variables into m clusters $C_1, C_2, \cdots, C_m$ that correspond to the number of basis functions. The centers $\boldsymbol{\mu}_j$ and width parameters h_j^2 are then determined by

$$\hat{\boldsymbol{\mu}}_j = \frac{1}{n_j}\sum_{\boldsymbol{x}_i \in C_j} \boldsymbol{x}_i, \qquad \hat{h}_j^2 = \frac{1}{n_j}\sum_{\boldsymbol{x}_i \in C_j} \|\boldsymbol{x}_i - \hat{\boldsymbol{\mu}}_j\|^2, \tag{3.21}$$

where n_j is the number of the observations that belong to the j-th cluster C_j. Substituting these estimates into the Gaussian basis function (3.20) gives us a set of m basis functions

$$\hat{\phi}_j(\boldsymbol{x}) = \exp\left(-\frac{\|\boldsymbol{x} - \hat{\boldsymbol{\mu}}_j\|^2}{2\hat{h}_j^2}\right), \qquad j = 1, 2, ..., m. \tag{3.22}$$

The nonlinear regression model based on the Gaussian basis functions is then given by

$$y_i = w_0 + \sum_{j=1}^{m} w_j \hat{\phi}_j(\boldsymbol{x}_i) + \varepsilon_i, \qquad i = 1, 2, \cdots, n. \tag{3.23}$$

As described in the next section, the key advantage of advance determination of the basis functions from the data on the predictor variables is that it facilitates estimation of the nonlinear regression model.

3.3 Basis Expansions

This section provides a unified approach to the modeling process (model specification, estimation, and evaluation) for linear regression, polynomial regression, spline and B-spline regression, radial basis functions, and other such procedures by incorporating a regression model based on basis functions.

3.3.1 Basis Function Expansions

For the p-dimensional predictor vector $\boldsymbol{x}$ and the response variable Y, as the regression functions approximating the true structure $E[Y|\boldsymbol{x}] = u(\boldsymbol{x})$, let us assume a linear combination of known functions $b_0(\boldsymbol{x}) \equiv 1$, $b_1(\boldsymbol{x})$, $b_2(\boldsymbol{x}), \cdots, b_m(\boldsymbol{x})$ called basis functions as follows:

$$u(\boldsymbol{x}; \boldsymbol{w}) = \sum_{j=0}^{m} w_j b_j(\boldsymbol{x}), \tag{3.24}$$

where $\boldsymbol{w} = (w_0, w_1, w_2, \cdots, w_m)^T$ is an $(m+1)$-dimensional unknown parameter vector (weight vector).

For example, we can express the linear regression model for p predictor variables $\{x_1, x_2, \cdots, x_p\}$, setting $b_j(\boldsymbol{x}) = x_j$ $(j = 1, 2, \cdots, p)$, as $\sum_{j=0}^{p} w_j b_j(\boldsymbol{x}) = w_0 + \sum_{j=1}^{p} w_j x_j$. We can also express the m-th order polynomial model of predictor variable x, setting $b_j(x) = x^j$, as $\sum_{j=0}^{m} w_j b_j(x) = w_0 + \sum_{j=1}^{m} w_j x^j$. The natural cubic spline given by (3.11) is expressed by the linear combination of the basis functions $1, x$, $d_1(x) - d_{m-1}(x), \cdots, d_{m-2}(x) - d_{m-1}(x)$, and all of the models based on the B-spline in (3.18) and the Gaussian basis function in (3.23) can be unified by (3.24), where, if no intercept is necessary, we set $b_0(\boldsymbol{x}) = 0$.

3.3.2 Model Estimation

Let us now consider fitting the regression model based on the basis expansion method to the observed data $\{(\boldsymbol{x}_i, y_i);\ i = 1, 2, \cdots, n\}$ given in the form

$$y_i = \sum_{j=0}^{m} w_j b_j(\boldsymbol{x}_i) + \varepsilon_i, \qquad i = 1, 2, \cdots, n, \tag{3.25}$$

where the error terms ε_i are mutually uncorrelated, and $E[\varepsilon_i] = 0$, $E[\varepsilon_i^2] = \sigma^2$. The n equations yielding this regression model can be written as follows:

$$\begin{pmatrix} y_1 \\ y_2 \\ \vdots \\ y_n \end{pmatrix} = \begin{bmatrix} 1 & b_1(\boldsymbol{x}_1) & b_2(\boldsymbol{x}_1) & \cdots & b_m(\boldsymbol{x}_1) \\ 1 & b_1(\boldsymbol{x}_2) & b_2(\boldsymbol{x}_2) & \cdots & b_m(\boldsymbol{x}_2) \\ \vdots & \vdots & \vdots & \ddots & \vdots \\ 1 & b_1(\boldsymbol{x}_n) & b_2(\boldsymbol{x}_n) & \cdots & b_m(\boldsymbol{x}_n) \end{bmatrix} \begin{pmatrix} w_0 \\ w_1 \\ w_2 \\ \vdots \\ w_m \end{pmatrix} + \begin{pmatrix} \varepsilon_1 \\ \varepsilon_2 \\ \vdots \\ \varepsilon_n \end{pmatrix}.$$

In vector and matrix notation, this equation is expressed as

$$\boldsymbol{y} = B\boldsymbol{w} + \boldsymbol{\varepsilon}, \tag{3.26}$$

where $\boldsymbol{y}$ is the n-dimensional observed values vector $\boldsymbol{y} = (y_1, y_2, \cdots, y_n)^T$, comprising the n data for the response variable Y and B is the $n \times (m+1)$ basis function matrix

$$B = \begin{bmatrix} 1 & b_1(\boldsymbol{x}_1) & b_2(\boldsymbol{x}_1) & \cdots & b_m(\boldsymbol{x}_1) \\ 1 & b_1(\boldsymbol{x}_2) & b_2(\boldsymbol{x}_2) & \cdots & b_m(\boldsymbol{x}_2) \\ \vdots & \vdots & \vdots & \ddots & \vdots \\ 1 & b_1(\boldsymbol{x}_n) & b_2(\boldsymbol{x}_n) & \cdots & b_m(\boldsymbol{x}_n) \end{bmatrix}, \tag{3.27}$$

$\boldsymbol{w} = (w_0, w_1, w_2, \cdots, w_m)^T$ is the $(m+1)$-dimensional weight vector and $\boldsymbol{\varepsilon} = (\varepsilon_1, \varepsilon_2, \cdots, \varepsilon_p)^T$ is the n-dimensional error vector.

(1) Least squares Comparison between the regression model based on the basis expansion in (3.26) and the linear regression model $\boldsymbol{y} = X\boldsymbol{\beta} + \boldsymbol{\varepsilon}$ in (2.15) reveals that the basis function matrix B corresponds to the design matrix X and the weight vector $\boldsymbol{w}$ corresponds to the regression coefficient vector $\boldsymbol{\beta}$. Accordingly, the least squares estimator of the weight vector $\boldsymbol{w}$ is given by minimizing the sum of squared errors

$$S(\boldsymbol{w}) = \boldsymbol{\varepsilon}^T \boldsymbol{\varepsilon} = (\boldsymbol{y} - B\boldsymbol{w})^T (\boldsymbol{y} - B\boldsymbol{w}). \tag{3.28}$$

Differentiating $S(\boldsymbol{w})$ with respect to $\boldsymbol{w}$ and setting this derivative equal to 0, we have

$$\hat{\boldsymbol{w}} = (B^T B)^{-1} B^T \boldsymbol{y}. \tag{3.29}$$

This yields the regression equation $y = \sum_{j=0}^{m} \hat{w}_j b_j(\boldsymbol{x})$ based on the basis function. The predicted value vector $\hat{\boldsymbol{y}}$ and the residual vector $\boldsymbol{e}$ are then given by

$$\begin{aligned} \hat{\boldsymbol{y}} &= B\hat{\boldsymbol{w}} = B(B^T B)^{-1} B^T \boldsymbol{y}, \\ \boldsymbol{e} &= \boldsymbol{y} - \hat{\boldsymbol{y}} = [I_n - B(B^T B)^{-1} B^T] \boldsymbol{y}. \end{aligned} \tag{3.30}$$

It may be seen that linear regression, polynomial regression, spline, and other models can therefore be estimated simply by finding the basis function matrix B from the basis functions and thus obtaining the estimated curve or estimated surface. Figure 3.9 shows the data fit of a nonlinear regression model based on a natural cubic spline basis function and a Gaussian basis function.

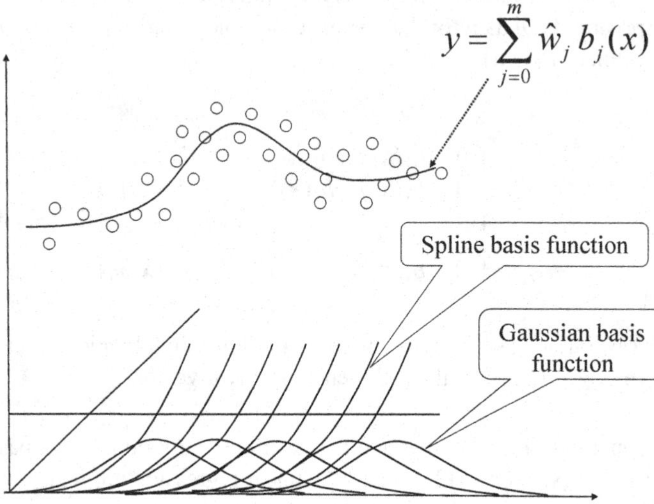

Figure 3.9 *Curve fitting; a nonlinear regression model based on a natural cubic spline basis function and a Gaussian basis function.*

(2) Maximum likelihood methods We next consider fitting a regression model based on basis expansion

$$y_i = \sum_{j=0}^{m} w_j b_j(\boldsymbol{x}_i) + \varepsilon_i, \qquad i = 1, 2, \cdots, n, \tag{3.31}$$

where ε_i $(i = 1, 2, \cdots, n)$ are independently distributed according to $N(0, \sigma^2)$, and thus assuming the *Gaussian nonlinear regression model*

$$\boldsymbol{y} = B\boldsymbol{w} + \boldsymbol{\varepsilon}, \qquad \boldsymbol{\varepsilon} \sim N_n(\boldsymbol{0}, \sigma^2 I_n), \tag{3.32}$$

where B is an $n \times (m + 1)$ basis function matrix defined by (3.27).

Since the observed values vector $\boldsymbol{y}$ follows an n-dimensional normal distribution with mean vector $B\boldsymbol{w}$ and variance-covariance matrix $\sigma^2 I_n$, its probability density function is given by

$$f(\boldsymbol{y}|X; \boldsymbol{w}, \sigma^2) = \frac{1}{(2\pi\sigma^2)^{n/2}} \exp\left\{-\frac{1}{2\sigma^2}(\boldsymbol{y} - B\boldsymbol{w})^T(\boldsymbol{y} - B\boldsymbol{w})\right\}, \tag{3.33}$$

where X is the n p-dimensional data concerning the predictor variables.

The log-likelihood function is then

$$\begin{aligned} \ell(\boldsymbol{w}, \sigma^2) &= \log f(\boldsymbol{y}|X; \boldsymbol{w}, \sigma^2) \\ &= -\frac{n}{2}\log(2\pi\sigma^2) - \frac{1}{2\sigma^2}(\boldsymbol{y} - B\boldsymbol{w})^T(\boldsymbol{y} - B\boldsymbol{w}) \end{aligned} \tag{3.34}$$

By differentiating $\ell(\boldsymbol{w}, \sigma^2)$ with respect to $\boldsymbol{w}$ and σ^2 and setting the result equal to zero, we have the likelihood equations

$$\begin{aligned} \frac{\partial \ell(\boldsymbol{w}, \sigma^2)}{\partial \boldsymbol{w}} &= \frac{1}{\sigma^2}\left(B^T\boldsymbol{y} - B^T B\boldsymbol{w}\right) = \boldsymbol{0}, \\ \frac{\partial \ell(\boldsymbol{w}, \sigma^2)}{\partial \sigma^2} &= -\frac{n}{2\sigma^2} + \frac{1}{2\sigma^4}(\boldsymbol{y} - B\boldsymbol{w})^T(\boldsymbol{y} - B\boldsymbol{w}) = 0. \end{aligned} \tag{3.35}$$

Then the maximum likelihood estimators of regression coefficient vector $\boldsymbol{w}$ and the error variance σ^2 are respectively given by

$$\hat{\boldsymbol{w}} = (B^T B)^{-1} B^T \boldsymbol{y}, \qquad \hat{\sigma}^2 = \frac{1}{n}(\boldsymbol{y} - B\hat{\boldsymbol{w}})^T(\boldsymbol{y} - B\hat{\boldsymbol{w}}). \tag{3.36}$$

We thus find that the maximum likelihood estimator of the weight vector of the Gaussian nonlinear regression model in (3.32) equals the least squares estimator.

(3) Other models In the nonlinear regression models we have considered the problem of fitting a smooth curve to data scattered on a two-dimensional plane or fitting a surface to data scattered in a space. We now consider a regression model that assumes a different nonlinear model $u_j(x_j; \boldsymbol{w}_j)$ for each of the variables x_j $(j = 1, 2, \cdots, p)$, and it is given by

$$y = u_1(x_1; \boldsymbol{w}_1) + u_2(x_2; \boldsymbol{w}_2) + \cdots + u_p(x_p; \boldsymbol{w}_p) + \varepsilon, \tag{3.37}$$

known as an *additive model*. We represent here $u_j(x_j; \boldsymbol{w}_j)$ by basis function expansion

$$u_j(x_j; \boldsymbol{w}_j) = \sum_{k=1}^{m_j} w_k^{(j)} b_k^{(j)}(x_j) = \boldsymbol{w}_j^T \boldsymbol{b}_j(x_j), \quad j = 1, \cdots, p, \tag{3.38}$$

where $\boldsymbol{w}_j = (w_1^{(j)}, w_2^{(j)}, \cdots, w_{m_j}^{(j)})^T$ and $\boldsymbol{b}_j(x_j) = (b_1^{(j)}(x_j), b_2^{(j)}(x_j), \cdots, b_{m_j}^{(j)}(x_j))^T$. Thus the additive model based on observed data $\{(y_i, x_{i1}, x_{i2},$

$\cdots, x_{ip}); i = 1, 2, \cdots, n\}$ for the response variable and p predictor variables may be written as

$$y_i = \boldsymbol{w}_1^T \boldsymbol{b}_1(x_{i1}) + \boldsymbol{w}_2^T \boldsymbol{b}_2(x_{i2}) \cdots + \boldsymbol{w}_p^T \boldsymbol{b}_p(x_{ip}) + \varepsilon_i, \quad i = 1, \cdots, n. \quad (3.39)$$

Just as in models based on the basis expansion method, the model can then be estimated either by least squares or by maximum likelihood.

Let us assume that the relationship between the response variable and the predictor variables can be explicated by a linear regression model if certain of the predictor variables are excluded. In many cases, observed data are nonlinear with respect to the time axis, and this therefore corresponds to the assumption of a model incorporating both predictor variables having a nonlinear structure and predictor variables having a linear structure. This type of model is known as a *semiparametric model* and is given by

$$y_i = \beta_0 + \beta_1 x_{i1} + \beta_2 x_{i2} + \cdots + \beta_{p-1} x_{i,p-1} + u_p(x_{ip}; \boldsymbol{w}_p) + \varepsilon_i \quad (3.40)$$

for $i = 1, 2, \cdots, n$, in which we assume a nonlinear structure $u_p(x_p; \boldsymbol{w}_p)$ for variable x_p. It may be seen that if we adopt a model based on basis function expansion for the presumably nonlinear variable x_p, then the model estimation can in the same way be performed by least squares or maximum likelihood. For details on additive and semiparametric models, see, for example, Hastie and Tibshirani (1990), Green and Silverman (1994), and Ruppert et al. (2003).

3.3.3 *Model Evaluation and Selection*

Models based on basis expansion, which unifies the expression of various regression models, all involve the problem of choosing an optimal model, in regard to variables in linear regression models, polynomial degree in polynomial regression models, and number of basis functions in spline and other nonlinear regression models. We shall regard this problem as a model selection and consider it in the context of evaluation and selection of nonlinear regression models.

Figure 3.10 shows cubic *B*-spline nonlinear regression models, each with a different number of basis functions, fitted to the data from the motorcycle crash experiment that is analyzed in Example 3.1 at the end of this subsection. The curves in this figure show the fit of cubic B-splines obtained with (a) 10, (b) 20, (c) 30, and (d) 40 basis functions.

Figure 3.10 illustrates the following important aspect of modeling with basis functions. It might be thought that simply increasing the number of basis functions is an effective means of fitting a model with many

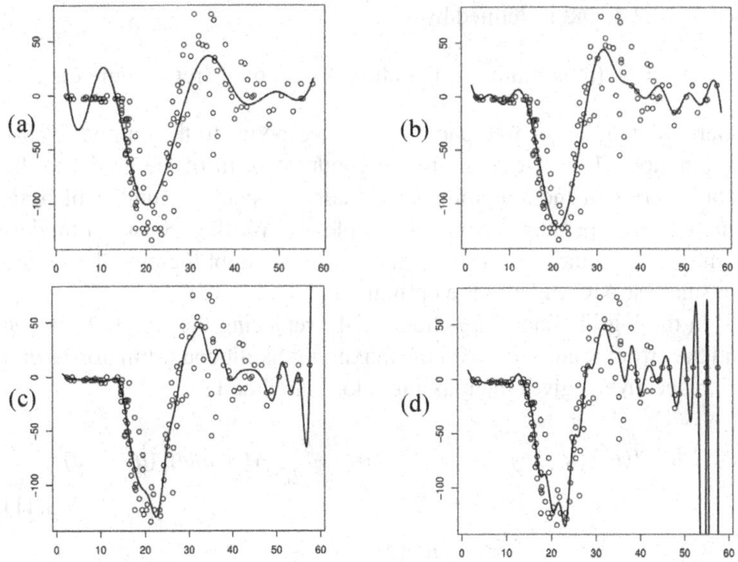

Figure 3.10 *Cubic B-spline nonlinear regression models, each with a different number of basis functions (a) 10, (b) 20, (c) 30, (d) 40, fitted to the motorcycle crash experiment data.*

parameters and thus explicating complex nonlinear structures in the analysis of phenomena. As the number of basis functions is increased, the model (the curve or surface) estimated by least squares or maximum likelihood draws closer to the data. The sum of squared residuals $n\hat{\sigma}^2 = (\boldsymbol{y} - \hat{\boldsymbol{y}})^T(\boldsymbol{y} - \hat{\boldsymbol{y}})$ accordingly moves toward zero, thus apparently improving the fit of the model to the data. As may be seen from this figure, however, with the very close approach of the curve to the data, the curve tends to undergo sharp local fluctuations, rendering the model ineffective for prediction. The variance of the parameter estimates also increases. The estimated model therefore lacks reliability and cannot be regarded as a good model for effective predictions. On the other hand, as may also be seen from the figure, an insufficient number of basis functions tends to result in a failure to properly elucidate the structure of the phenomenon.

In this light, it may therefore be useful to consider the choice of the number of basis functions as a model selection problem, using the criteria AIC and BIC for model evaluation and selection. It may be recalled that the AIC was used for variable selection in linear regression models in

Section 2.2.4, and is defined by

$$\text{AIC} = -2(\text{maximum log-likelihood}) + 2(\text{no. of free parameters}),$$

where the number of free parameters corresponds to the number of basis functions. Here we measure the goodness of fit of the model to the data in terms of the maximum log-likelihood, and the number of basis functions as a penalty for model complexity. We thus construct models with different numbers of basis functions and adopt the model yielding the smallest AIC value as the optimum model.

In the log-likelihood function (3.34), replacing the weight vector $\hat{\boldsymbol{w}}$ and the error variance $\hat{\sigma}^2$ with the maximum likelihood estimators $\hat{\boldsymbol{w}}$ and $\hat{\sigma}^2$, respectively, gives the maximum log-likelihood

$$\begin{aligned}\log f(\boldsymbol{y}|X;\hat{\boldsymbol{w}},\hat{\sigma}^2) &= -\frac{n}{2}\log(2\pi\hat{\sigma}^2) - \frac{1}{2\hat{\sigma}^2}(\boldsymbol{y} - B\hat{\boldsymbol{w}})^T(\boldsymbol{y} - B\hat{\boldsymbol{w}}) \\ &= -\frac{n}{2}\log(2\pi\hat{\sigma}^2) - \frac{n}{2},\end{aligned} \tag{3.41}$$

the first term in AIC for the Gaussian nonlinear regression model. From this equation, it may be seen that the maximum log-likelihood increases as the sum of squared residuals $n\hat{\sigma}^2$ decreases. The number of free parameters in the model is in total $(m + 2)$, since it includes the number of basis functions m together with the error variance for the model and the added intercept w_0. The AIC for the Gaussian nonlinear regression model is generally given by

$$\begin{aligned}\text{AIC} &= -2\left\{\log f(\boldsymbol{y}|X;\hat{\boldsymbol{w}},\hat{\sigma}^2) - (m+2)\right\} \\ &= n\log(2\pi\hat{\sigma}^2) + n + 2(m+2).\end{aligned} \tag{3.42}$$

Among the models having different numbers of basis functions, the one yielding the smallest AIC value is selected as the optimum model. Chapter 5 will provide a more detailed description of the various model evaluation and selection criteria, including the AIC, and their derivation.

Example 3.1 (*B*-spline regression model fitting) The motorcycle crash experiment data shown in Figure 3.11 (see also Figure 3.2) represent the measured acceleration y (in terms of g, the acceleration due to gravity) of a crash-dummy's head at time x (in milliseconds) from the instant of impact. We fitted a cubic B-spline nonlinear regression model in (3.18) to

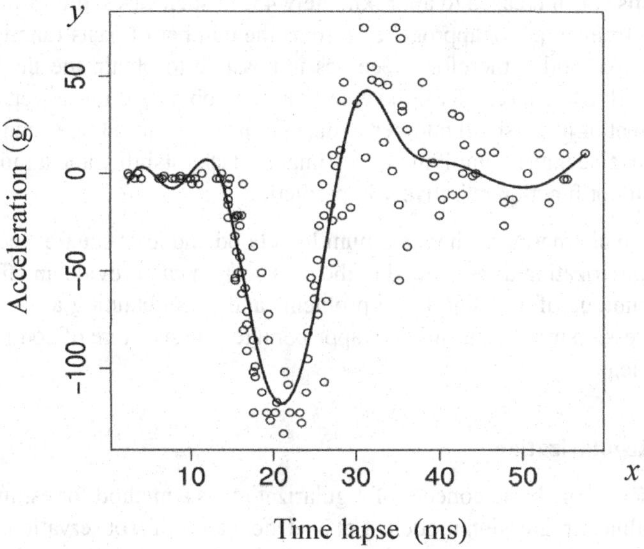

Figure 3.11 *The cubic B-spline nonlinear regression model* $y = \sum_{j=1}^{13} \hat{w}_j b_j(x)$. *The model is estimated by maximum likelihood and selected the number of basis functions by* AIC.

these data. We estimated the model by maximum likelihood and selected the number of basis functions by minimization of AIC in (3.42). The number of basis functions selected on this basis was 13. The estimated regression curve $y = \sum_{j=1}^{13} \hat{w}_j b_j(x)$ is shown in the figure.

As may be seen from the figure, the fit of this nonlinear regression model to the data is rather poor in some regions, and the model might therefore be regarded as falling short of an acceptable explication of the structure of the phenomenon under investigation. The number of parameters in AIC represents the model complexity. As suggested by this figure, a limit exists in the fitting of a model to data that represent a complex structure in regard to the degree of explication of model complexity depending on the number of basis functions. In some cases, increasing the number of basis functions may lead to an inability to calculate the inverse matrix $(B^T B)^{-1}$ in $\hat{\boldsymbol{w}} = (B^T B)^{-1} B^T \boldsymbol{y}$, which is the maximum likelihood estimator for the weight vector in (3.36).

This actually occurred in fitting to the data shown in the figure, as it became impossible to perform the calculation when the number of basis functions was increased to approximately 45. In such cases, the value of the determinant $|B^T B|$ approaches zero as the number of basis functions is increased and it therefore becomes impossible to obtain the inverse matrix $(B^T B)^{-1}$. Even if the inverse matrix is obtained, moreover, replacement of just a small number of data points may cause large changes in the inverse matrix, implying the estimate $\hat{\boldsymbol{w}}$ lacks stability and a model that will not function effectively in prediction.

As an alternative to the maximum likelihood and least squares methods, regularization as described in the next section can provide an effective technique of avoiding these problems and thus obtaining a nonlinear regression model that suitably approximates the structure of complex phenomena.

3.4 Regularization

We describe the basic concept of regularization as a method for estimating nonlinear regression models. Let us denote a set of n observations of p-dimensional vector $\boldsymbol{x}$ of predictor variables and response variable y as $\{(\boldsymbol{x}_i, y_i);\ i = 1, 2, \cdots, n\}$, where $\boldsymbol{x}_i = (x_{i1}, x_{i2}, \cdots, x_{ip})^T$, and attempt to fit a nonlinear regression model based on basis functions

$$y_i = \sum_{j=1}^{m} w_j b_j(\boldsymbol{x}_i) + \varepsilon_i \qquad i = 1, 2, \cdots, n. \tag{3.43}$$

In vector and matrix notation, the model is expressed as

$$\boldsymbol{y} = B\boldsymbol{w} + \boldsymbol{\varepsilon}, \tag{3.44}$$

where $\boldsymbol{y}$ is an n-dimensional observed values vector, $\boldsymbol{w}$ is an m-dimensional parameter vector, and B is an $n \times m$ basis function matrix defined by

$$B = \begin{bmatrix} b_1(\boldsymbol{x}_1) & b_2(\boldsymbol{x}_1) & \cdots & b_m(\boldsymbol{x}_1) \\ b_1(\boldsymbol{x}_2) & b_2(\boldsymbol{x}_2) & \cdots & b_m(\boldsymbol{x}_2) \\ \vdots & \vdots & \ddots & \vdots \\ b_1(\boldsymbol{x}_n) & b_2(\boldsymbol{x}_n) & \cdots & b_m(\boldsymbol{x}_n) \end{bmatrix}. \tag{3.45}$$

In estimating m-dimensional parameter vector (weight vector) $\boldsymbol{w}$, which represents the basis function coefficients, we shall describe the estimation methods including the additional constraints.

3.4.1 Regularized Least Squares

We assume that the error terms ε_i $(i = 1, 2, \cdots, n)$ in the nonlinear regression model (3.43) are mutually uncorrelated, and that $E[\varepsilon_i] = 0$, $E[\varepsilon_i^2] = \sigma^2$. The least squares estimator for the model parameter is given by $\hat{\boldsymbol{w}} = (B^T B)^{-1} B^T \boldsymbol{y}$ as shown in (3.29) of Section 3.3.

Increasing the number of basis functions, and thus the number of parameters, may induce over-fitting of the model to the data and instability in the estimate and in some cases make it impossible to calculate the inverse matrix $(B^T B)^{-1}$ in the least squares estimator $\hat{\boldsymbol{w}}$. In such cases, by imposing the penalty $R(\boldsymbol{w})$, which increases with increasing complexity, on the term representing the decreasing sum of squared errors as the model complexity increases, we attempt to minimize the function

$$S_\gamma(\boldsymbol{w}) = (\boldsymbol{y} - B\boldsymbol{w})^T(\boldsymbol{y} - B\boldsymbol{w}) + \gamma R(\boldsymbol{w}). \tag{3.46}$$

The γ (> 0) is known as the *regularization parameter* or *smoothing parameter*, and together with adjusting the fit of the model and the smoothness of its curve, it serves the role of contributing to stabilization of the estimator. The penalty term $R(\boldsymbol{w})$ is also known as the *regularization term*, and this estimation method is known as the *regularized least squares method* or *penalized least squares method.*

Figure 3.12 illustrates a role of the penalty term $\gamma R(\boldsymbol{w})$, with model complexity (e.g., degree of polynomial or number of basis functions) increasing along the x-axis. It shows the decrease in the first term $S_e(\boldsymbol{w})$ and the increase in the second term (the penalty term) $R(\boldsymbol{w})$ with two different weights $\gamma_1 < \gamma_2$ that occur with decreasing model complexity, together with the curve representing the sum of the first and second terms with each weight, thus representing the two equations

$$S_{\gamma_1}(\boldsymbol{w}) = S_e(\boldsymbol{w}) + \gamma_1 R(\boldsymbol{w}), \qquad S_{\gamma_2}(\boldsymbol{w}) = S_e(\boldsymbol{w}) + \gamma_2 R(\boldsymbol{w}). \tag{3.47}$$

As shown in this figure, changing the weight in the second term by the regularization parameter γ changes $S_\gamma(\boldsymbol{w})$ continuously, thus enabling continuous adjustment of the model complexity.

In many cases, the penalty term $R(\boldsymbol{w})$ can be expressed as the quadratic form $\boldsymbol{w}^T K \boldsymbol{w}$ of parameter vector $\boldsymbol{w}$, in which K is an $m \times m$ non-negative definite matrix, and an identity matrix or second-order difference matrix is then used as follows:

$$K = I_m, \quad K = D_2^T D_2 \text{ with } D_2 = \begin{bmatrix} 1 & -2 & 1 & 0 & \cdots & 0 \\ 0 & 1 & -2 & 1 & \ddots & \vdots \\ \vdots & \ddots & \ddots & \ddots & \ddots & 0 \\ 0 & \cdots & 0 & 1 & -2 & 1 \end{bmatrix}. \tag{3.48}$$

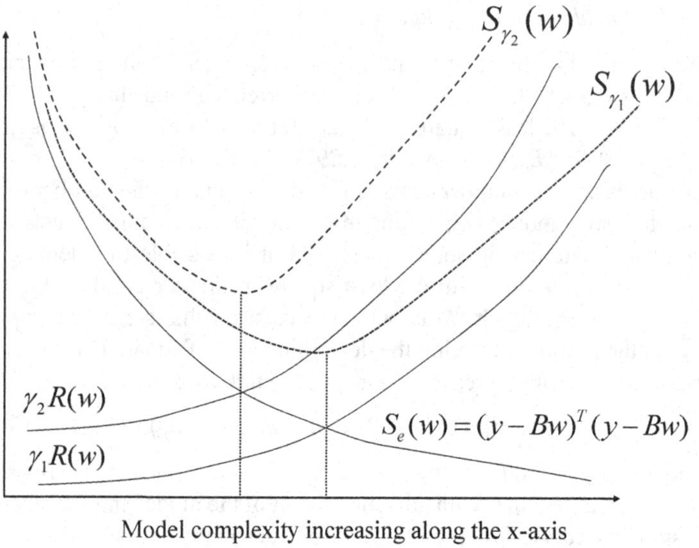

Figure 3.12 *The role of the penalty term: Changing the weight in the second term by the regularization parameter γ changes $S_\gamma(\boldsymbol{w})$ continuously, thus enabling continuous adjustment of the model complexity.*

With the penalty term thus represented by a quadratic form, the regularized least squares estimator of $\boldsymbol{w}$ is given as the solution that minimizes

$$S_\gamma(\boldsymbol{w}) = (\boldsymbol{y} - B\boldsymbol{w})^T(\boldsymbol{y} - B\boldsymbol{w}) + \gamma \boldsymbol{w}^T K \boldsymbol{w}. \tag{3.49}$$

By differentiating each side of (3.49) with respect to the parameter vector $\boldsymbol{w}$ and setting the derivative equal to 0, we have

$$\frac{\partial S_\gamma(\boldsymbol{w})}{\partial \boldsymbol{w}} = -2B^T\boldsymbol{y} + 2B^T B\boldsymbol{w} + 2\gamma K\boldsymbol{w} = \boldsymbol{0}. \tag{3.50}$$

Solving the resulting equation, the *regularized least squares estimator* of the parameter vector $\boldsymbol{w}$ is then given by

$$\hat{\boldsymbol{w}} = (B^T B + \gamma K)^{-1} B^T \boldsymbol{y}. \tag{3.51}$$

In particular, if we let $K = I_m$, the m-dimensional identity matrix, we have $\hat{\boldsymbol{w}} = (B^T B + \gamma I_m)^{-1} B^T \boldsymbol{y}$ and thus obtain the ridge estimator described in Section 2.3.1 for multicollinearity in linear regression models.

Even in cases where the inverse matrix $(B^T B)^{-1}$ in the least squares estimator cannot be calculated, the solution can be obtained by applying the regularized least squares method. With an appropriate value for the regularization parameter, this also contributes to stabilization of the estimator. For example, if for simplicity we let $B^T B$ be a 2×2 matrix having 1 as each of its components, the value of its determinant becomes 0 and an inverse matrix cannot be obtained. But if we add γ (> 0) to the 1 of each diagonal element, the value of the determinant is then $(1 + \gamma)^2 - 1$ and it becomes possible to calculate the inverse matrix.

3.4.2 Regularized Maximum Likelihood Method

Suppose that the random errors ε_i ($i = 1, 2, \cdots, n$) in the nonlinear regression model (3.43) are independently and identically distributed according to $N(0, \sigma^2)$. The log-likelihood function of the nonlinear regression model with Gaussian noise is then given by

$$\begin{aligned} \ell(\boldsymbol{\theta}) &= \sum_{i=1}^{n} \log f(y_i|\boldsymbol{x}_i; \boldsymbol{w}, \sigma^2) \\ &= -\frac{n}{2}\log(2\pi\sigma^2) - \frac{1}{2\sigma^2}(\boldsymbol{y} - B\boldsymbol{w})^T(\boldsymbol{y} - B\boldsymbol{w}), \end{aligned} \tag{3.52}$$

where $\boldsymbol{\theta} = (\boldsymbol{w}^T, \sigma^2)^T$. The maximum likelihood estimators of $\boldsymbol{w}$ and σ^2 obtained by maximizing this log-likelihood function are $\hat{\boldsymbol{w}} = (B^T B)^{-1} B^T \boldsymbol{y}$ and $\hat{\sigma}^2 = (\boldsymbol{y} - B\hat{\boldsymbol{w}})^T(\boldsymbol{y} - B\hat{\boldsymbol{w}})/n$. By substituting these estimators into the log-likelihood function (3.52), we obtain the maximum log-likelihood

$$\ell(\hat{\boldsymbol{\theta}}) = -\frac{n}{2}\log(2\pi) - \frac{n}{2}\log\hat{\sigma}^2 - \frac{n}{2}. \tag{3.53}$$

In modeling a phenomenon involving a complex nonlinear structure, as described in the previous subsection, it is necessary to fit a model having a large number of parameters, but in model estimation by maximum likelihood, the log-likelihood function increases as the model approximating the structure nears the data. This occurs because the term $-\log\hat{\sigma}^2$ in the maximum log-likelihood (3.53) becomes larger, since the maximum likelihood estimate of the variance corresponds to the sum of squared residuals, which decreases as the model becomes more complex.

In Section 3.3.3, we attempted to resolve this problem by adjusting the number of parameters of the model through selection of the number of basis functions based on minimization of AIC, as a method of controlling the variation in the curve (or surface). When the structure of the phenomenon generating the data is complex, however, it may not be possible to sufficiently approximate the structure in this way. We therefore estimate the parameters by maximizing the following objective function imposed the penalty term $R(\boldsymbol{w})$ on the log-likelihood function

$$\ell_\lambda(\boldsymbol{\theta}) = \sum_{i=1}^{n} \log f(y_i|\boldsymbol{x}_i; \boldsymbol{\theta}) - \frac{\lambda}{2} R(\boldsymbol{w}), \tag{3.54}$$

where λ (> 0) is a regularization or smoothing parameter that serves the role of adjusting the model fit and smoothing the curve. This function is known as the *regularized log-likelihood function* or *penalized log-likelihood function,* and the method of parameter estimation by its maximization is known as the *regularized maximum likelihood method* or *penalized maximum likelihood method.*

For a nonlinear regression model with Gaussian noise, the regularized log-likelihood function with a regularization term expressed in quadratic form is given by

$$\ell_\lambda(\boldsymbol{\theta}) = -\frac{n}{2}\log(2\pi\sigma^2) - \frac{1}{2\sigma^2}(\boldsymbol{y} - B\boldsymbol{w})^T(\boldsymbol{y} - B\boldsymbol{w}) - \frac{\lambda}{2}\boldsymbol{w}^T K\boldsymbol{w}. \tag{3.55}$$

By differentiating this equation with respect to the parameter vector $\boldsymbol{w}$ and the error variance σ^2, and setting the resulting derivatives equal to 0, we can then obtain the likelihood equations

$$\begin{aligned} \frac{\partial \ell_\lambda(\boldsymbol{\theta})}{\partial \boldsymbol{w}} &= \frac{1}{\sigma^2}(B^T\boldsymbol{y} - B^T B\boldsymbol{y}) - \lambda K\boldsymbol{w} = \boldsymbol{0}, \\ \frac{\partial \ell_\lambda(\boldsymbol{w}, \sigma^2)}{\partial \sigma^2} &= -\frac{n}{2\sigma^2} + \frac{1}{2\sigma^4}(\boldsymbol{y} - B\boldsymbol{w})^T(\boldsymbol{y} - B\boldsymbol{w}) = 0. \end{aligned} \tag{3.56}$$

Accordingly, the *regularized maximum likelihood estimators* for $\boldsymbol{w}$ and σ^2 are, respectively, given by

$$\hat{\boldsymbol{w}} = (B^T B + \lambda\hat{\sigma}^2 K)^{-1}B^T\boldsymbol{y}, \qquad \hat{\sigma}^2 = \frac{1}{n}(\boldsymbol{y} - \hat{\boldsymbol{y}})^T(\boldsymbol{y} - \hat{\boldsymbol{y}}), \tag{3.57}$$

where $\hat{\boldsymbol{y}} = B\hat{\boldsymbol{w}} = B(B^T B + \lambda\hat{\sigma}^2 K)^{-1}B^T\boldsymbol{y}$.

Because the estimator $\hat{\boldsymbol{w}}$ is dependent on the variance estimator $\hat{\sigma}^2$, in practice it is calculated as follows. We first let $\beta = \lambda\hat{\sigma}^2$ and obtain $\hat{\boldsymbol{w}} = (B^T B + \beta_0 K)^{-1}B^T\boldsymbol{y}$ for the given $\beta = \beta_0$. After obtaining the variance estimator $\hat{\sigma}^2$, we then calculate the value of the smoothing parameter as $\lambda = \beta_0/\hat{\sigma}^2$. The selection of the smoothing parameter will be described in the next section.

With the assumption of a normal distribution $N(0, \sigma^2)$ for the random error, the regularized maximum likelihood method for estimating the parameter vector $\boldsymbol{w}$ is equivalent to the regularized least squares method. The regularized log-likelihood function in (3.55) may actually be rewritten as

$$\ell_\lambda(\boldsymbol{\theta}) = -\frac{n}{2}\log(2\pi\sigma^2) - \frac{1}{2\sigma^2}\left\{(\boldsymbol{y} - B\boldsymbol{w})^T(\boldsymbol{y} - B\boldsymbol{w}) + \lambda\sigma^2\boldsymbol{w}^T K\boldsymbol{w}\right\}.$$

It may therefore be seen that, if we let $\lambda\sigma^2 = \gamma$, the maximization of the regularized log-likelihood function with Gaussian noise is equivalent to maximization of the regularized least squares $S_\gamma(\boldsymbol{w})$ in (3.49).

3.4.3 Model Evaluation and Selection

Model estimation by regularized least squares and regularized maximum likelihood with respect to each of the continuously changing smoothing parameters entails the construction of an infinite number of models. Appropriate smoothing parameter selection is therefore essential in the modeling process.

Figure 3.13 illustrates the effect of a smoothing parameter on the estimated curve in the case of model estimation by regularized maximum likelihood. As shown in the figure, the curve closely follows the data points when λ is small and tends toward a straight line when λ is large. This tendency is inherent to the composition of the regularized log-likelihood function $\ell_\lambda(\boldsymbol{\theta}) = \ell(\boldsymbol{\theta}) - \lambda R(\boldsymbol{w})$. When λ is quite small, $\ell(\boldsymbol{\theta})$ becomes the governing term in the log-likelihood function and the curve tends to pass near the data points even when this requires large variations in curvature. Conversely, increasing λ increases the value of the second term $\lambda R(\boldsymbol{w})$, which has the effect of reducing the value of $R(\boldsymbol{w})$, and the curve therefore tends to become linear.

It is for this reason that an increase in the regularization term $R(\boldsymbol{w})$ corresponds to an increase in the complexity of the model. A quadratic regularization term is used to capture model complexity based on the concept of curvature (the second derivative of a fitted function), thus utilizing the inherent tendency for curve linearization with diminishing curvature. For example, let $u(x)$ be a twice continuously differentiable function in the interval $[a, b]$. The complexity of a model $u(x)$ can be measured by the total curvature

$$R = \int_a^b \{u''(t)\}^2\, dt. \tag{3.58}$$

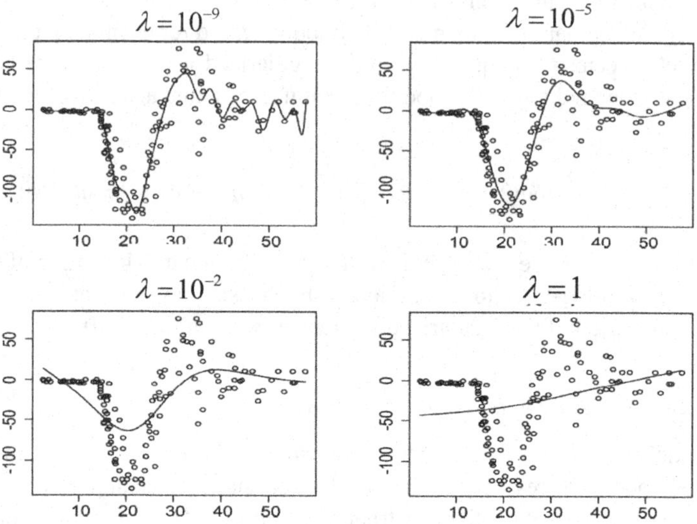

Figure 3.13 *The effect of a smoothing parameter λ: The curves are estimated by the regularized maximum likelihood method for various values of λ.*

The regularized least squares method is then given by

$$\sum_{i=1}^{n} \{y_i - u(x_i)\}^2 + \gamma \int_a^b \{u''(t)\}^2 \, dt. \tag{3.59}$$

The following methods are available, among others, for evaluation of models of this nature that are dependent on the regularization parameter λ or γ.

(1) Effective degrees of freedom In regression models based on B-splines and other basis expansions, the model complexity depends on the number of basis functions as well as on the smoothing parameter. The problem of selecting the parameter value for the desired smoothing may therefore be viewed as a model selection, and model evaluation criteria AIC or BIC may also be applied. These criteria cannot be directly applied, however, for models estimated by regularized maximum likelihood, because the assessment of the model complexity would not be based on the concept of the number of free parameters. In such cases, the smoothing parameter may instead be considered as a means of assessing the complexity in the following manner.

Let us consider the estimation of a nonlinear regression model $\boldsymbol{y} = B\boldsymbol{w} + \boldsymbol{\varepsilon}$ by regularized maximum likelihood using $R(\boldsymbol{w}) = \boldsymbol{w}^T K \boldsymbol{w}$. The regularized maximum likelihood estimate of the weight vector $\boldsymbol{w}$ is $\hat{\boldsymbol{w}} = (B^T B + \lambda\hat{\sigma}^2 K)^{-1} B^T \boldsymbol{y}$, and the predicted value vector is given by $\hat{\boldsymbol{y}} = B\hat{\boldsymbol{w}} = H(\lambda, m)\boldsymbol{y}$. Here, we define

$$H(\lambda, m) = B(B^T B + \lambda\hat{\sigma}^2 K)^{-1} B^T, \tag{3.60}$$

where B is an $n \times m$ basis function matrix defined by (3.45) and $\hat{\sigma}^2 = (\boldsymbol{y} - \hat{\boldsymbol{y}})^T(\boldsymbol{y} - \hat{\boldsymbol{y}})/n$. The $n \times n$ matrix $H(\lambda, m)$ is generally referred to as the *hat matrix* or *smoother matrix* for curve (or surface) estimation; H transforms the observed value vector $\boldsymbol{y}$ into the predicted value vector $\hat{\boldsymbol{y}}$.

As may be seen from (3.60), if $\lambda = 0$ the smoother matrix then becomes $\mathrm{tr}\{B(B^T B)^{-1} B^T\} = \mathrm{tr}\{(B^T B)^{-1} B^T B\} = m$, the number of basis functions and accordingly the number of free parameters. The complexity of a model controlled by the smoothing parameter is thus defined in terms of H as

$$\mathrm{tr}H(\lambda, m) = \mathrm{tr}\left\{B(B^T B + \lambda\hat{\sigma}^2 K)^{-1} B^T\right\}, \tag{3.61}$$

which is referred to as the *effective degrees of freedom* (Hastie and Tibshirani, 1990). By formally replacing the number of free parameters in AIC by the effective degrees of freedom, we thus obtain the criterion for evaluating a model estimated by regularized maximum likelihood as follows:

$$\mathrm{AIC_M} = n(\log 2\pi + 1) + n \log \hat{\sigma}^2 + 2\mathrm{tr}H(\lambda, m). \tag{3.62}$$

The model selection consists of finding the optimum model in terms of λ and m that minimizes the criterion $\mathrm{AIC_M}$.

The use of the trace of the hat matrix as the effective degrees of freedom has been investigated in smoothing methods (Wahba, 1990) and generalized additive models (Hastie and Tibshirani, 1990). Ye (1998) developed a concept of the effective degrees of freedom that is applicable to complex modeling procedures.

As described in Chapter 5, the AIC was developed for evaluating models estimated by maximum likelihood, and the use of the $\mathrm{AIC_M}$ for models estimated by regularized maximum likelihood therefore involves a problem in terms of its theoretical justification. Konishi and Kitagawa (1996) proposed the generalized information criterion (GIC) for evaluating models estimated by a broad range of methods, including maximum likelihood and regularized maximum likelihood. An extended version of the Bayesian information criterion (BIC) for the evaluation of models based on regularization has also been proposed by Konishi et al. (2004). These criteria are described in Konishi and Kitagawa (2008).

(2) Cross-validation Cross-validation may be used for the estimation of the prediction error (see Section 5.1.2), and applied in selecting the smoothing parameter and the number of basis functions in regularization. Cross-validation consists of evaluating the error by separating the observed data into two groups and using the data in one for model estimation and the data in the other for model evaluation, essentially as follows.

In general, let us represent a spline, B-spline, or other such model as $u(\boldsymbol{x}; \boldsymbol{w})$. After specifying the value of a smoothing parameter and the number of basis functions, we begin by excluding the i-th data $(\boldsymbol{x}_i, y_i)$ from among the n observed data and estimating the weight vector $\boldsymbol{w}$ based on the remaining $(n-1)$ data; the resulting weight vector is denoted $\hat{\boldsymbol{w}}^{(-i)}$. We next compute the residual $y_i - u(\boldsymbol{x}_i; \hat{\boldsymbol{w}}^{(-i)})$ at the excluded data point $(\boldsymbol{x}_i, y_i)$ in the estimated regression model $y = u(\boldsymbol{x}; \hat{\boldsymbol{w}}^{(-i)})$. This process is repeated for all of the data to obtain

$$\mathrm{CV}(\lambda, m) = \frac{1}{n} \sum_{i=1}^{n} \left\{ y_i - u(\boldsymbol{x}_i; \hat{\boldsymbol{w}}^{(-i)}) \right\}^2. \tag{3.63}$$

The CV values are obtained for various values of a smoothing parameter and numbers of basis functions, and the model that minimizes the CV value is selected as the optimal model.

In cross-validation, the n iterations of the estimation process with successive exclusion of each data point in turn become unnecessary if the predicted value vector $\hat{\boldsymbol{y}}$ can be given by $\hat{\boldsymbol{y}} = H(\lambda, m)\boldsymbol{y}$, in which the smoother matrix $H(\lambda, m)$ is not dependent on the data vector $\boldsymbol{y}$, thus facilitating efficient computation (see Section 5.1.2). In such cases, if we designate the i-th diagonal element of matrix $H(\lambda, m)$ as $h_{ii}(\lambda, m)$ we can then rewrite (3.63) as

$$\mathrm{CV}(\lambda, m) = \frac{1}{n} \sum_{i=1}^{n} \left\{ \frac{y_i - u(\boldsymbol{x}_i; \hat{\boldsymbol{w}})}{1 - h_{ii}(\lambda, m)} \right\}^2. \tag{3.64}$$

(See Konishi and Kitagawa, 2008, p. 243). If we replace $1-h_{ii}(\lambda, m)$ in the denominator by its mean value $1 - n^{-1}\mathrm{tr}H(\lambda, m)$, we then have the *generalized cross-validation* (Craven and Wahba, 1979)

$$\mathrm{GCV}(\lambda, m) = \frac{1}{n}\sum_{i=1}^{n}\left\{\frac{y_i - u(\boldsymbol{x}_i; \hat{\boldsymbol{w}})}{1 - n^{-1}\mathrm{tr}H(\lambda, m)}\right\}^2. \tag{3.65}$$

The generalized cross-validation GCV can be rewritten as follows:

$$\begin{aligned} n\log \mathrm{GCV} &= n\log\left[\frac{1}{n}\sum_{i=1}^{n}\{y_i - u(\boldsymbol{x}_i; \hat{\boldsymbol{w}})\}^2\right] - 2n\log\left\{1 - \frac{1}{n}\mathrm{tr}H(\lambda, m)\right\} \\ &\approx n\log\hat{\sigma}^2 + 2\mathrm{tr}H(\lambda, m), \end{aligned} \tag{3.66}$$

where $\hat{\sigma}^2$ is the residual sum of squares (RSS) divided by the number of data points n. In its final form, obtained by Taylor's expansion, this equation embodying elimination of the constant terms that are not dependent on the model is clearly equivalent to the information criterion $\mathrm{AIC_M}$ in (3.62) with its modified degrees of freedom.

Exercises

3.1 Verify that the following function satisfies the conditions for a cubic spline at knot t_2, that is, the first and second derivatives of $u(x; \boldsymbol{\theta})$ at t_2 are continuous:

$$u(x; \boldsymbol{\theta}) = \beta_0 + \beta_1 x + \beta_2 x^2 + \beta_3 x^3 + \sum_{i=1}^{3}\theta_i (x - t_i)_+^3,$$

where $\boldsymbol{\theta} = (\beta_0, \beta_1, \beta_2, \beta_3, \theta_1, \theta_2, \theta_3)^T$ and $(x - t_i)_+ = \max\{0, x - t_i\}$ (see (3.10)).

3.2 Show that the following piecewise function is a natural cubic spline with knots $-1, 0$ and 1:

$$f(x) = \begin{cases} 2x + 4, & -2 \le x \le -1 \\ -x^3 - 3x^2 - x + 3, & -1 \le x \le 0 \\ x^3 - 3x^2 - x + 3, & 0 \le x \le 1 \\ -4x + 4, & 1 \le x \le 2. \end{cases}$$

3.3 Verify that the following function satisfies the conditions for a natural cubic spline (see (3.11)):

$$u(x; \boldsymbol{\theta}) = \beta_0 + \beta_1 x + \sum_{i=1}^{m-2}\theta_i\{d_i(x) - d_{m-1}(x)\},$$

where $\boldsymbol{\theta} = (\beta_0, \beta_1, \theta_1, \theta_2, \cdots, \theta_{m-2})^T$ and

$$d_i(x) = \frac{(x - t_i)_+^3 - (x - t_m)_+^3}{t_m - t_i}.$$

3.4 Derive the least squares estimator of the weight vector $\boldsymbol{w}$ of the nonlinear regression model given by (3.26).

3.5 Derive the maximum likelihood estimators of the weight vector $\boldsymbol{w}$ and the error variance σ^2 of the Gaussian nonlinear regression model given by (3.32).

3.6 Show that the maximum likelihood estimator of the weight vector of the Gaussian nonlinear regression model based on the basis expansion in (3.32) equals the least squares estimator.

Chapter 4

Logistic Regression Models

The objectives of linear and nonlinear regression models are the understanding, elucidation, prediction, and control of phenomena by linking a number of factors that may influence their outcome. This inherently raises the question of how to construct models from data-based information for probabilistic assessment of unknown risks posed by phenomena arising out of various factors. For example, in considering several factors that might lead to the occurrence of a given disease, the question is how to model the probability of a disease occurrence (the risk probability) from the levels of those factors. If we can assess the risk of its occurrence probabilistically, this may lead to future prevention. In the process of modeling this type of risk, moreover, it may be possible to isolate and identify the causal factors.

In this chapter, we describe the basic concept of probabilistic assessment of risk arising out of various factors using linear and nonlinear logistic regression models.

4.1 Risk Prediction Models

Let us first consider the construction of a model in relation to one type of stimulus by analysis of whether individuals have reacted at various levels of stimulus to predict the risk of response to given levels.

4.1.1 Modeling for Proportional Data

In cases in which it is possible to observe the response to different stimulus levels in many individuals in terms of the proportion of the individuals responding, it is desirable to build a model of the correspondence between the stimulus level and the response rate, and predict the risk of response at future stimulus levels. Let us consider a case that involved the exposure to aphids on the underside of plant leaves to five insecticides differing in their concentration of a chemical having an insecticidal effect and find a model that expresses this insecticidal effect.

Table 4.1 *Stimulus levels and the proportion of individuals responded.*

Stimulus level (x)	0.4150	0.5797	0.7076	0.8865	1.0086
No. of individuals	50	48	46	49	50
Response number	6	16	24	42	44
Response rate (y)	0.120	0.333	0.522	0.857	0.880

Table 4.1 shows the number and proportion of individuals (y) that died among around 50 aphids at five incrementally increasing concentrations of the insecticide, taken as the stimulus levels (x), following its dispersal on the aphids (Chatterjee et al., 1999). We will construct a model to ascertain how the response rate (lethality) changes with the incrementally increasing stimulus level.

In the framework of a regression model, the stimulus level x would be taken as the predictor variable and the rate y as the response variable. The present model differs from a regression model in that, although the values of the predictor variable x are a range of real numbers, those of the response variable y are limited to the range $0 < y < 1$ because they represent probability ratios. Figure 4.1 shows a plot of the graduated stimulus levels shown in Table 4.1 along the x axis and the response rate along the y axis. Our objective is to construct from this data a model that outputs the y values in the range 0 to 1 for the input x values as

$$y = f(x), \qquad 0 < y < 1, \quad -\infty < x < +\infty. \tag{4.1}$$

As neither a linear model nor a polynomial model can provide a fitted model that outputs risk probability values limited to the range $0 < y < 1$, it is necessary to consider a new type of model.

In general, a model that outputs values of y in the interval (0, 1) for real-number inputs x, here respectively representing the response rate (y) to the stimulus level (x), is a *logistic regression model* of the form

$$y = \frac{\exp(\beta_0 + \beta_1 x)}{1 + \exp(\beta_0 + \beta_1 x)}, \qquad 0 < y < 1, \quad -\infty < x < +\infty. \tag{4.2}$$

The logistic regression model is a monotone function that expresses either of two curves, as illustrated in Figure 4.2, depending on whether the coefficient parameter β_1 of the predictor variable x is positive or negative. It may alternatively take the form

$$y = \frac{1}{1 + \exp\{-(\beta_0 + \beta_1 x)\}}. \tag{4.3}$$

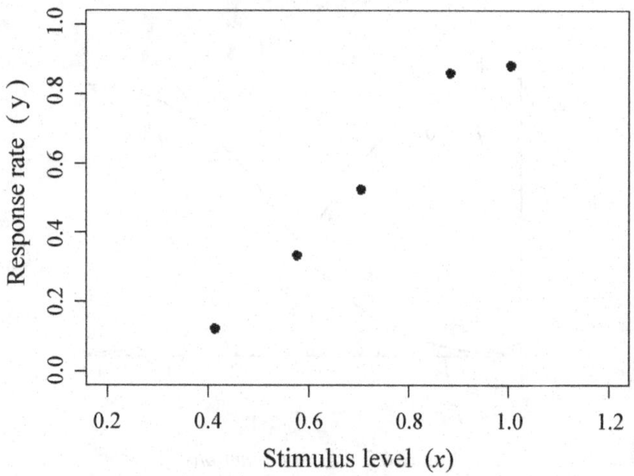

Figure 4.1 *Plot of the graduated stimulus levels shown in Table 4.1 along the x axis and the response rate along the y axis.*

which is obtained by multiplying the denominator and the numerator in (4.2) by $\exp\{-(\beta_0 + \beta_1 x)\}$. If y is then converted as follows by what is called the *logit transformation*, the function becomes linear in x.

$$\log \frac{y}{1-y} = \beta_0 + \beta_1 x. \tag{4.4}$$

By fitting the logistic regression model to the observed data shown in Table 4.1 for the relation between the stimulus level x and the response rate y, we obtain

$$y = \frac{\exp(-4.85 + 7.167x)}{1 + \exp(-4.85 + 7.167x)}, \tag{4.5}$$

(see Figure 4.3). The parameters of this model were estimated by maximum likelihood, which is described in Section 4.2. As shown in the figure, the estimated logistic regression model (4.5) yields a value of 0.677 as the stimulus level resulting in death for 50% of the total number and 1.1 as that resulting in death for 95%.

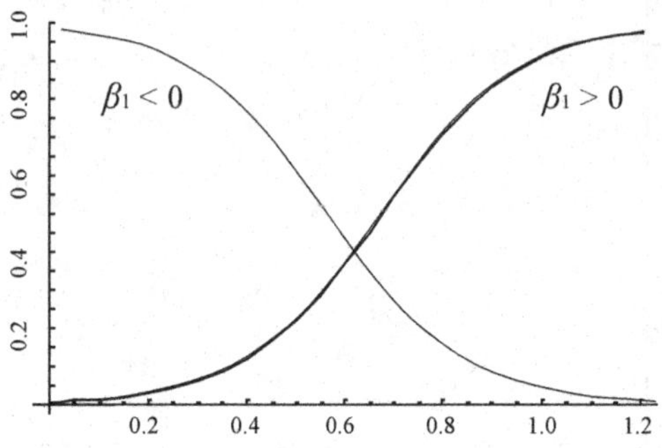

Figure 4.2 *Logistic functions.*

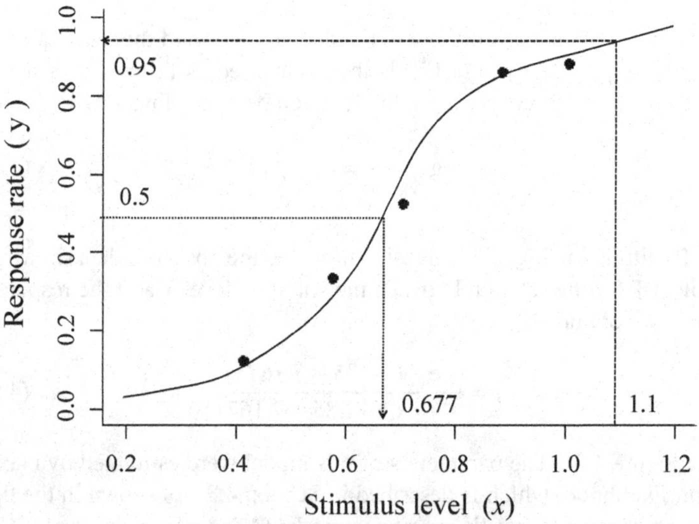

Figure 4.3 *Fitting the logistic regression model to the observed data shown in Table 4.1 for the relation between the stimulus level x and the response rate y.*

4.1.2 Binary Response Data

In the previous section, we fitted a logistic regression model for a group of numerous individuals at each of several stimulus levels to find the proportion responding to the stimulus. We next address the question of how to proceed in cases in which such a ratio cannot be obtained. For example, in modeling the relation between the level of a certain component in the blood and the incidence of a disease, in many cases it is impossible to find large groups of people with the same blood level values and investigate the proportion in which the disease has occurred. What is available instead is only binary data (i.e., 0 or 1) representing occurrence or non-occurrence of the disease (the response) at particular laboratory test values (stimulus levels). For a predictor variable x that takes real-number values in this way, and with observed data in the binary form of response or non-response as the variable y, the objective of modeling becomes *binary response data analysis*. Let us consider the following example as a basis for discussion of risk modeling from observed binary data obtained at various stimulus levels.

In this example, we will construct a model to predict the risk of calcium oxalate crystals presence in the body from the specific gravity of urine, which can be readily determined by medical examination. Such crystals are a cause of kidney and ureteral stone formation. The data shown here are the measured values of urine specific gravity, taken from a report by Andrews and Herzberg (1985, p. 249) showing the measured values obtained from six tests for urine factors thought to cause crystal formation in clinical examinations yielding 77 data. A value of 0 was assigned if no calcium oxalate crystals were found in the urine, and a value of 1 was assigned if they were found. In Section 4.2.1 below, multiple-risk model analysis is applied to the calcium oxalate crystal data obtained from the six tests.

In Figure 4.4, the data on presence and non-presence of the crystals are plotted along the vertical axis as $y = 0$ for the 44 individuals exhibiting their non-presence and $y = 1$ for the 33 exhibiting their presence, and the values found for their urine specific gravity are plotted along the x axis. This figure suggests, in relative terms, that the probability of calcium oxalate crystal presence tends to increase with increasing urine specific gravity. If we can model the relation between the specific gravity value and the probability of crystal presence from this $\{0, 1\}$ binary data, then it may become possible to predict the risk probability for individuals, such as a prediction of 0.82 crystal presence probability in an individual having urine with a specific gravity of 1.026.

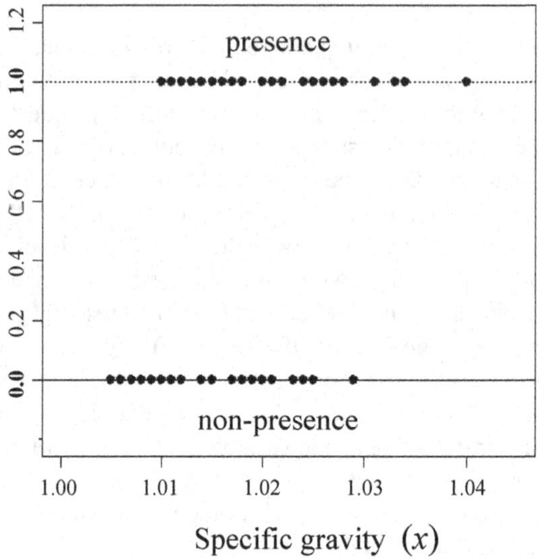

Figure 4.4 *The data on presence and non-presence of the crystals are plotted along the vertical axis as* $y = 0$ *for the 44 individuals exhibiting their non-presence and* $y = 1$ *for the 33 exhibiting their presence. The x axis takes the values of their urine specific gravity.*

In general, if we assign the values $y = 1$ and $y = 0$ to individuals showing response and non-response, respectively, at level x (stimulus levels), we then have the following n-set of data relative to the binary {0, 1} values.

$$(x_1, y_1),\ (x_2, y_2),\ \cdots,\ (x_n, y_n), \qquad y_i = \begin{cases} 1 & \text{response} \\ 0 & \text{non-response.} \end{cases} \tag{4.6}$$

We will construct a model on the basis of this type of binary response data for estimation of the response probability π at given level x.

For the random variable Y representing whether response has occurred, the probability at stimulus level x may then be expressed as $P(Y = 1|x) = \pi$, and the non-response probability as $P(Y = 0|x) = 1 - \pi$. As the model linking the factor x causing this response and the response probability π, let us assume the logistic regression model

$$\pi = \frac{\exp(\beta_0 + \beta_1 x)}{1 + \exp(\beta_0 + \beta_1 x)}, \qquad 0 < \pi < 1, \quad -\infty < x < +\infty. \tag{4.7}$$

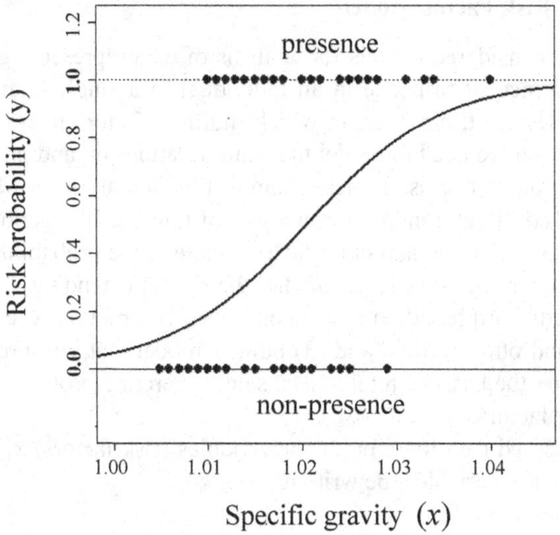

Figure 4.5 *The fitted logistic regression model for the 77 set of data expressing observed urine specific gravity and presence or non-presence of calcium oxalate crystals.*

By model estimation based on the observed data (4.6), we can obtain a model that outputs values in the interval (0, 1) for given levels x and can thus be applied to risk prediction. For the 77 set of data shown in Figure 4.4 expressing observed urine specific gravity and presence or non-presence of calcium oxalate crystals, by applying the logistic regression model using the maximum likelihood method described in Section 4.2.1, we obtain

$$\hat{\pi} = \frac{\exp(-142.57 + 139.71x)}{1 + \exp(-142.57 + 139.71x)}, \qquad 0 < \hat{\pi} < 1. \tag{4.8}$$

The fitted logistic regression model is represented by the curve shown in Figure 4.5. The probability of the presence of calcium oxalate crystals in the body increases with increasing urine specific gravity. This risk prediction model can be applied to probabilistic risk assessment for individuals.

4.2 Multiple Risk Factor Models

So far we have considered models for analysis of data representing the presence or absence of response in an individual to a single stimulus type. Many cases exist, however, in which multiple factors interact to induce risk, and so we need to model their interrelationship and predict risk probability on that basis. Typical examples include those in which we wish to (1) construct a model for analysis of how the interaction of blood pressure, cholesterol, and other factors relate to the probability of occurrence of a certain disease; (2) predict the risk in extending a loan or issuing a credit card based on age, income, family structure, years of employment, and other factors; and (3) build a model that will predict corporate risk on the basis of total assets, sales, operating profit, equity ratio, and other factors.

Let the observed data for p predictor variables (risk factors) x_1, x_2, $\cdots$, x_p and response variable y be written

$$\Big\{(x_{i1}, x_{i2}, \cdots, x_{ip}),\ y_i;\ i = 1, 2, \cdots, n\Big\},\quad y_i = \begin{cases} 1 & \text{response} \\ 0 & \text{non-response,} \end{cases} \tag{4.9}$$

where x_{ij} $(i = 1, 2, \cdots, n, j = 1, 2, \cdots, p)$ is the data for the j-th predictor variable of the i-th individual. If response occurs when the i-th individual is exposed to stimulus comprising p risk factors, we take the data as $y_i = 1$, and if there is no response, we take it as $y_i = 0$.

As in the previous section, if we use the random variable Y representing response or non-response, then the respective response and non-response probabilities for the multiple risk factors x_1, $\cdots$, x_p may be expressed as

$$P(Y = 1|x_1, \cdots, x_p) = \pi, \qquad P(Y = 0|x_1, \cdots, x_p) = 1 - \pi. \tag{4.10}$$

The multiple risk factors and the response probabilities are linked, moreover, by

$$\begin{aligned} \pi &= \frac{\exp(\beta_0 + \beta_1 x_1 + \beta_2 x_2 + \cdots + \beta_p x_p)}{1 + \exp(\beta_0 + \beta_1 x_1 + \beta_2 x_2 + \cdots + \beta_p x_p)} \\ &= \frac{\exp(\boldsymbol{\beta}^T \boldsymbol{x})}{1 + \exp(\boldsymbol{\beta}^T \boldsymbol{x})}, \end{aligned} \tag{4.11}$$

where $\boldsymbol{\beta} = (\beta_0, \beta_1, \cdots, \beta_p)^T$ and $\boldsymbol{x} = (1, x_1, x_2, \cdots, x_p)^T$. This is referred to as a *multiple logistic regression model*. By logit transformation in the same manner as in (4.4), this becomes

$$\log \frac{\pi}{1-\pi} = \beta_0 + \beta_1 x_1 + \beta_2 x_2 + \cdots + \beta_p x_p = \boldsymbol{\beta}^T \boldsymbol{x}, \qquad (4.12)$$

which expresses the linear combination of the predictor variables representing the multiple risk factors. In the next section, we discuss the estimation of model parameters by maximum likelihood with the binary response expressed in the form of the probability distribution.

4.2.1 Model Estimation

We consider the use of maximum likelihood for parameter estimation in the logistic regression model. The first question is then how to represent a phenomenon for which the outcomes are response or non-response given in binary form in a probability distribution model. Let us begin by considering the toss of a coin that has a π probability of coming up heads, which may be thought of as a value that can be determined only by an infinite number of trials. We represent the outcome of each trial by the random variable Y, which is assigned a value of 1 if the coin comes up heads and 0 if it comes up tails. We may express this as $\mathrm{P}(Y = 1) = \pi$, $\mathrm{P}(Y = 0) = 1 - \pi$. The random variable Y thus takes a value of either 0 or 1 and is therefore a discrete random variable with a probability distribution

$$f(y|\pi) = \pi^y (1-\pi)^{1-y}, \qquad y = 0, 1, \qquad (4.13)$$

which is the *Bernoulli distribution*. We consider modeling based on the data in (4.9) using this distribution.

The observed data for the i-th individual consists of the possible values y_i of the binary response random variable Y_i representing response or non-response for the multiple risk factors $(x_{i1}, x_{i2}, \cdots, x_{ip})$. If we take π_i as the true response rate for the i-th multiple risk factors, the probability distribution of Y_i is then given by

$$f(y_i|\pi_i) = \pi_i^{y_i} (1-\pi_i)^{1-y_i}, \qquad y_i = 0, 1, \quad i = 1, 2, \cdots, n. \qquad (4.14)$$

The likelihood function based on $y_1, y_2, \cdots, y_n$ is accordingly

$$L(\pi_1, \pi_2, \cdots, \pi_n) = \prod_{i=1}^{n} f(y_i|\pi_i) = \prod_{i=1}^{n} \pi_i^{y_i} (1-\pi_i)^{1-y_i}. \qquad (4.15)$$

By substituting the logistic regression model (4.11), which links the multiple risk factors that influence the response rate π_i of the i-th individual, into (4.15), we have the likelihood function of the $(p+1)$-dimensional parameter vector $\boldsymbol{\beta}$

$$\begin{aligned} L(\boldsymbol{\beta}) &= \prod_{i=1}^{n} \pi_i^{y_i}(1-\pi_i)^{1-y_i} \\ &= \prod_{i=1}^{n} \left(\frac{\pi_i}{1-\pi_i}\right)^{y_i}(1-\pi_i) \qquad (4.16) \\ &= \prod_{i=1}^{n} \left[\exp(y_i\boldsymbol{\beta}^T\boldsymbol{x}_i)\left\{\frac{1}{1+\exp(\boldsymbol{\beta}^T\boldsymbol{x}_i)}\right\}\right], \end{aligned}$$

where $\boldsymbol{x}_i = (1, x_{i1}, x_{i2}, \cdots, x_{ip})^T$. Accordingly, the log-likelihood function for the parameter vector $\boldsymbol{\beta}$ of the logistic regression model is

$$\ell(\boldsymbol{\beta}) = \log L(\boldsymbol{\beta}) = \sum_{i=1}^{n} y_i\boldsymbol{\beta}^T\boldsymbol{x}_i - \sum_{i=1}^{n} \log\{1+\exp(\boldsymbol{\beta}^T\boldsymbol{x}_i)\}. \quad (4.17)$$

The optimization process with respect to unknown parameter vector $\boldsymbol{\beta}$ is nonlinear, and the equation does not have an explicit solution. The maximum likelihood estimate, $\hat{\boldsymbol{\beta}}$, in this case may be obtained using a numerical optimization technique, such as the Newton-Raphson method.

The first and second derivatives of the log-likelihood function $\ell(\boldsymbol{\beta})$ with respect to $\boldsymbol{\beta}$ are given by

$$\frac{\partial \ell(\boldsymbol{\beta})}{\partial \boldsymbol{\beta}} = \sum_{i=1}^{n} \{y_i - \pi_i\}\boldsymbol{x}_i = X^T\Lambda\mathbf{1}_n, \quad (4.18)$$

$$\frac{\partial^2 \ell(\boldsymbol{\beta})}{\partial \boldsymbol{\beta}\partial \boldsymbol{\beta}^T} = -\sum_{i=1}^{n} \pi_i(1-\pi_i)\boldsymbol{x}_i\boldsymbol{x}_i^T = -X^T\Pi(I_n - \Pi)X, \quad (4.19)$$

where $X = (\boldsymbol{x}_1, \boldsymbol{x}_2, \cdots, \boldsymbol{x}_n)^T$ is an $n \times (p+1)$ matrix, I_n is an $n \times n$ identity matrix, $\mathbf{1}_n$ is an n-dimensional vector, the elements of which are all 1, and Λ and Π are $n \times n$ diagonal matrices defined as

$$\Lambda = \begin{pmatrix} y_1-\pi_1 & 0 & \cdots & 0 \\ 0 & y_2-\pi_2 & \cdots & 0 \\ \vdots & \vdots & \ddots & \vdots \\ 0 & 0 & \cdots & y_n-\pi_n \end{pmatrix}, \quad \Pi = \begin{pmatrix} \pi_1 & 0 & \cdots & 0 \\ 0 & \pi_2 & \cdots & 0 \\ \vdots & \vdots & \ddots & \vdots \\ 0 & 0 & \cdots & \pi_n \end{pmatrix}.$$

Starting from an initial value, we numerically obtain a solution using the following update formula:

$$\boldsymbol{\beta}^{new} = \boldsymbol{\beta}^{old} + \left[E\left\{ -\frac{\partial^2 \ell(\boldsymbol{\beta}^{old})}{\partial \boldsymbol{\beta} \partial \boldsymbol{\beta}^T} \right\} \right]^{-1} \frac{\partial \ell(\boldsymbol{\beta}^{old})}{\partial \boldsymbol{\beta}}. \tag{4.20}$$

This update formula is referred to as *Fisher's scoring algorithm* (Nelder and Wedderburn, 1972; Green and Silverman, 1994), and the $(r + 1)$st estimator, $\hat{\boldsymbol{\beta}}^{(r+1)}$, is updated by

$$\hat{\boldsymbol{\beta}}^{(r+1)} = \left\{ X^T \Pi^{(r)} (I_n - \Pi^{(r)}) X \right\}^{-1} X^T \Pi^{(r)} (I_n - \Pi^{(r)}) \boldsymbol{\xi}^{(r)},$$

where $\boldsymbol{\xi}^{(r)} = X\boldsymbol{\beta}^{(r)} + \{\Pi^{(r)}(I_n - \Pi^{(r)})\}^{-1}(\boldsymbol{y} - \Pi^{(r)}\mathbf{1}_n)$ and $\Pi^{(r)}$ is an $n \times n$ diagonal matrix having $\pi_i = \exp(\hat{\boldsymbol{\beta}}^{(r)T} \boldsymbol{x}_i)/\{1 + \exp(\hat{\boldsymbol{\beta}}^{(r)T} \boldsymbol{x}_i)\}$ for the rth estimator $\hat{\boldsymbol{\beta}}^{(r)}$ in the i-th diagonal element. Thus, by substituting the estimator $\hat{\boldsymbol{\beta}}$ determined by the numerical optimization procedure into (4.11), we have the estimated logistic regression model

$$\hat{\pi}(\boldsymbol{x}) = \frac{\exp(\hat{\boldsymbol{\beta}}^T \boldsymbol{x})}{1 + \exp(\hat{\boldsymbol{\beta}}^T \boldsymbol{x})}, \tag{4.21}$$

which is used to predict risk probability for the multiple risk factors.

Example 4.1 (Probability of the presence of calcium oxalate crystal) The presence of calcium oxalate crystals in the body leads to the formation of kidney or uretral stones, but detection of their presence requires a thorough medical examination. We consider the construction of a simple method of risk diagnosis with a model comprising six properties of urine that are thought to be factors in crystal formation.

The six-dimensional data that are available (Andrews and Herzberg, 1985, p. 249) are the results of thorough examination in which calcium oxalate crystals were found to be present in the urine of 33 of the 77 individuals and not present in that of 44. The six urine properties that were measured for each individual were specific gravity x_1, pH x_2, osmolality (mOsm) x_3, conductivity x_4, urea concentration x_5, and calcium concentration (CALC) x_6. The label variable Y is represented as $y_i = 1;\ i = 1, 2, \cdots, 33$ for the individuals in which the crystals are present and as $y_i = 0;\ i = 1, 2, \cdots, 44$ for those in which it was absent.

Estimating the logistic regression model in (4.11) by the maximum likelihood method yields

$$\hat{\pi}(\boldsymbol{x}) = \frac{\exp(\hat{\beta}_0 + \hat{\beta}_1 x_1 + \hat{\beta}_2 x_2 + \hat{\beta}_3 x_3 + \hat{\beta}_4 x_4 + \hat{\beta}_5 x_5 + \hat{\beta}_6 x_6)}{1 + \exp(\hat{\beta}_0 + \hat{\beta}_1 x_1 + \hat{\beta}_2 x_2 + \hat{\beta}_3 x_3 + \hat{\beta}_4 x_4 + \hat{\beta}_5 x_5 + \hat{\beta}_6 x_6)},$$

where

$$\hat{\beta}_0 + \hat{\beta}_1 x_1 + \hat{\beta}_2 x_2 + \hat{\beta}_3 x_3 + \hat{\beta}_4 x_4 + \hat{\beta}_5 x_5 + \hat{\beta}_6 x_6 \qquad (4.22)$$

$$= -355.34 + 355.94x_1 - 0.5x_2 + 0.02x_3 - 0.43x_4 + 0.03x_5 + 0.78x_6.$$

If this risk prediction model is applied, for example, to the data $\boldsymbol{x}$ = (1.017, 5.74, 577, 20, 296, 4.49), then the logistic regression model (4.22) yields a value of 1.342 and the estimate of risk probability is $\hat{\pi}(\boldsymbol{x})$ = exp(1.342)/{1 + exp(1.342)} = 0.794, thus indicating a fairly high probability of the presence of calcium oxalate crystals in the body.

4.2.2 Model Evaluation and Selection

In constructing a multiple logistic regression model for the modeling of multiple risk factors and risk probabilities, a key question is what risk factors to include in the model to obtain optimum risk prediction. Utilization of the AIC (Akaike information criterion) as defined by (5.31) in Section 5.2.2 provides an answer to this question.

The AIC is generally defined by

$$\text{AIC} = -2(\text{maximum log-likelihood}) + 2(\text{no. of free parameters}).$$

We first replace the parameter vector $\boldsymbol{\beta}$ in the log-likelihood function (4.17) with the maximum likelihood estimator $\hat{\boldsymbol{\beta}}$, to obtain the maximum log-likelihood $\ell(\hat{\boldsymbol{\beta}})$. The number of free parameters in the model is $(p + 1)$, which is the number of multiple risk factors p together with one intercept. Then the AIC for evaluating the logistic regression model estimated by the maximum likelihood method is given by

$$\begin{aligned} \text{AIC} &= -2\ell(\hat{\boldsymbol{\beta}}) + 2(p+1) \\ &= -2\sum_{i=1}^{n} y_i \hat{\boldsymbol{\beta}}^T \boldsymbol{x}_i + 2\sum_{i=1}^{n} \log\{1 + \exp(\hat{\boldsymbol{\beta}}^T \boldsymbol{x}_i)\} + 2(p+1). \end{aligned} \qquad (4.23)$$

We select as the optimum model the one with the combination of risk factors for which the n predictor variables yield the smallest AIC value.

4.3 Nonlinear Logistic Regression Models

The logistic regression models that we have considered to this point all suppose monotonically increasing or decreasing response rates for

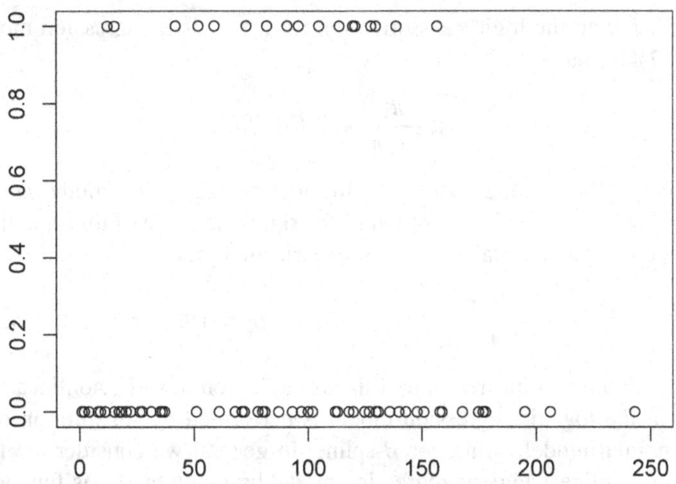

Figure 4.6 *Plot of post-operative kyphosis occurrence along* $Y = 1$ *and non-occurrence along* $Y = 0$ *versus the age (x; in months) of 83 patients.*

changing stimulus levels. In some cases, however, this assumption does not hold and it is therefore necessary to consider model nonlinearization.

Figure 4.6 represents one such case. It is a plot of post-operative kyphosis occurrence along $Y = 1$ and non-occurrence along $Y = 0$ versus the age (x; in months) of 83 patients at the time of laminectomy, a corrective spinal surgery for kyphosis, which is a severe curving of the spine (Hastie and Tibshirani, 1990, p. 301). The objective, in this case, is to predict the timing of laminectomy that will result in a smallest probability of post-operative kyphosis occurrence $P(Y = 1|x) = \pi(x)$, where $\pi(x)$ is the probability of post-operative occurrence relative to age in months at the time of surgery x.

If the probability of post-operative kyphosis is taken as monotonic with respect to age, then logistic regression model (4.7) is appropriate. If the estimated model shows an increase in occurrence probability with increasing age, then surgery should be performed during early childhood, and conversely, if the probability decreases with increasing age, then surgery should be performed after the completion of body growth. It is not clear from the data plotted in Figure 4.6, however, that the relation between probability and age is monotonic. In this case, it is therefore necessary to consider the fit of a logistic regression model for nonlinearity.

By taking the logit transformation for the logistic regression model in (4.7), we have

$$\log \frac{\pi(x)}{1-\pi(x)} = \beta_0 + \beta_1 x. \tag{4.24}$$

Conceptually, nonlinearization of the logistic regression model essentially consists of nonlinearization of the right-hand side of this equation. For a cubic polynomial model, nonlinearization yields

$$\log \frac{\pi(x)}{1-\pi(x)} = \beta_0 + \beta_1 x + \beta_2 x^2 + \beta_3 x^3. \tag{4.25}$$

As in the nonlinearization of linear regression models, nonlinearization of the logistic regression model is performed by substitution of a polynomial model, spline, or B-spline. In general we consider the following nonlinear logistic regression model based on the basis functions $b_0(x) \equiv 1, b_1(x), b_2(x), \cdots, b_m(x)$

$$\log \frac{\pi(x)}{1-\pi(x)} = \sum_{j=0}^{m} w_j b_j(x) = \boldsymbol{w}^T \boldsymbol{b}(x), \tag{4.26}$$

where $\boldsymbol{b}(x) = (b_0(x), b_1(x), b_2(x), \cdots, b_m(x))^T$ and $\boldsymbol{w} = (w_0, w_1, w_2, \cdots, w_m)^T$ is an $(m+1)$-dimensional parameter vector. Then the *nonlinear logistic regression model* linking multiple risk factors and a risk probability can be given by

$$\pi(x) = \frac{\exp\left\{\sum_{j=0}^{m} w_j b_j(x)\right\}}{1 + \exp\left\{\sum_{j=0}^{m} w_j b_j(x)\right\}} = \frac{\exp\left\{\boldsymbol{w}^T \boldsymbol{b}(x)\right\}}{1 + \exp\left\{\boldsymbol{w}^T \boldsymbol{b}(x)\right\}}. \tag{4.27}$$

Nonlinear logistic regression models can be constructed using the Gaussian basis function described in Section 3.2.3 as the basis function. Logistic regression models can also be applied to linear and nonlinear discriminant analysis by Bayes' theorem. This logistic discrimination will be described in Section 7.3.

4.3.1 Model Estimation

The nonlinear logistic regression model (4.27) based on the observed data $\{(x_i, y_i);\ i = 1, 2, \cdots, n\}$ is estimated by the method of maximum likelihood. Let $y_1, \cdots, y_n$ be an independent sequence of binary random variables taking values of 0 and 1 with conditional probabilities

$$P(Y = 1|x_i) = \pi(x_i) \quad \text{and} \quad P(Y = 0|x_i) = 1 - \pi(x_i). \qquad (4.28)$$

The y_i is a discrete random variable with Bernoulli distribution

$$f(y_i|x_i; \boldsymbol{w}) = \pi(x_i)^{y_i} \{1 - \pi(x_i)\}^{1-y_i}, \quad y_i = 0, 1. \qquad (4.29)$$

Hence the log-likelihood function for y_i in terms of $\boldsymbol{w} = (w_0, \cdots, w_m)^T$ is

$$\begin{aligned} \ell(\boldsymbol{w}) &= \sum_{i=1}^{n} y_i \log \pi_i + \sum_{i=1}^{n} (1 - y_i) \log(1 - \pi_i) \\ &= \sum_{i=1}^{n} y_i \boldsymbol{w}^T \boldsymbol{b}(x_i) - \sum_{i=1}^{n} \log\left[1 + \exp\{\boldsymbol{w}^T \boldsymbol{b}(x_i)\}\right]. \end{aligned} \qquad (4.30)$$

The solution $\boldsymbol{w} = \hat{\boldsymbol{w}}$, which maximizes $\ell(\boldsymbol{w})$, is obtained using the numerical optimization method described in Section 4.2.1. In the estimation process, the following substitutions are made in (4.18) and (4.19):

$$\boldsymbol{\beta} \Rightarrow \boldsymbol{w}, \qquad X \Rightarrow B,$$

$$\pi_i = \frac{\exp(\boldsymbol{\beta}^T \boldsymbol{x}_i)}{1 + \exp(\boldsymbol{\beta}^T \boldsymbol{x}_i)} \Rightarrow \pi_i = \frac{\exp\left\{\boldsymbol{w}^T \boldsymbol{b}(x_i)\right\}}{1 + \exp\{\boldsymbol{w}^T \boldsymbol{b}(x_i)\}},$$

where B is an $n \times (m + 1)$ basis function matrix $B = (\boldsymbol{b}(x_1), \boldsymbol{b}(x_2), \cdots, \boldsymbol{b}(x_n))^T$. Substituting the estimator $\hat{\boldsymbol{w}}_\lambda$ obtained by the numerical optimization method into (4.27), we have the nonlinear logistic regression model

$$\hat{\pi}(x) = \frac{\exp\left\{\hat{\boldsymbol{w}}^T \boldsymbol{b}(x)\right\}}{1 + \exp\left\{\hat{\boldsymbol{w}}^T \boldsymbol{b}(x)\right\}}. \qquad (4.31)$$

4.3.2 Model Evaluation and Selection

When addressing data containing a complex nonlinear structure, we need to construct a model that provides flexibility in describing the structure. One approach for this issue is to capture the structure by selecting the number of basis functions included in (4.26). It may therefore be useful to consider selection of the number of basis functions by applying model evaluation and selection criteria.

Replacing the parameter vector $\boldsymbol{w}$ in the log-likelihood function (4.30) with the maximum likelihood estimator $\hat{\boldsymbol{w}}$ yields

$$\ell(\hat{\boldsymbol{w}}) = \sum_{i=1}^{n} y_i \hat{\boldsymbol{w}}^T \boldsymbol{b}(x_i) - \sum_{i=1}^{n} \log\left[1 + \exp\{\hat{\boldsymbol{w}}^T \boldsymbol{b}(x_i)\}\right]. \tag{4.32}$$

The number of free parameters in the model is $(m+1)$ for $w_0, w_1, \cdots, w_m$. Then the AIC for evaluating the nonlinear logistic regression model estimated by the maximum likelihood method is given by

$$\begin{aligned} \text{AIC} &= -2\ell(\hat{\boldsymbol{w}}) + 2(m+1) \qquad (4.33) \\ &= -2\sum_{i=1}^{n} y_i \hat{\boldsymbol{w}}^T \boldsymbol{b}(x_i) + 2\sum_{i=1}^{n} \log\left[1 + \exp\{\hat{\boldsymbol{w}}^T \boldsymbol{b}(x_i)\}\right] + 2(m+1). \end{aligned}$$

Out of the statistical models constructed by the various values m, the number of basis functions, the optimal model is selected by minimizing the information criterion AIC.

Example 4.2 (Probability of occurrence of kyphosis) Figure 4.7 shows a plot of data for 83 patients who received laminectomy, in terms of their age (x, in months) at the time of operation, and $Y = 1$ if the patient developed kyphosis and $Y = 0$ otherwise. As the figure indicates, the probability of onset is not necessarily monotone with respect to age expressed in months. Therefore, we fit the following nonlinear logistic regression model based on polynomials

$$\pi(x) = \frac{\exp\left(\beta_0 + \beta_1 x + \beta_2 x^2 + \cdots + \beta_p x^p\right)}{1 + \exp\left(\beta_0 + \beta_1 x + \beta_2 x^2 + \cdots + \beta_p x^p\right)}. \tag{4.34}$$

By applying the AIC in (4.33), we selected the model with 2nd order polynomial (AIC = 84.22). The corresponding logistic curve is given by

$$\hat{\pi}(x) = \frac{\exp(-3.0346 + 0.0558x - 0.0003x^2)}{1 + \exp(-3.0346 + 0.0558x - 0.0003x^2)}. \tag{4.35}$$

The curve in Figure 4.7 represents the estimated curve. It can be seen from the estimated logistic curve that while the rate of onset increases with the patient's age in months at the time of surgery, a peak occurs at approximately 100 months, and the rate of onset begins to decrease thereafter.

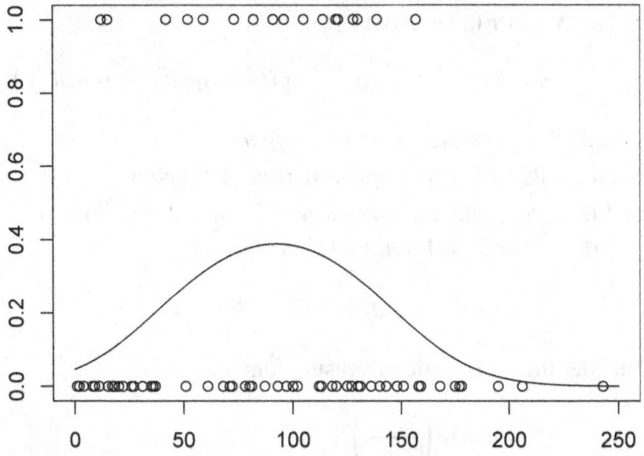

Figure 4.7 *Fitting the polynomial-based nonlinear logisitic regression model to the kyphosis data.*

Exercises

4.1 Consider the logistic model

$$y = f(x) = \frac{\exp(\beta_0 + \beta_1 x)}{1 + \exp(\beta_0 + \beta_1 x)}.$$

(a) Show that $f(x)$ is a monotonic function.

(b) Finding the inverse function of $f(x)$, derive the logit transformation in (4.4).

4.2 Show that the log-likelihood function for the parameter vector $\boldsymbol{\beta}$ of the multiple logistic regression model in (4.11) is given by (4.17).

4.3 Show that the first and second derivatives of the log-likelihood function $\ell(\boldsymbol{\beta})$ in (4.17) with respect to $\boldsymbol{\beta}$ are given by (4.18) and (4.19), respectively.

4.4 Consider the function

$$f(t) = \beta_1 \exp\left\{(\beta_0 + \beta_1 t) - \exp(\beta_0 + \beta_1 t)\right\}, \quad (\beta_1 > 0).$$

Let

$$y = F(x) = \int_{-\infty}^{x} f(t)dt, \qquad -\infty < x < \infty.$$

(a) Show that $F(x)$ is given by

$$y = F(x) = 1 - \exp\{-\exp(\beta_0 + \beta_1 x)\}, \qquad 0 < y < 1,$$

called the *complementary log-log model.*

(b) Show that $F(x)$ is a strictly increasing function.

(c) Show that the inverse linearizing transformation $g(y)$ (e.g., $g(y) = \beta_0 + \beta_1 x$) for $y = F(x)$ is given by

$$g(y) = \log\{-\log(1 - y)\}.$$

4.5 Consider the monotonic increasing function

$$\begin{aligned} y &= \Phi\left(\frac{x-\mu}{\sigma}\right) \\ &= \frac{1}{\sqrt{2\pi}\sigma}\int_{-\infty}^{x} \exp\left\{-\frac{1}{2}\left(\frac{t-\mu}{\sigma}\right)^2\right\} dt, \end{aligned}$$

where $\Phi(z)$ is the standard normal distribution function. Then, taking $-\mu/\sigma = \beta_0$ and $\sigma^{-1} = \beta_1$, the inverse transformation $\Phi^{-1}(y)$ yields a linear model $\beta_0 + \beta_1 x$, called the *probit model.*

Chapter 5

Model Evaluation and Selection

The purpose of statistical modeling is to construct a model that approximates the structure of a phenomenon as accurately as possible through the use of available data. In practice, model evaluation and selection are central issues, and a crucial aspect is selecting the optimum model from a set of candidate models. It requires the use of appropriate criteria to evaluate the *goodness* or *badness* of the constructed models. In this chapter, we describe the basic concepts and outline the derivation of several criteria that have been widely applied in the model evaluation and selection process.

In Section 5.1, we consider criteria derived as an estimator of prediction error, error ascertained from a predictive point of view, as the difference between the estimated model and the true structure of phenomena. In Section 5.2, we discuss the information criterion AIC, which was derived as an asymptotic approximate estimator of the Kullback-Leibler discrepancy measure. In Chapters 2 through 4, we utilized the AIC in evaluating and selecting linear and nonlinear regression models. Here, we consider the AIC in the broader context of model evaluation, as a concept for model evaluation in a general framework, bearing in mind its applicability not only to regression models but also to many other problems.

In Section 5.3, we outline the derivation of the BIC, the Bayesian information criterion, which is based on the posterior probability of models as determined by Bayes' theorem. The discussion also extends to model inference and model averaging (Burnham and Anderson, 2002) based on information from multiple models and their relative importance.

5.1 Criteria Based on Prediction Errors

In this section, we begin by defining prediction error, which represents the error from the perspective of accuracy in prediction, and then discuss the evaluation criteria derived as its estimator.

5.1.1 Prediction Errors

Suppose that we have n set of independent data $\{(y_i, \boldsymbol{x}_i);\ i = 1, 2, \cdots, n\}$ observed for response variable y and p predictor variables $\boldsymbol{x} = (x_1, x_2, \cdots, x_p)^T$. Consider the regression model

$$y_i = u(\boldsymbol{x}_i; \boldsymbol{\beta}) + \varepsilon_i, \qquad i = 1, 2, \cdots, n, \tag{5.1}$$

where $\boldsymbol{\beta}$ is a parameter vector. Let $y = u(\boldsymbol{x}; \hat{\boldsymbol{\beta}})$ be the fitted model. The *residual sum of squares* (RSS)

$$\text{RSS} = \sum_{i=1}^{n} \left\{ y_i - u(\boldsymbol{x}_i; \hat{\boldsymbol{\beta}}) \right\}^2 \tag{5.2}$$

may be used as a measure of the goodness of fit of a regression model to the observed data. It is not effective, however, as a model evaluation criterion for variable selection or order selection. This may be verified by considering the following polynomial model for a predictor variable x

$$u(x; \boldsymbol{\beta}) = \beta_0 + \beta_1 x + \beta_2 x^2 + \cdots + \beta_p x^p. \tag{5.3}$$

The curves in Figure 5.1 represent the fitting of 3rd-, 8th-, and 12th-order polynomial models to 15 data points. As shown by this figure, increasing the degree of the polynomial increases the closeness of the curve to the data, and effectively reduces the RSS. Ultimately, in this case, the 14th-order polynomial model (i.e., of degree one less than the number of data points) passes through all of the data points and reduces the RSS to 0. In this way, the RSS generally results in selection of the highest-degree polynomial and thus the most complex model, and therefore does not function effectively as a criterion for variable selection or order selection.

This shows that a predictive perspective is needed for the model evaluation. More specifically, to evaluate the goodness of a model that has been constructed on the basis of observed data (*training data*), it is necessary to use data (*test data*) that have been obtained independently from those data. For this purpose, the goodness of a model $y = u(\boldsymbol{x}; \hat{\boldsymbol{\beta}})$ that has been estimated on the basis of observed data is evaluated by applying the *predictive sum of squares* (PSS), rather than the RSS, to data z_i obtained independently from the observed data at points $\boldsymbol{x}_i$, as follows:

$$\text{PSS} = \sum_{i=1}^{n} \left\{ z_i - u(\boldsymbol{x}_i; \hat{\boldsymbol{\beta}}) \right\}^2. \tag{5.4}$$

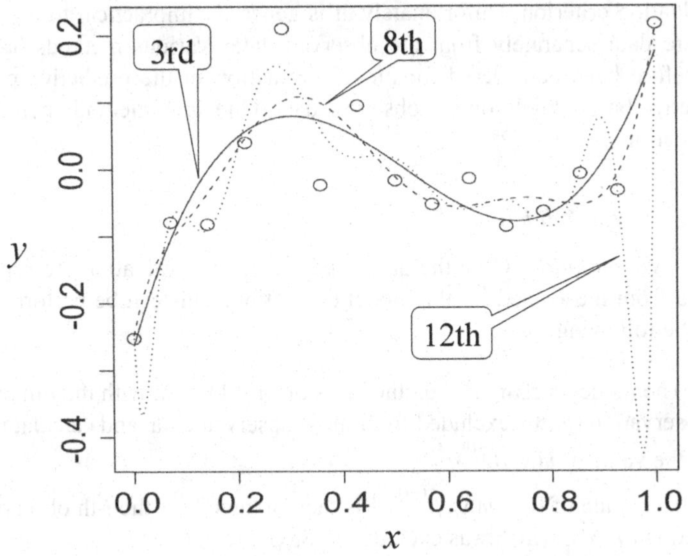

Figure 5.1 *Fitting of 3rd-, 8th-, and 12th-order polynomial models to 15 data points.*

It should also be noted that RSS generally yields values that are smaller than those of PSS and thus overestimates the goodness of an estimated model. This is important, as the underlying objective of evaluation is to determine which of the constructed models will best predict future phenomena, and not simply the one that best fits the observed data.

Equation (5.4) represents the error that will occur with a single future data set. On the other hand, the error that will occur when data sets of the same size are repeatedly obtained is defined as follows:

$$\text{PSE} = \sum_{i=1}^{n} E\left[\{Z_i - u(\boldsymbol{x}_i; \hat{\boldsymbol{\beta}})\}^2\right]. \tag{5.5}$$

The *predictive mean squared error* (PSE) is a measure of the difference that occurs on average in data $Z_i = z_i$ obtained randomly at points $\boldsymbol{x}_i$ independently of the observed data, and it is essential to find an estimator for this measure.

Using the RSS as a PSE estimator would simply mean reusing the same data y_i that were used in estimating the model as a substitute for future data z_i, and the RSS would thus not function effectively as a model evaluation criterion. Unfortunately, it is generally impractical to obtain future data separately from the observed data. Various methods have therefore been considered for model evaluation in the predictive perspective based solely on the observed data. One such method is cross-validation.

5.1.2 Cross-Validation

In *cross-validation* (CV), the data used for model evaluation are separated from those used for the model estimation. This can be performed by the following steps:

(1) The model is estimated on the basis of $(n-1)$ data, with the i-th observation $(y_i, \boldsymbol{x}_i)$ excluded from the n observed data, and calculating the value of $u(\boldsymbol{x}; \hat{\boldsymbol{\beta}}^{(-i)})$.

(2) The value of $\{y_i - u(\boldsymbol{x}_i; \hat{\boldsymbol{\beta}}^{(-i)})\}^2$ is then obtained for the i-th observation $(y_i, \boldsymbol{x}_i)$, which was excluded in Step 1.

(3) Steps 1 and 2 are repeated for all $i \in \{1, 2, \cdots, n\}$, and the resulting

$$\mathrm{CV} = \frac{1}{n} \sum_{i=1}^{n} \left\{ y_i - u(\boldsymbol{x}_i; \hat{\boldsymbol{\beta}}^{(-i)}) \right\}^2 \tag{5.6}$$

is taken as the criterion for assessment of the goodness of the estimated model based on the observed data.

This process is known as *leave-one-out cross-validation.* In polynomial modeling, the CV is computed for various polynomial degrees, and the model yielding the smallest CV is selected as the optimum model. Similarly, in linear regression modeling the model yielding the smallest CV is selected from among the different models obtained in correspondence with the combination of predictor variables. Theoretical derivation has shown, moreover, that the CV is an estimator of the PSE as defined in (5.5) (e.g., Konishi and Kitagawa, 2008, p. 241).

K-fold cross-validation In CV, the n observed data are generally partitioned into k data sets $\{\chi_1, \chi_2, \cdots, \chi_k\}$ with approximately the same number of data in each set. The model is estimated using the $(k-1)$ data sets remaining after exclusion of the i-th data set χ_i, and then evaluated using the excluded data set χ_i. This process is performed for $i = 1, \cdots, k$, in this order, and the resulting average value is taken as the estimated value of the PSE. This is known as *K-fold cross-validation.*

Generalized cross-validation Computing time may become a problem in executing CV for model evaluation for large-scale, large numbers of data. It can be substantially reduced in cases where the predicted value vector $\hat{\boldsymbol{y}}$ can be given as $\hat{\boldsymbol{y}} = H\boldsymbol{y}$ for the matrix H that does not depend on the observation vector $\boldsymbol{y}$, since it then becomes unnecessary to perform the estimation process n times (once for each excluded observation). Matrix H is known as the *hat matrix*, as it maps observed data $\boldsymbol{y}$ to prediction values $\hat{\boldsymbol{y}}$. It is also known as the *smoother matrix* (see Section 3.4.3) because of its application to curve (surface) estimations such as those of nonlinear regression models with basis functions.

One example of the hat matrix occurs in the predicted value of the linear regression model given by (2.24) as $\hat{\boldsymbol{y}} = X\hat{\boldsymbol{\beta}} = X(X^TX)^{-1}X^T\boldsymbol{y}$, in which $H = X(X^TX)^{-1}X^T$ is the hat matrix and is thus not dependent on data $\boldsymbol{y}$. Others occur in the predicted value $\hat{\boldsymbol{y}} = X\hat{\boldsymbol{\beta}} = X(X^TX+\lambda I_n)^{-1}X^T\boldsymbol{y}$ of the ridge estimator in Section 2.3.1 and in the predicted value $\hat{\boldsymbol{y}} = B\hat{\boldsymbol{w}} = B(B^TB + \gamma K)^{-1}B^T\boldsymbol{y}$ for the nonlinear regression model (3.26) (see (3.51) for the parameter estimation). The hat matrices are then $H(\lambda) = X(X^TX + \lambda I_n)^{-1}X^T$ and $H(\lambda, m) = B(B^TB + \lambda\hat{\sigma}^2K)^{-1}B^T$, respectively.

With hat matrix $H = (h_{ij})$, (5.6) can be expressed as

$$\mathrm{CV} = \frac{1}{n}\sum_{i=1}^{n}\left\{\frac{y_i - u(\boldsymbol{x}_i;\hat{\boldsymbol{\beta}})}{1 - h_{ii}}\right\}^2, \tag{5.7}$$

where h_{ii} is the i-th diagonal element of the hat matrix. If we replace the term $1 - h_{ii}$ in the denominator with the mean value $1 - n^{-1}\mathrm{tr}H$, moreover, we then have the *generalized cross-validation* (Craven and Wahba, 1979) written as follows:

$$\mathrm{GCV} = \frac{\displaystyle\sum_{i=1}^{n}\left\{y_i - u(\boldsymbol{x}_i;\hat{\boldsymbol{\beta}})\right\}^2}{n\left\{1 - \dfrac{1}{n}\mathrm{tr}H\right\}^2}. \tag{5.8}$$

This equation eliminates the need to execute n iterations with one observation excluded in each iteration and therefore facilitates efficient computation. For an explanation of why the replacement can be performed using the hat matrix in this manner, see Green and Silverman (1994) and Konishi and Kitagawa (2008, p. 243).

5.1.3 Mallows' C_p

One of the model evaluation criteria based on prediction error is Mallows' C_p (Mallows, 1973), which is particularly used for variable selection in regression modeling. This criterion was derived under the assumption that the probabilistic structure of the specified model is different from the true probabilistic structure that generates data in the framework of a linear regression model. It is assumed that the expectation and the variance covariance matrix of the n-dimensional observation vector $\boldsymbol{y} = (y_1, y_2, \cdots, y_n)^T$ for response variable y are

$$E[\boldsymbol{y}] = \boldsymbol{\mu}, \qquad \mathrm{cov}(\boldsymbol{y}) = E[(\boldsymbol{y} - \boldsymbol{\mu})(\boldsymbol{y} - \boldsymbol{\mu})^T] = \omega^2 I_n, \tag{5.9}$$

respectively. We then estimate the true expectation $\boldsymbol{\mu}$, using the linear regression model

$$\boldsymbol{y} = X\boldsymbol{\beta} + \boldsymbol{\varepsilon}, \qquad E[\boldsymbol{\varepsilon}] = \boldsymbol{0}, \quad \mathrm{cov}(\boldsymbol{\varepsilon}) = \sigma^2 I_n, \tag{5.10}$$

where $\boldsymbol{\beta} = (\beta_0, \beta_1, \cdots, \beta_p)^T$, $\boldsymbol{\varepsilon} = (\varepsilon_1, \cdots, \varepsilon_n)^T$ and X is an $n \times (p+1)$ design matrix formed from the data for the predictor variables. The expectation and the variance-covariance matrix of the observation vector $\boldsymbol{y}$ under this linear regression model are, respectively,

$$E[\boldsymbol{y}] = X\boldsymbol{\beta}, \quad \mathrm{cov}(\boldsymbol{y}) = E[(\boldsymbol{y} - X\boldsymbol{\beta})(\boldsymbol{y} - X\boldsymbol{\beta})^T] = \sigma^2 I_n. \tag{5.11}$$

Comparison of (5.9) and (5.11) shows that the objective is estimation of the true structure $\boldsymbol{\mu}$ via the assumed model under the assumption that the variance (ω^2) of the observed data is different from the one (σ^2) in the linear regression model.

Using the least squares estimator $\hat{\boldsymbol{\beta}} = (X^T X)^{-1} X^T \boldsymbol{y}$ of the regression coefficient vector $\boldsymbol{\beta}$, $\boldsymbol{\mu}$ is estimated by

$$\hat{\boldsymbol{\mu}} = X\hat{\boldsymbol{\beta}} = X(X^T X)^{-1} X^T \boldsymbol{y} \equiv H\boldsymbol{y}. \tag{5.12}$$

The goodness of this estimator $\hat{\boldsymbol{\mu}}$ is measured by the mean square error

$$\Delta_p = E[(\hat{\boldsymbol{\mu}} - \boldsymbol{\mu})^T (\hat{\boldsymbol{\mu}} - \boldsymbol{\mu})]. \tag{5.13}$$

It follows from (5.9) and (5.12) that the expectation of the estimator $\hat{\boldsymbol{\mu}}$ is

$$E[\hat{\boldsymbol{\mu}}] = X(X^T X)^{-1} X^T E[\boldsymbol{y}] \equiv H\boldsymbol{\mu}. \tag{5.14}$$

Hence the mean square error Δ_p can be expressed as

$$\Delta_p = E[(\hat{\boldsymbol{\mu}} - \boldsymbol{\mu})^T (\hat{\boldsymbol{\mu}} - \boldsymbol{\mu})]$$

$$
\begin{aligned}
&= E\left[\{Hy - H\mu - (I_n - H)\mu\}^T \{Hy - H\mu - (I_n - H)\mu\}\right] \\
&= E\left[(y - \mu)^T H(y - \mu)\right] + \mu^T (I_n - H)\mu \\
&= \text{tr}\left\{HE[(y - \mu)(y - \mu)^T]\right\} + \mu^T (I_n - H)\mu \\
&= (p + 1)\omega^2 + \mu^T (I_n - H)\mu. \qquad (5.15)
\end{aligned}
$$

Here, since H and $I_n - H$ are idempotent matrices, we used the relationships $H^2 = H$, $(I_n - H)^2 = I_n - H$, $H(I_n - H) = \mathbf{0}$ and tr$H$ = tr$\{X(X^T X)^{-1} X^T\}$ = tr$I_{p+1} = p + 1$, tr$(I_n - H) = n - p - 1$.

In (5.15) the first term, $(p+1)\omega^2$, increases as the number of parameters increases. The second term, $\mu^T (I_n - H)\mu$, is the sum of squared biases of the estimator $\hat{\mu}$, and decreases as the number of parameters increases. If an estimator of Δ_p is available, then it can be used as a criterion for model evaluation.

The expectation of the residual sum of squares can be calculated as

$$
\begin{aligned}
&E[(y - \hat{y})^T (y - \hat{y})] \\
&\quad = E[(y - Hy)^T (y - Hy)] \\
&\quad = E[\{(I_n - H)(y - \mu) + (I_n - H)\mu\}^T \{(I_n - H)(y - \mu) + (I_n - H)\mu\}] \\
&\quad = E[(y - \mu)^T (I_n - H)(y - \mu)] + \mu^T (I_n - H)\mu \\
&\quad = \text{tr}\left\{(I_n - H)E[(y - \mu)(y - \mu)^T]\right\} + \mu^T (I_n - H)\mu \\
&\quad = (n - p - 1)\omega^2 + \mu^T (I_n - H)\mu. \qquad (5.16)
\end{aligned}
$$

Comparison between (5.15) and (5.16) reveals that, if ω^2 is assumed known, then the unbiased estimator of Δ_p is given by

$$
\hat{\Delta}_p = (y - \hat{y})^T (y - \hat{y}) + \{2(p + 1) - n\}\omega^2. \qquad (5.17)
$$

By dividing both sides of the above equation by the estimator $\hat{\omega}^2$ of ω^2, we obtain Mallows' C_p criterion as an estimator of Δ_p in the form

$$
C_p = \frac{(y - \hat{y})^T (y - \hat{y})}{\hat{\omega}^2} + \{2(p + 1) - n\}. \qquad (5.18)
$$

It may be seen that model preferability increases as the C_p criterion decreases. The estimator $\hat{\omega}^2$ is usually represented by the unbiased estimator of error variance in the most complex model. In a linear regression model, for example, it is represented by the unbiased estimator of error variance in the model including all of the predictor variables.

For order selection in a time series autoregressive model, the *final prediction error* (FPE; Akaike, 1969) is given as the PSE estimator. For a linear regression model $y = \hat{\beta}_0 + \hat{\beta}_1 x_1 + \cdots + \hat{\beta}_p x_p$ estimated by least squares, it is given as

$$\text{FPE} = \frac{n+p+1}{n(n-p-1)}(\boldsymbol{y} - \hat{\boldsymbol{y}})^T(\boldsymbol{y} - \hat{\boldsymbol{y}}). \tag{5.19}$$

Moreover, FPE can be rewritten as

$$\begin{aligned} n \log \text{FPE} &= n \log \left\{ \frac{n+p+1}{n-(p+1)} \right\} + n \log \frac{1}{n}(\boldsymbol{y} - \hat{\boldsymbol{y}})^T(\boldsymbol{y} - \hat{\boldsymbol{y}}) \\ &\approx n \log \left\{ 1 + \frac{2}{n}(p+1) \right\} + n \log \frac{1}{n}(\boldsymbol{y} - \hat{\boldsymbol{y}})^T(\boldsymbol{y} - \hat{\boldsymbol{y}}) \qquad (5.20) \\ &\approx 2(p+1) + n \log \frac{1}{n}(\boldsymbol{y} - \hat{\boldsymbol{y}})^T(\boldsymbol{y} - \hat{\boldsymbol{y}}). \end{aligned}$$

In contrast to $\text{RSS} = (\boldsymbol{y} - \hat{\boldsymbol{y}})^T(\boldsymbol{y} - \hat{\boldsymbol{y}})$, which decreases with increasing model complexity, both Mallows' C_p criterion and the FPE criterion thus include the number of model parameters and thereby penalize model complexity. It may also be noted that they yield equations equivalent to that of the AIC in (5.43) for the Gaussian linear regression model in Example 5.4 of Section 5.2.2, which shows that they are closely related to AIC.

5.2 Information Criteria

In its prediction of the value at each data point $\boldsymbol{x}_i$ by $\hat{y}_i = u(\boldsymbol{x}_i; \hat{\boldsymbol{\beta}})$ for a model $y = u(\boldsymbol{x}; \hat{\boldsymbol{\beta}})$ estimated on the basis of observed data, the PSE defined by (5.5) is a measure of the distance from future data z_i on average, and each criterion described in the previous section may be essentially regarded as an estimate of this distance. The basic idea behind information criteria, however, consists of expressing the estimated model in a probability distribution and using the Kullback-Leibler information (Kullback-Leibler, 1951) to measure the distance from the true probability distribution generating the data. By its nature, this measurement is not just an assessment of the goodness of fit of the model but one that is performed from the perspective of prediction, i.e., from the perspective of whether the model can be used to predict future data.

In this section, we discuss the basic concept of information criteria with consideration for evaluation of various statistical models expressed by probability distribution, as well as regression models.

5.2.1 Kullback-Leibler Information

Let us begin by considering the closeness of a probability distribution model estimated by maximum likelihood to the true probability distribution generating the data. For this purpose, we assume that the data $y_1, \cdots, y_n$ are generated in accordance with the density function $g(y)$ or the probability distribution function $G(y)$. Let $\mathcal{F} = \{f(y|\boldsymbol{\theta});\ \boldsymbol{\theta} \in \Theta \subset R^p\}$ be a probability distribution model specified to approximate the true distribution $g(y)$ that generated the data. If the true probability distribution $g(y)$ is contained in the specified probability distribution model, then there exists a $\boldsymbol{\theta}_0 \in \Theta$ such that $g(y) = f(y|\boldsymbol{\theta}_0)$. One example of this is the case of a normal distribution model $N(\mu, \sigma^2)$ where the probability distribution that generates the data is also a normal distribution $N(\mu_0, \sigma_0^2)$.

We estimate the probability distribution model $f(y|\boldsymbol{\theta})$ by maximum likelihood and set the estimate $\hat{\boldsymbol{\theta}} = (\hat{\theta}_1, \hat{\theta}_2, \cdots, \hat{\theta}_p)^T$. Then by substituting the estimates for the parameters of the specified probability distribution model, we approximate the true probability distribution $g(y)$ by $f(y|\hat{\boldsymbol{\theta}})$. The model thus expressed by a probability distribution $f(y|\hat{\boldsymbol{\theta}})$ is known as a *statistical model*. A simple example of a statistical model is a probability distribution model $N(\overline{y}, s^2)$ in which the sample mean $\overline{y}$ and sample variance s^2, as maximum likelihood estimates, are substituted for the normal distribution model $N(\mu, \sigma^2)$ ((5.32) in Example 5.1).

Let us first consider statistical models within the framework of regression modeling. Suppose that the data $y_1, y_2, \cdots, y_n$ for the response variable y are mutually independent and have been obtained in accordance with the probability distribution $g(y|\boldsymbol{x})$, which is a conditional distribution of y given data points $\boldsymbol{x}$ for the predictor variables. To approximate this true distribution, we assume a probability distribution model $f(y|\boldsymbol{x}; \boldsymbol{\theta})$ characterized by parameters, and attempt to approximate $g(y|\boldsymbol{x})$ with a statistical model $f(y|\boldsymbol{x}; \hat{\boldsymbol{\theta}})$ having maximum likelihood estimators $\hat{\boldsymbol{\theta}}$ substituted for the model parameters $\boldsymbol{\theta}$. Equation (5.40) in Example 5.4 is a statistical model for a Gaussian linear regression model expressed by a probability distribution model. Our purpose here is to evaluate the goodness or badness of the statistical model $f(y|\boldsymbol{x}; \hat{\boldsymbol{\theta}})$. We describe the basic concept for derivation of an information-theoretic criterion from the standpoint of making a prediction.

We represent as the estimator $\hat{\boldsymbol{\theta}} = \hat{\boldsymbol{\theta}}(\boldsymbol{y})$ dependent on the data $\boldsymbol{y}$ for a case in which it is necessary to elucidate the difference from future data. To evaluate the models from the perspective of prediction, we assess the expected goodness or badness of fit for the estimated model $f(z|\boldsymbol{x}; \boldsymbol{\theta})$ when it is used to predict the independent future data $Z = z$ taken at random from $g(z|\boldsymbol{x})$. To measure the closeness of these two distributions, we use the *Kullback-Leibler information* (K-L information)

$$I\{g(z); f(z|\hat{\boldsymbol{\theta}})\} = E_G\left[\log \frac{g(Z)}{f(Z|\hat{\boldsymbol{\theta}})}\right]$$

$$= \begin{cases} \displaystyle\int_{-\infty}^{\infty} \log\left\{\frac{g(z)}{f(z|\hat{\boldsymbol{\theta}})}\right\} g(z)dz & \text{continuous models,} \\ \displaystyle\sum_{i=1}^{\infty} g(z_i) \log \frac{g(z_i)}{f(z_i|\hat{\boldsymbol{\theta}})} & \text{discrete models,} \end{cases} \tag{5.21}$$

where the expectation is taken with respect to the unknown true distribution $g(z)$ by fixing $\hat{\boldsymbol{\theta}}=\hat{\boldsymbol{\theta}}(\boldsymbol{y})$.

In the discrete model approach, the procedure may be regarded as first measuring the difference between $g(z_i)$ and $f(z_i|\hat{\boldsymbol{\theta}})$ at z_i by $\log\{g(z_i)/f(z_i|\hat{\boldsymbol{\theta}})\}$, adding the probability $g(z_i)$, which occurs at z_i as a weight, and measuring the difference between the distributions on this basis. Accordingly, if the true distribution and the estimated model are distant at those places where the occurrence probability of data is high, the K-L information is then also large. The same applies for continuous models.

The K-L information has the following properties:

$$\text{(i) } I\{g(z); f(z|\hat{\boldsymbol{\theta}})\} \geq 0, \quad \text{(ii) } I\{g(z); f(z|\hat{\boldsymbol{\theta}})\} = 0 \iff g(z) = f(z|\hat{\boldsymbol{\theta}}).$$

This implies that a smaller K-L information indicates closer proximity of the model $f(z|\hat{\boldsymbol{\theta}})$ to the probability distribution $g(z)$ generating the data. In comparing multiple models, accordingly, the model yielding the smallest K-L information is the one that is selected. As may be seen when K-L information is rewritten

$$I\{g(z); f(z|\hat{\boldsymbol{\theta}})\} = E_G\left[\log \frac{g(Z)}{f(Z|\hat{\boldsymbol{\theta}})}\right]$$

$$= E_G[\log g(Z)] - E_G[\log f(Z|\hat{\boldsymbol{\theta}})], \tag{5.22}$$

the first term $E_G[\log g(Z)]$ on the far right side is always constant, independent of the individual model, and the closeness of the model to the true distribution thus increases with

$$E_G\left[\log f(Z|\hat{\boldsymbol{\theta}})\right] = \int \log f(z|\hat{\boldsymbol{\theta}}) g(z)dz, \tag{5.23}$$

which is known as the *expected log-likelihood* of the estimated model $f(z|\hat{\theta})$. It must be noted, however, that this expected log-likelihood cannot be directly obtained as it is dependent on the unknown probability distribution $g(z)$ that has generated the data. For this reason, the problem of constructing the information criterion is ultimately one of finding an effective estimator for this expected log-likelihood.

5.2.2 *Information Criterion AIC*

The expected log-likelihood in (5.23) depends on the unknown probability distribution $g(z)$ generating the data, which raises the possibility of estimating this unknown probability distribution from the observed data. One such estimator is

$$\hat{g}(z) = \frac{1}{n}, \qquad z = y_1, y_2, \cdots, y_n, \tag{5.24}$$

a discrete probability distribution known as the *empirical distribution* in which an equal probability of $1/n$ is given to each of the n observations. This yields, as an estimator of the expected log-likelihood,

$$\begin{aligned} E_{\hat{G}}\left[\log f(Z|\hat{\theta})\right] &= \log f(y_1|\hat{\theta}(\boldsymbol{y}))\hat{g}(y_1) + \cdots + \log f(y_n|\hat{\theta}(\boldsymbol{y}))\hat{g}(y_n) \\ &= \frac{1}{n}\sum_{i=1}^{n} \log f(y_i|\hat{\theta}(\boldsymbol{y})). \end{aligned} \tag{5.25}$$

This represents the log-likelihood of the estimated model $f(z|\hat{\theta}(\boldsymbol{y}))$ or the maximum log-likelihood

$$\ell(\hat{\theta}) = \sum_{i=1}^{n} \log f(y_i|\hat{\theta}(\boldsymbol{y})) = \log f(\boldsymbol{y}|\hat{\theta}(\boldsymbol{y})). \tag{5.26}$$

It should be noted that the estimator of the expected log-likelihood $E_G[\log f(Z|\hat{\theta})]$ is here $\ell(\hat{\theta})/n$, and the log-likelihood $\ell(\hat{\theta})$ is the estimator of $nE_G[\log f(Z|\hat{\theta})]$.

The log-likelihood given by (5.26) is an estimator of the expected log-likelihood. However, the log-likelihood was obtained by estimating the expected log-likelihood by reusing the observed data $\boldsymbol{y}$ that were initially used to estimate the model in place of the future data. The use of the same data twice for estimating the parameters and for estimating the evaluation measure (the expected log-likelihood) of the goodness of the estimated model gives rise to the bias

$$\log f(\boldsymbol{y}|\hat{\theta}(\boldsymbol{y})) - nE_{G(z)}\left[\log f(Z|\hat{\theta}(\boldsymbol{y}))\right]. \tag{5.27}$$

The above equation represents the bias for specific data $\boldsymbol{y}$, but the bias that may be expected when data sets of the same size are repeatedly extracted may be given as

$$\text{bias}(G) = E_{G(\boldsymbol{y})}\left[\log f(\boldsymbol{Y}|\hat{\boldsymbol{\theta}}(\boldsymbol{Y})) - nE_{G(z)}\left[\log f(Z|\hat{\boldsymbol{\theta}}(\boldsymbol{Y}))\right]\right], \quad (5.28)$$

where the expectation is taken with respect to the joint distribution of $\boldsymbol{Y}$. Hence by correcting the bias of the log-likelihood, an estimator of the expected log-likelihood is given by

$$\log f(\boldsymbol{y}|\hat{\boldsymbol{\theta}}) - (\text{estimator for bias}(G)). \quad (5.29)$$

The general form of the *information criterion* (IC) can be constructed by evaluating the bias and correcting for the bias of the log-likelihood as follows:

$$\begin{aligned} \text{IC} &= -2(\text{log-likelihood of the estimated model} - \text{bias estimator}) \\ &= -2\log f(\boldsymbol{y}|\hat{\boldsymbol{\theta}}) + 2(\text{estimator for bias}(G)). \end{aligned} \quad (5.30)$$

In general, the bias in (5.28) can take various forms depending on the relationship between the true distribution generating the data and the specified model and on the estimation method employed to construct a statistical model.

The *AIC*, the information criterion proposed by Akaike (1973, 1974), is a criterion for evaluating models estimated by maximum likelihood, and is given by the following equation:

$$\begin{aligned} \text{AIC} &= -2(\text{maximum log-likelihood}) + 2(\text{no. of free parameters}) \\ &= -2\log f(\boldsymbol{y}|\hat{\boldsymbol{\theta}}) + 2(\text{number of free parameters}). \end{aligned} \quad (5.31)$$

This implies that the bias of the log-likelihood of the estimated model may be approximated by the number of free parameters contained in the model. In what is known as *AIC minimization*, the model that minimizes the AIC is selected as the optimum model from all the models under comparison. A detailed derivation of the information criterion will be given in the next section.

We consider several illustrations of the information criterion AIC, in Examples 5.1 through 5.4, and then sort out the relationship between the maximum log-likelihood and the number of free parameters in a statistical model.

Example 5.1 (AIC in normal distribution model) Assume a normal distribution model $N(\mu, \sigma^2)$ as the probability distribution for approximating a histogram constructed from the data $y_1, y_2, \cdots, y_n$. The statistical model is

$$f(y|\hat{\boldsymbol{\theta}}) = \frac{1}{\sqrt{2\pi s^2}} \exp\left\{-\frac{(y-\overline{y})^2}{2s^2}\right\}, \tag{5.32}$$

in which the parameters μ and σ^2 have been replaced by the sample mean $\overline{y} = \sum_{i=1}^n y_i/n$ and the sample variance $s^2 = \sum_{i=1}^n (y_i - \overline{y})^2/n$, respectively. The maximum log-likelihood is

$$\begin{aligned} \ell(\hat{\boldsymbol{\theta}}) = \sum_{i=1}^n \log f(y_i|\hat{\boldsymbol{\theta}}) &= -\frac{n}{2}\log(2\pi s^2) - \frac{n}{2s^2}\frac{1}{n}\sum_{i=1}^n (y_i - \overline{y})^2 \\ &= -\frac{n}{2}\log(2\pi s^2) - \frac{n}{2} \end{aligned} \tag{5.33}$$

and the model contains two free parameters μ and σ^2. From (5.31), the AIC is then given as

$$\text{AIC} = n\log(2\pi s^2) + n + 2 \times 2. \tag{5.34}$$

Example 5.2 (AIC in polynomial regression model order selection) For response variable y and predictor variable x, assume a polynomial regression model with Gaussian noise $N(0, \sigma^2)$

$$y = \beta_0 + \beta_1 x + \beta_2 x^2 + \cdots + \beta_p x^p + \varepsilon = \boldsymbol{\beta}^T \boldsymbol{x} + \varepsilon, \tag{5.35}$$

where $\boldsymbol{\beta} = (\beta_0, \beta_1, \cdots, \beta_p)^T$ and $\boldsymbol{x} = (1, x, x^2, \cdots, x^p)^T$. The specified probability distribution model is then a normal distribution $N(\boldsymbol{\beta}^T \boldsymbol{x}, \sigma^2)$ with mean $\boldsymbol{\beta}^T \boldsymbol{x}$ and variance σ^2.

For the n observations $\{(y_i, x_i);\ i = 1, 2, \cdots, n\}$, the maximum likelihood estimates of the regression coefficient vector $\boldsymbol{\beta}$ and the error variance are respectively given by $\hat{\boldsymbol{\beta}} = (X^T X)^{-1} X \boldsymbol{y}$ and $\hat{\sigma}^2 = \sum_{i=1}^n (y_i - \hat{\boldsymbol{\beta}}^T \boldsymbol{x}_i)^2/n$, where $\boldsymbol{y} = (y_1, y_2, \cdots, y_n)^T$, $\boldsymbol{x}_i = (1, x_i, x_i^2, \cdots, x_i^p)^T$ and $X = (\boldsymbol{x}_1, \boldsymbol{x}_2, \cdots, \boldsymbol{x}_n)^T$. The statistical model is then

$$f(y|\boldsymbol{x}; \hat{\boldsymbol{\theta}}) = \frac{1}{\sqrt{2\pi\hat{\sigma}^2}} \exp\left\{-\frac{1}{2\hat{\sigma}^2}(y - \hat{\boldsymbol{\beta}}^T \boldsymbol{x})^2\right\} \tag{5.36}$$

where $\hat{\boldsymbol{\theta}} = (\hat{\beta}_0, \hat{\beta}_1, \cdots, \hat{\beta}_p, \hat{\sigma}^2)^T$. The log-likelihood of this statistical model (maximum log-likelihood) is

$$\sum_{i=1}^{n} \log f(y_i|x_i;\hat{\theta}) = -\frac{n}{2}\log(2\pi\hat{\sigma}^2) - \frac{n}{2\hat{\sigma}^2}\frac{1}{n}\sum_{i=1}^{n}(y_i - \hat{\beta}^T x_i)^2$$
$$= -\frac{n}{2}\log(2\pi\hat{\sigma}^2) - \frac{n}{2}. \tag{5.37}$$

Note that the number of free parameters in the model is $(p + 2)$, which is the sum of the $(p + 1)$ number of regression coefficients in the p-th order polynomial plus 1 for the error variance. The AIC for the polynomial regression model is therefore given by

$$\text{AIC} = n\log(2\pi\hat{\sigma}^2) + n + 2(p + 2). \tag{5.38}$$

The polynomial model in the order that minimizes the value of AIC is selected as the optimum model.

Example 5.3 (Numerical example for polynomial regression model) Figure 5.2 shows the fitting of a linear model (dashed line), a 2nd-order polynomial model (solid line), and an 8th-order polynomial model (dotted line) to 20 data observed for predictor variable x and response variable y. Table 5.1 shows the RSS, prediction error estimate by CV, and AIC for the fits of the polynomial models of order 1 through 9. Although the RSS value decreases with further increases in the polynomial order, the values obtained by CV and AIC are both smallest with the 2nd-order polynomial model, and that model is therefore selected. As exemplified in the figure, a model that is too simple fails to reflect the structure of the target phenomenon, and a complex model of excessively high degree overfits the observed data and incorporates more error than necessary, and both would therefore be ineffective for the prediction of future phenomenon.

Example 5.4 (AIC for Gaussian linear regression model) Let us model the relationship between the response variable y and the predictor variables $x_1, x_2, \cdots, x_p$ with Gaussian linear regression model

$$y = \boldsymbol{\beta}^T \boldsymbol{x} + \varepsilon, \qquad \varepsilon \sim N(0, \sigma^2), \tag{5.39}$$

where $\boldsymbol{\beta} = (\beta_0, \beta_1, \cdots, \beta_p)^T$ and $\boldsymbol{x} = (1, x_1, x_2, \cdots, x_p)^T$. The specified probability distribution model is then a normal distribution with mean $\boldsymbol{\beta}^T\boldsymbol{x}$ and variance σ^2. The maximum likelihood estimates of the regression coefficient vector and the error variance, as found in Section 2.2.2

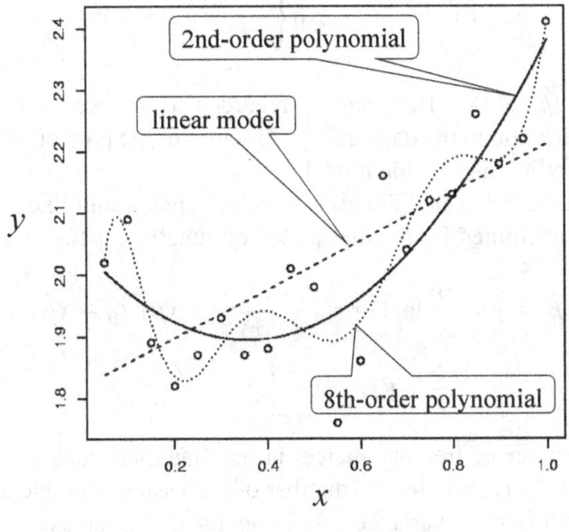

Figure 5.2 *Fitting a linear model (dashed line), a 2nd-order polynomial model (solid line), and an 8th-order polynomial model (dotted line) to 20 data.*

Table 5.1 *Comparison of the values of RSS, CV, and AIC for fitting the polynomial models of order 1 through 9.*

order	RSS	CV	AIC
1	0.0144	0.0185	-22.009
2	0.0068	**0.0092**	**-35.003**
3	0.0065	0.0099	-33.912
4	0.0064	0.0121	-32.111
5	0.0064	0.0177	-30.116
6	0.0062	0.0421	-29.034
7	0.0048	0.0284	-32.163
8	0.0044	0.0280	-31.934
9	0.0043	0.0283	-30.363

(2), are given by $\hat{\boldsymbol{\beta}} = (X^T X)^{-1} X^T \boldsymbol{y}$ and $\hat{\sigma}^2 = (\boldsymbol{y} - X\hat{\boldsymbol{\beta}})^T(\boldsymbol{y} - X\hat{\boldsymbol{\beta}})/n$, respectively. The statistical model is therefore the probability distribution

$$f(y|\boldsymbol{x}; \hat{\boldsymbol{\theta}}) = \frac{1}{\sqrt{2\pi\hat{\sigma}^2}} \exp\left\{-\frac{1}{2\hat{\sigma}^2}(y - \hat{\boldsymbol{\beta}}^T \boldsymbol{x})^2\right\}, \tag{5.40}$$

where $\hat{\boldsymbol{\theta}} = (\hat{\boldsymbol{\beta}}^T, \hat{\sigma}^2)^T$. This may be regarded as an expression of the stochastic variation in the data and the structure of the phenomenon, in a single probability distribution model.

The maximum log-likelihood, in which the maximum likelihood estimates are substituted for the log-likelihood function, is then

$$\begin{aligned} \ell(\hat{\boldsymbol{\beta}}, \hat{\sigma}^2) &= -\frac{n}{2}\log(2\pi\hat{\sigma}^2) - \frac{1}{2\hat{\sigma}^2}(\boldsymbol{y} - X\hat{\boldsymbol{\beta}})^T(\boldsymbol{y} - X\hat{\boldsymbol{\beta}}) \\ &= -\frac{n}{2}\log(2\pi\hat{\sigma}^2) - \frac{n}{2}. \end{aligned} \tag{5.41}$$

Since the number of free parameters in the Gaussian linear regression model is $(p + 2)$, $(p + 1)$ for the number of regression coefficients and 1 for the number of error variances, AIC may then be given as

$$\text{AIC} = n\log(2\pi\hat{\sigma}^2) + n + 2(p + 2). \tag{5.42}$$

The predictor variable combination that minimizes the value of this AIC is taken as the optimum model.

Maximum log-likelihood and number of free parameters

Both the AIC of polynomial regression model in (5.38) and the AIC of linear regression model in (5.42) can be rewritten as

$$\begin{aligned} \text{AIC} &= -2\log f(\boldsymbol{y}|X; \hat{\boldsymbol{\theta}}) + 2(p + 2) \\ &= n\log(2\pi) + n + n\log\hat{\sigma}^2 + 2(p + 2). \end{aligned} \tag{5.43}$$

We consider only terms that affect the model comparison, fixing the term $n\log(2\pi) + n$. Then we find that $n\log\hat{\sigma}^2$, which represents the model fit to the data, decreases with increasing model complexity and thus tends to lower AIC, whereas increasing the number of model parameters, and thus $(p+2)$, has the opposite effect, clearly showing the role of the number of parameters as a penalty in AIC.

Complex models characterized by a large number of parameters generally tend to yield a relatively good fit to observed data, but models of excessive complexity do not function effectively in prediction of future phenomena. For optimum model selection based on information contained in observed data from the perspective of effective prediction, it is therefore necessary to control both goodness of model fit to the data and model complexity.

5.2.3 *Derivation of Information Criteria*

An information criterion was essentially constructed as an estimator of the Kullback-Leibler information between the true distribution $g(z)$ generating the data $\boldsymbol{y}$ and the estimated model $f(z|\hat{\boldsymbol{\theta}}(\boldsymbol{y}))$ for a future observation z that might be obtained on the same random structure. We recall that the information criterion in (5.30) is obtained by correcting the bias

$$\text{bias}(G) = E_{G(\boldsymbol{y})}\left[\log f(\boldsymbol{Y}|\hat{\boldsymbol{\theta}}(\boldsymbol{Y})) - nE_{G(z)}\left[\log f(Z|\hat{\boldsymbol{\theta}}(\boldsymbol{Y}))\right]\right], \quad (5.44)$$

when the expected log-likelihood $nE_{G(z)}\left[\log f(Z|\hat{\boldsymbol{\theta}}(\boldsymbol{Y}))\right]$ is estimated by the log-likelihood $\log f(\boldsymbol{Y}|\hat{\boldsymbol{\theta}}(\boldsymbol{Y}))$. Here $E_{G(\boldsymbol{y})}$ denotes the expectation with respect to the joint distribution of a random sample $\boldsymbol{Y}$, and $E_{G(z)}$ represents the expectation with respect to the true distribution $G(z)$. We derive the bias correction term (5.44) for the log-likelihood of a statistical model estimated by the methods of maximum likelihood.

The maximum likelihood estimator $\hat{\boldsymbol{\theta}}$ is given as the solution of the likelihood equation

$$\frac{\partial \ell(\boldsymbol{\theta})}{\partial \boldsymbol{\theta}} = \sum_{i=1}^{n} \frac{\partial}{\partial \boldsymbol{\theta}} \log f(Y_i|\boldsymbol{\theta}) = \boldsymbol{0}. \quad (5.45)$$

Let $\boldsymbol{\theta}_0$ be a solution of the equation

$$E_{G(z)}\left[\frac{\partial}{\partial \boldsymbol{\theta}} \log f(Z|\boldsymbol{\theta})\right] = \int g(z)\frac{\partial}{\partial \boldsymbol{\theta}} \log f(z|\boldsymbol{\theta})dz = \boldsymbol{0}. \quad (5.46)$$

Then it can be shown that the maximum likelihood estimator $\hat{\boldsymbol{\theta}}$ converges in probability to $\boldsymbol{\theta}_0$ as n tends to infinity.

We first decompose the bias in (5.44) as follows:

$$\begin{aligned} &E_{G(\boldsymbol{y})}\left[\log f(\boldsymbol{Y}|\hat{\boldsymbol{\theta}}(\boldsymbol{Y})) - nE_{G(z)}\left[\log f(Z|\hat{\boldsymbol{\theta}}(\boldsymbol{Y}))\right]\right] \\ &\quad = E_{G(\boldsymbol{y})}\left[\log f(\boldsymbol{Y}|\hat{\boldsymbol{\theta}}(\boldsymbol{Y})) - \log f(\boldsymbol{Y}|\boldsymbol{\theta}_0)\right] \\ &\qquad + E_{G(\boldsymbol{y})}\left[\log f(\boldsymbol{Y}|\boldsymbol{\theta}_0) - nE_{G(z)}\left[\log f(Z|\boldsymbol{\theta}_0)\right]\right] \end{aligned} \quad (5.47)$$

$$+E_{G(\boldsymbol{y})}\left[nE_{G(z)}\left[\log f(Z|\boldsymbol{\theta}_0)\right] - nE_{G(z)}\left[\log f(Z|\hat{\boldsymbol{\theta}}(\boldsymbol{Y}))\right]\right]$$
$$= D_1 + D_2 + D_3.$$

Note that $\hat{\boldsymbol{\theta}} = \hat{\boldsymbol{\theta}}(\boldsymbol{Y})$ depends on the random sample $\boldsymbol{Y}$ drawn from the true distribution $G(y)$. In the next step, we calculate separately the three expectations D_1, D_2 and D_3.

(1) Expectation D_2: Since D_2 does not contain an estimator, it can easily be seen that

$$\begin{aligned} D_2 &= E_{G(\boldsymbol{y})}\left[\log f(\boldsymbol{Y}|\boldsymbol{\theta}_0) - nE_{G(z)}\left[\log f(Z|\boldsymbol{\theta}_0)\right]\right] \\ &= E_{G(\boldsymbol{y})}\left[\sum_{i=1}^{n} \log f(Y_i|\boldsymbol{\theta}_0)\right] - nE_{G(z)}\left[\log f(Z|\boldsymbol{\theta}_0)\right] \\ &= 0. \end{aligned} \tag{5.48}$$

This implies that although D_2 varies randomly depending on the data, its expectation is 0.

(2) Expectation D_3: Noting that D_3 depends on the maximum likelihood estimator $\hat{\boldsymbol{\theta}}$, we write

$$\eta(\hat{\boldsymbol{\theta}}) \equiv E_{G(z)}\left[\log f(Z|\hat{\boldsymbol{\theta}})\right]. \tag{5.49}$$

By expanding $\eta(\hat{\boldsymbol{\theta}})$ in a Taylor series around $\boldsymbol{\theta}_0$ given as a solution of (5.46), we have

$$\begin{aligned} \eta(\hat{\boldsymbol{\theta}}) &= \eta(\boldsymbol{\theta}_0) + \sum_{i=1}^{p}(\hat{\theta}_i - \theta_i^{(0)})\frac{\partial \eta(\boldsymbol{\theta}_0)}{\partial \theta_i} \\ &\quad + \frac{1}{2}\sum_{i=1}^{p}\sum_{j=1}^{p}(\hat{\theta}_i - \theta_i^{(0)})(\hat{\theta}_j - \theta_j^{(0)})\frac{\partial^2 \eta(\boldsymbol{\theta}_0)}{\partial \theta_i \partial \theta_j} + \cdots, \end{aligned} \tag{5.50}$$

where $\hat{\boldsymbol{\theta}} = (\hat{\theta}_1, \hat{\theta}_2, \cdots, \hat{\theta}_p)^T$ and $\boldsymbol{\theta}_0 = (\theta_1^{(0)}, \theta_2^{(0)}, \cdots, \theta_p^{(0)})^T$. From the fact that $\boldsymbol{\theta}_0$ is a solution of (5.46), it holds that

$$\frac{\partial \eta(\boldsymbol{\theta}_0)}{\partial \theta_i} = E_{G(z)}\left[\frac{\partial}{\partial \theta_i}\log f(Z|\boldsymbol{\theta})\bigg|_{\boldsymbol{\theta}_0}\right] = 0, \quad i = 1, 2, \cdots, p, \tag{5.51}$$

where $|_{\boldsymbol{\theta}_0}$ is the value of the partial derivative at $\boldsymbol{\theta} = \boldsymbol{\theta}_0$. Hence, $\eta(\hat{\boldsymbol{\theta}})$ in (5.50) can be approximated as

$$\eta(\hat{\boldsymbol{\theta}}) = \eta(\boldsymbol{\theta}_0) - \frac{1}{2}(\hat{\boldsymbol{\theta}} - \boldsymbol{\theta}_0)^T J(\boldsymbol{\theta}_0)(\hat{\boldsymbol{\theta}} - \boldsymbol{\theta}_0), \tag{5.52}$$

where $J(\boldsymbol{\theta}_0)$ is the $p \times p$ matrix given by

$$J(\boldsymbol{\theta}_0) = -E_{G(z)}\left[\left. \frac{\partial^2 \log f(z|\boldsymbol{\theta})}{\partial\boldsymbol{\theta}\partial\boldsymbol{\theta}^T} \right|_{\boldsymbol{\theta}_0} \right] = -\int g(z) \left. \frac{\partial^2 \log f(Z|\boldsymbol{\theta})}{\partial\boldsymbol{\theta}\partial\boldsymbol{\theta}^T} \right|_{\boldsymbol{\theta}_0} dz. \tag{5.53}$$

Here (a, b)-th element of $J(\boldsymbol{\theta}_0)$ is

$$j_{ab} = -E_{G(z)}\left[\left. \frac{\partial^2 \log f(Z|\boldsymbol{\theta})}{\partial\theta_a\partial\theta_b} \right|_{\boldsymbol{\theta}_0} \right] = -\int g(z) \left. \frac{\partial^2 \log f(z|\boldsymbol{\theta})}{\partial\theta_a\partial\theta_b} \right|_{\boldsymbol{\theta}_0} dz. \tag{5.54}$$

Then, since D_3 is the expectation of $\eta(\boldsymbol{\theta}_0) - \eta(\hat{\boldsymbol{\theta}})$ with respect to $G(\boldsymbol{y})$, it follows from (5.52) that D_3 can be approximately calculated as

$$\begin{aligned} D_3 &= E_{G(\boldsymbol{y})}\left[nE_{G(z)}\left[\log f(Z|\boldsymbol{\theta}_0)\right] - nE_{G(z)}\left[\log f(Z|\hat{\boldsymbol{\theta}})\right]\right] \\ &= \frac{n}{2} E_{G(\boldsymbol{y})}\left[(\hat{\boldsymbol{\theta}} - \boldsymbol{\theta}_0)^T J(\boldsymbol{\theta}_0)(\hat{\boldsymbol{\theta}} - \boldsymbol{\theta}_0) \right] \\ &= \frac{n}{2} E_{G(\boldsymbol{y})}\left[\operatorname{tr}\left\{ J(\boldsymbol{\theta}_0)(\hat{\boldsymbol{\theta}} - \boldsymbol{\theta}_0)(\hat{\boldsymbol{\theta}} - \boldsymbol{\theta}_0)^T \right\} \right] \\ &= \frac{n}{2} \operatorname{tr}\left\{ J(\boldsymbol{\theta}_0) E_{G(\boldsymbol{y})}\left[(\hat{\boldsymbol{\theta}} - \boldsymbol{\theta}_0)(\hat{\boldsymbol{\theta}} - \boldsymbol{\theta}_0)^T \right] \right\}. \end{aligned} \tag{5.55}$$

It is known (see, e.g., Konishi and Kitagawa, 2008, p. 47) that for the maximum likelihood estimator $\hat{\boldsymbol{\theta}}$, the distribution of $\sqrt{n}(\hat{\boldsymbol{\theta}}_n - \boldsymbol{\theta}_0)$ converges in law to the p-dimensional normal distribution with mean vector $\mathbf{0}$ and the variance-covariance matrix $J^{-1}(\boldsymbol{\theta}_0)I(\boldsymbol{\theta}_0)J^{-1}(\boldsymbol{\theta}_0)$ as n tends to infinity, that is,

$$\sqrt{n}(\hat{\boldsymbol{\theta}}_n - \boldsymbol{\theta}_0) \rightarrow N_p\left(\mathbf{0}, J^{-1}(\boldsymbol{\theta}_0)I(\boldsymbol{\theta}_0)J^{-1}(\boldsymbol{\theta}_0)\right) \quad \text{as } n \rightarrow +\infty, \tag{5.56}$$

where the $p \times p$ matrices $I(\boldsymbol{\theta}_0)$ and $J(\boldsymbol{\theta}_0)$ evaluated at $\boldsymbol{\theta} = \boldsymbol{\theta}_0$ are respectively given by

$$I(\boldsymbol{\theta}) = \int g(z) \frac{\partial \log f(z|\boldsymbol{\theta})}{\partial\boldsymbol{\theta}} \frac{\partial \log f(z|\boldsymbol{\theta})}{\partial\boldsymbol{\theta}^T} dz, \tag{5.57}$$

$$J(\boldsymbol{\theta}) = -\int g(z) \frac{\partial^2 \log f(z|\boldsymbol{\theta})}{\partial\boldsymbol{\theta}\partial\boldsymbol{\theta}^T} dz. \tag{5.58}$$

By substituting the asymptotic variance-covariance matrix

$$E_{G(\boldsymbol{y})}\left[(\hat{\boldsymbol{\theta}} - \boldsymbol{\theta}_0)(\hat{\boldsymbol{\theta}} - \boldsymbol{\theta}_0)^T \right] = \frac{1}{n} J(\boldsymbol{\theta}_0)^{-1} I(\boldsymbol{\theta}_0) J(\boldsymbol{\theta}_0)^{-1} \tag{5.59}$$

into (5.55), we have

$$D_3 = \frac{1}{2}\text{tr}\left\{I(\boldsymbol{\theta}_0)J(\boldsymbol{\theta}_0)^{-1}\right\}. \tag{5.60}$$

(3) Expectation D_1: Letting $\ell(\boldsymbol{\theta}) = \log f(\boldsymbol{Y}|\boldsymbol{\theta})$ and expanding $\ell(\boldsymbol{\theta})$ in a Taylor series around the maximum likelihood estimator $\hat{\boldsymbol{\theta}}$ yield

$$\ell(\boldsymbol{\theta}) = \ell(\hat{\boldsymbol{\theta}}) + (\boldsymbol{\theta} - \hat{\boldsymbol{\theta}})^T \frac{\partial \ell(\hat{\boldsymbol{\theta}})}{\partial \boldsymbol{\theta}} + \frac{1}{2}(\boldsymbol{\theta} - \hat{\boldsymbol{\theta}})^T \frac{\partial^2 \ell(\hat{\boldsymbol{\theta}})}{\partial \boldsymbol{\theta} \partial \boldsymbol{\theta}^T}(\boldsymbol{\theta} - \hat{\boldsymbol{\theta}}) + \cdots. \tag{5.61}$$

Here $\partial \ell(\hat{\boldsymbol{\theta}})/\partial \boldsymbol{\theta} = \mathbf{0}$ for the maximum likelihood estimator given as a solution of the likelihood equation $\partial \ell(\boldsymbol{\theta})/\partial \boldsymbol{\theta} = \mathbf{0}$. By applying the law of large numbers, we see that

$$-\frac{1}{n}\frac{\partial^2 \ell(\boldsymbol{\theta}_0)}{\partial \boldsymbol{\theta} \partial \boldsymbol{\theta}^T} = -\frac{1}{n}\sum_{i=1}^{n} \frac{\partial^2}{\partial \boldsymbol{\theta} \partial \boldsymbol{\theta}^T} \log f(y_i|\boldsymbol{\theta})\bigg|_{\boldsymbol{\theta}_0} \rightarrow J(\boldsymbol{\theta}_0), \tag{5.62}$$

as n tends to infinity. Using these results, the equation in (5.61) can be approximated as

$$\ell(\boldsymbol{\theta}_0) - \ell(\hat{\boldsymbol{\theta}}) \approx -\frac{n}{2}(\boldsymbol{\theta}_0 - \hat{\boldsymbol{\theta}})^T J(\boldsymbol{\theta}_0)(\boldsymbol{\theta}_0 - \hat{\boldsymbol{\theta}}). \tag{5.63}$$

Then, taking expectation and using the result in (5.59), we have

$$\begin{aligned} D_1 &= E_{G(\boldsymbol{y})}\left[\log f(\boldsymbol{Y}|\hat{\boldsymbol{\theta}}(\boldsymbol{Y})) - \log f(\boldsymbol{Y}|\boldsymbol{\theta}_0)\right] \\ &= \frac{n}{2}E_{G(\boldsymbol{y})}\left[(\boldsymbol{\theta}_0 - \hat{\boldsymbol{\theta}})^T J(\boldsymbol{\theta}_0)(\boldsymbol{\theta}_0 - \hat{\boldsymbol{\theta}})\right] \\ &= \frac{n}{2}E_{G(\boldsymbol{y})}\left[\text{tr}\left\{J(\boldsymbol{\theta}_0)(\boldsymbol{\theta}_0 - \hat{\boldsymbol{\theta}})(\boldsymbol{\theta}_0 - \hat{\boldsymbol{\theta}})^T\right\}\right] \\ &= \frac{n}{2}\text{tr}\left\{J(\boldsymbol{\theta}_0)E_{G(\boldsymbol{y})}[(\hat{\boldsymbol{\theta}} - \boldsymbol{\theta}_0)(\hat{\boldsymbol{\theta}} - \boldsymbol{\theta}_0)^T]\right\} \\ &= \frac{1}{2}\text{tr}\left\{I(\boldsymbol{\theta}_0)J(\boldsymbol{\theta}_0)^{-1}\right\}. \end{aligned} \tag{5.64}$$

Finally, combining the results in (5.48), (5.60), and (5.64), we obtain the bias (5.44) for the log-likelihood in estimating the expected log-likelihood as follows:

$$\begin{aligned} b(G) &= D_1 + D_2 + D_3 \\ &= \frac{1}{2}\text{tr}\left\{I(\boldsymbol{\theta}_0)J(\boldsymbol{\theta}_0)^{-1}\right\} + 0 + \frac{1}{2}\text{tr}\left\{I(\boldsymbol{\theta}_0)J(\boldsymbol{\theta}_0)^{-1}\right\} \\ &= \text{tr}\left\{I(\boldsymbol{\theta}_0)J(\boldsymbol{\theta}_0)^{-1}\right\}, \end{aligned} \tag{5.65}$$

where $I(\boldsymbol{\theta}_0)$ and $J(\boldsymbol{\theta}_0)$ are respectively given by (5.57) and (5.58).

Information criterion AIC

Let us now assume that the parametric model is $\{f(y|\boldsymbol{\theta}); \boldsymbol{\theta} \in \Theta \subset R^p\}$, and that the true distribution $g(y)$ is contained in the specified parametric model, that is, there exists a $\boldsymbol{\theta}_0 \in \Theta$ such that $g(y) = f(y|\boldsymbol{\theta}_0)$. Under the assumption, the equality $I(\boldsymbol{\theta}_0) = J(\boldsymbol{\theta}_0)$ holds for the $p \times p$ matrices $J(\boldsymbol{\theta}_0)$ and $I(\boldsymbol{\theta}_0)$ given in the bias (5.65). Then the bias of the log-likelihood of $f(y|\hat{\boldsymbol{\theta}})$ in estimating the expected log-likelihood is asymptotically given by

$$
\begin{aligned}
E_{G(\boldsymbol{y})} & \left[\log f(\boldsymbol{Y}|\hat{\boldsymbol{\theta}}) - n E_{G(z)} \left[\log f(Z|\hat{\boldsymbol{\theta}}) \right] \right] \\
& = \operatorname{tr}\left\{ I(\boldsymbol{\theta}_0) J(\boldsymbol{\theta}_0)^{-1} \right\} = \operatorname{tr}(I_p) = p, \qquad (5.66)
\end{aligned}
$$

where I_p is the identity matrix of dimension p. Hence, by correcting the asymptotic bias p of the log-likelihood, we obtain AIC

$$
\text{AIC} = -2 \log f(\boldsymbol{y} \mid \hat{\boldsymbol{\theta}}) + 2p. \qquad (5.67)
$$

We see that AIC is an asymptotic approximate estimator of the Kullback-Leibler information discrepancy between the estimated model and the true distribution generating the data. It was derived under the assumptions that model estimation is by maximum likelihood, and this is carried out in a parametric family of distributions including the true distribution. The AIC does not require any analytical derivation of the bias correction terms for individual problems and does not depend on the unknown probability distribution G, which removes fluctuations due to the estimation of the bias. Further, Akaike (1974) states that if the true distribution that generated the data exists near the specified parametric model, the bias associated with the log-likelihood of the estimated model can be approximated by the number of free parameters. These attributes make the AIC a highly flexible technique from a practical standpoint.

Although direct application of AIC is limited to the models with parameters estimated by the maximum likelihood methods, Akaike's basic idea of bias correction for the log-likelihood can be applied to a wider class of models defined by statistical functionals (Konishi and Kitagawa, 1996, 2008). Another direction of the extension is based on the bootstrapping. In Ishiguro et al. (1997), the bias of the log-likelihood was estimated by using the bootstrap methods. The theory of the bootstrap bias correction for the log-likelihood was further developed in Kitagawa and Konishi (2010).

Burnham and Anderson (2002) provided a nice review and explanation of the use of AIC in the model selection and evaluation problems (see also Linhart and Zucchini, 1986; Sakamoto et al., 1986; Bozdogan, 1987; Kitagawa and Gersch, 1996; Akaike and Kitagawa, 1998; McQuarrie and Tsai, 1998; Konishi, 1999, 2002; Clarke et al., 2009; Kitagawa, 2010). Burnham and Anderson (2002) also discussed modeling philosophy and perspectives on model selection from an information-theoretic point of view, focusing on AIC.

Finite correction of Gaussian linear regression models

If the true probability distribution that generated the data is included in the specified Gaussian linear regression model, then we obtain the following information criterion (Sugiura, 1978)

$$\text{AIC}_\text{C} = n\left\{\log(2\pi\hat{\sigma}^2) + 1\right\} + 2\frac{n(p+2)}{n-p-3}, \tag{5.68}$$

incorporating exact log-likelihood bias correction. That is, with the assumption that the true distribution is included in the Gaussian linear regression model, the bias of the log-likelihood can be evaluated exactly using the properties of the normal distribution, and it is given by

$$E_G\left[\log f(\boldsymbol{Y}|\hat{\boldsymbol{\theta}}(\boldsymbol{Y})) - \int \log f(z|\hat{\boldsymbol{\theta}}(\boldsymbol{Y}))g(z)dz\right] = \frac{n(p+2)}{n-p-3}, \tag{5.69}$$

where the expectation is taken over the joint normal distribution of $\boldsymbol{Y}$.

The exact bias correction term in (5.69) can be expanded in powers of order $1/n$ as

$$\frac{n(p+2)}{n-p-3} = (p+2)\left\{1 + \frac{1}{n}(p+3) + \frac{1}{n^2}(p+3)^2 + \cdots\right\}. \tag{5.70}$$

The leading factor on the right-hand side, $(p+2)$, is an asymptotic bias, and this example thus shows that AIC corrects the asymptotic bias for the log-likelihood of the estimated model.

Hurvich and Tsai (1989, 1991) have obtained similar results for time series models and discussed the effectiveness of exact bias correction. For specific models and methods of estimation, it is thus possible to perform exact bias correction by utilizing the characteristics of the normal distribution. In a general framework, however, the discussion becomes more difficult.

5.2.4 Multimodel Inference

In contrast to model selection, which involves the selection of a single model that best approximates the probabilistic data structure from among a set of candidate models, *multimodel inference* (Burnham and Anderson, 2002) involves inference from a model set using the relative importance or certainty of multiple constructed models as weights. One form of multimodel inference is *model averaging*. It may essentially be regarded as a process of assigning large weights to good models, obtaining the mean value of the regression coefficients of the multiple models, and then constructing a single model on this basis. In contrast to the Akaike (1978, 1979) concept, which relates the relative plausibility of models based on the difference of the AIC value of each model from the minimum AIC value in likelihood, the concept of Burnham and Anderson (2002) relates the relative certainty by weights in model averaging, which may be formally described as follows.

We consider the model set constructed from the data $\boldsymbol{y}$ as $\{f_i(\boldsymbol{y}|\hat{\theta}_i);\ i = 1, \cdots, r\}$. Let AIC_i be the AIC of the individual models, and AIC_{min} the smallest AIC. For each model, we then have

$$\Delta\text{AIC}_i = \text{AIC}_i - \text{AIC}_{min}, \qquad i = 1, \cdots, r, \tag{5.71}$$

which is taken as an expression of the certainty (likelihood) of that model based on the smallest AIC. This difference in AIC is standardized by

$$w_i = \frac{\exp\left(-\frac{1}{2}\Delta\text{AIC}_i\right)}{\displaystyle\sum_{k=1}^{r} \exp\left(-\frac{1}{2}\Delta\text{AIC}_k\right)}, \tag{5.72}$$

and used as the weight in model averaging. Since $\sum_{i=1}^{r} w_i = 1$, the weight w_i serves as an indicator of relative certainty for each model in the set. The weighting here is based on AIC, but it may also be based on other criteria, as shown for BIC in the next section.

Weighting defined in this manner may, for example, be used to estimate the regression coefficient β_j of the predictor variable x_j in a linear regression model, as

$$\hat{\bar{\beta}}_j = \sum_{i=1}^{r} w_i I_j(M_i)\hat{\beta}_{j,i} \bigg/ \sum_{i=1}^{r} w_i I_j(M_i), \tag{5.73}$$

where we let $I_j(M_i)=1$ if the predictor variable x_j is in the model M_i and $I_j(M_i)=0$ if it is not, and take $\hat{\beta}_{j,i}$ as the estimate of the regression coefficient $\beta_{j,i}$ of model M_i.

Inference based on model averaging, with the relative certainty of each model in the candidate model set ordered by weighting based on the difference between its information criterion value and the minimum information criterion value, seems to hold the potential for application to a broad range of problem types, but further research will be necessary on the theoretical underpinnings.

Other approaches besides model averaging can be applied to resolve the uncertainty in model selection. For example, one such approach focuses on assessment of the significance of the difference between two models in AIC or assessment of uncertainty by construction of a reliability set (Linhart, 1988; Shimodaira, 1997). Another is based on *bootstrap selection probability*, with repeated extraction of bootstrap samples from observed data and iteration of model selection; inference proceeds from the frequency of selection for each model (see Appendix A.3).

5.3 Bayesian Model Evaluation Criterion

Like the AIC, the Bayesian information criterion (BIC) proposed by Schwarz (1978) is a criterion for evaluation of models estimated by maximum likelihood. Its construction is based on the posterior probability of the model as an application of Bayes' theorem, which is described in more detail in Chapter 7. In this section, we consider the basic concept of its derivation.

5.3.1 Posterior Probability and BIC

In its basic concept, the BIC consists of selecting as the optimum model the one that yields the highest posterior probability under Bayes' theorem. The derivation of this criterion is essentially as follows.

We begin by setting the r candidate models as $M_1, M_2, \cdots, M_r$, and characterizing each model M_i by the probability distribution $f_i(y|\boldsymbol{\theta}_i)$ ($\boldsymbol{\theta}_i \in \Theta_i \subset R^{p_i}$) and the prior distribution $\pi_i(\boldsymbol{\theta}_i)$ of the parameter vector $\boldsymbol{\theta}_i$, as

$$\text{Model } M_i: \; f_i(y|\boldsymbol{\theta}_i), \quad \pi_i(\boldsymbol{\theta}_i), \quad \boldsymbol{\theta}_i \in \Theta_i \subset R^{p_i}, \quad i = 1, 2, \cdots, r. \tag{5.74}$$

The r probability distribution models that have been estimated contain unknown parameter vectors $\boldsymbol{\theta}_i$ of differing dimensions. We have previously replaced the unknown parameter vectors by estimators $\hat{\boldsymbol{\theta}}_i$, and constructed an information criterion for evaluating the goodness of the statistical models $f_i(y|\hat{\boldsymbol{\theta}}_i)$. In the Bayesian approach, in contrast, the target of the evaluation with n observed data $\boldsymbol{y}$ is the following distribution obtained by integrating $f_i(y|\hat{\boldsymbol{\theta}}_i)$ over the prior distribution $\pi_i(\boldsymbol{\theta}_i)$ of model parameters

$$p_i(\boldsymbol{y}) = \int f_i(\boldsymbol{y}|\boldsymbol{\theta}_i)\pi_i(\boldsymbol{\theta}_i)d\boldsymbol{\theta}_i. \tag{5.75}$$

The $p_i(\boldsymbol{y})$ represents the likelihood when the data have been observed and thus the certainty (plausibility) of the data's observation with model M_i, and is known as the *marginal likelihood* or the *marginal distribution*.

If we designate as $P(M_i)$ the prior probability that the i-th model will occur, the posterior probability of the i-th model is then given by Bayes' theorem as

$$P(M_i|\boldsymbol{y}) = \frac{p_i(\boldsymbol{y})P(M_i)}{\sum_{j=1}^{r} p_j(\boldsymbol{y})P(M_j)}, \qquad i = 1, 2, \cdots, r. \tag{5.76}$$

This posterior probability represents the probability that, when the data $\boldsymbol{y}$ are observed, they will have originated in the i-th model. It follows that if one model is to be selected from among r models, it is most naturally the one that exhibits the highest posterior probability. Since the denominator in (5.76) is the same for all of the models, moreover, the selected model will be the one that maximizes the numerator $p_i(\boldsymbol{y})P(M_i)$. When the prior probability $P(M_i)$ is the same for all of the models, furthermore, the selected model will be the one that maximizes the marginal likelihood $p_i(\boldsymbol{y})$ of the data. If we can express the marginal likelihood represented by the integral of (5.75) in a form that is practical and easy to use, there will be no need to obtain the integral for each problem and, just as with the AIC, it can thus be used as a general model evaluation criterion.

The BIC proposed by Schwarz (1978) was obtained by approximating the integral in (5.75) using Laplace method of integration, which will be discussed in a later section, and is usually applied in the form of the natural logarithm multiplied by -2 and thus as

$$\begin{aligned} -2\log p_i(\boldsymbol{y}) &= -2\log\left\{\int f_i(\boldsymbol{y}|\boldsymbol{\theta}_i)\pi_i(\boldsymbol{\theta}_i)d\boldsymbol{\theta}_i\right\} \\ &\approx -2\log f_i(\boldsymbol{y}|\hat{\boldsymbol{\theta}}_i) + p_i \log n, \end{aligned} \tag{5.77}$$

where $\hat{\boldsymbol{\theta}}_i$ is a maximum likelihood estimate of a p_i-dimensional parameter vector $\boldsymbol{\theta}_i$ included in $f_i(y|\boldsymbol{\theta}_i)$. The BIC for evaluating statistical models estimated by maximum likelihood is given by

$$\begin{aligned}\text{BIC} &= -2(\text{maximum log-likelihood}) + \log n\,(\text{no. of free parameters})\\ &= -2\log f(\boldsymbol{y}|\hat{\boldsymbol{\theta}}) + \log n\,(\text{number of free parameters}). \qquad (5.78)\end{aligned}$$

The model that minimizes BIC is selected as the optimum model.

In the absence of observed data, the equality of all of the models in prior probability implies that all of them may be selected with the same probability. Once data are observed, however, the posterior probability of each model can be calculated from Bayes' theorem. Then, even though the same prior probability $P(M_i)$ is assumed in (5.76) for all of the models, the posterior probability $P(M_i|\boldsymbol{y})$ incorporating the information gained from the data resolves the comparison between the models and thus identifies the model that generates the data.

5.3.2 Derivation of the BIC

The marginal likelihood (5.75) of the data $\boldsymbol{y}$ can be approximated by using the *Laplace approximation for integrals* (Barndorff-Nielsen and Cox, 1989, p. 169). In this description, we omit the index i and the marginal likelihood is expressed as

$$p(\boldsymbol{y}) = \int f(\boldsymbol{y}|\boldsymbol{\theta})\pi(\boldsymbol{\theta})d\boldsymbol{\theta}, \qquad (5.79)$$

where $\boldsymbol{\theta}$ is a p-dimensional parameter vector. This equation can be rewritten as

$$p(\boldsymbol{y}) = \int \exp\{\log f(\boldsymbol{y}|\boldsymbol{\theta})\}\,\pi(\boldsymbol{\theta})d\boldsymbol{\theta} = \int \exp\{\ell(\boldsymbol{\theta})\}\,\pi(\boldsymbol{\theta}), \qquad (5.80)$$

where $\ell(\boldsymbol{\theta})$ is the log-likelihood function $\ell(\boldsymbol{\theta}) = \log f(\boldsymbol{y}|\boldsymbol{\theta})$.

The Laplace approximation takes advantage of the fact that when the number n of observations is sufficiently large, the integrand is concentrated in a neighborhood of the mode of $\ell(\boldsymbol{\theta})$, or in this case, in a neighborhood of the maximum likelihood estimator $\hat{\boldsymbol{\theta}}$, and that the value of the integral depends on the behavior of the function in this neighborhood. By applying Laplace's method of integration, we approximate the marginal likelihood defined by (5.79), and then derive the Bayesian information criterion BIC.

For the maximum likelihood estimator $\hat{\boldsymbol{\theta}}$, the Taylor expansion of the log-likelihood function $\ell(\boldsymbol{\theta})$ around $\hat{\boldsymbol{\theta}}$ is given by

$$\ell(\boldsymbol{\theta}) = \ell(\hat{\boldsymbol{\theta}}) - \frac{n}{2}(\boldsymbol{\theta} - \hat{\boldsymbol{\theta}})^T J(\hat{\boldsymbol{\theta}})(\boldsymbol{\theta} - \hat{\boldsymbol{\theta}}) + \cdots, \qquad (5.81)$$

where

$$J(\hat{\theta}) = -\frac{1}{n}\frac{\partial^2 \ell(\theta)}{\partial\theta\partial\theta^T}\bigg|_{\theta=\hat{\theta}} = -\frac{1}{n}\frac{\partial^2 \log f(y|\theta)}{\partial\theta\partial\theta^T}\bigg|_{\theta=\hat{\theta}}. \tag{5.82}$$

Similarly, we expand the prior distribution $\pi(\theta)$ in a Taylor series around the maximum likelihood estimator $\hat{\theta}$ as

$$\pi(\theta) = \pi(\hat{\theta}) + (\theta - \hat{\theta})^T \frac{\partial \pi(\theta)}{\partial\theta}\bigg|_{\theta=\hat{\theta}} + \cdots. \tag{5.83}$$

Substituting (5.81) and (5.83) into (5.80) and arranging the results leads to the following approximation of the marginal likelihood

$$\begin{aligned} p(y) &= \int \exp\left\{\ell(\hat{\theta}) - \frac{n}{2}(\theta - \hat{\theta})^T J(\hat{\theta})(\theta - \hat{\theta}) + \cdots\right\} \\ &\qquad \times \left\{\pi(\hat{\theta}) + (\theta - \hat{\theta})^T \frac{\partial \pi(\theta)}{\partial\theta}\bigg|_{\theta=\hat{\theta}} + \cdots\right\} d\theta \qquad (5.84) \\ &\approx \exp\left\{\ell(\hat{\theta})\right\}\pi(\hat{\theta}) \int \exp\left\{-\frac{n}{2}(\theta - \hat{\theta})^T J(\hat{\theta})(\theta - \hat{\theta})\right\} d\theta. \end{aligned}$$

Here, we used the fact that $\hat{\theta}$ converges to θ in probability with order $\hat{\theta} - \theta = O_p(n^{-1/2})$, and also that the following equation holds:

$$\int (\theta - \hat{\theta}) \exp\left\{-\frac{n}{2}(\theta - \hat{\theta})^T J(\hat{\theta})(\theta - \hat{\theta})\right\} d\theta = \mathbf{0}, \tag{5.85}$$

which can be considered as the expectation of the random variable $\theta - \hat{\theta}$ distributed as the multivariate normal distribution with mean vector $\mathbf{0}$.

In (5.84), integrating with respect to the parameter vector θ yields

$$\int \exp\left\{-\frac{n}{2}(\theta - \hat{\theta})^T J(\hat{\theta})(\theta - \hat{\theta})\right\} d\theta = (2\pi)^{p/2} n^{-p/2} |J(\hat{\theta})|^{-1/2}, \tag{5.86}$$

since the integrand is the density function of the p-dimensional normal distribution with mean vector $\hat{\theta}$ and variance-covariance matrix $J^{-1}(\hat{\theta})/n$. Consequently, when the sample size n is large, the marginal likelihood defined by (5.79) can be approximated as

$$p(y) \approx \exp\left\{\ell(\hat{\theta})\right\}\pi(\hat{\theta})(2\pi)^{p/2} n^{-p/2} |J(\hat{\theta})|^{-1/2}. \tag{5.87}$$

Taking the logarithm of this expression and multiplying it by -2, we have

$$\begin{aligned} -2\log p(y) &= -2\log\left\{\int f(y|\theta)\pi(\theta)d\theta\right\} \qquad (5.88) \\ &\approx -2\log f(y|\hat{\theta}) + p\log n + \log|J(\hat{\theta})| - p\log 2\pi - 2\log \pi(\hat{\theta}). \end{aligned}$$

Then the following model evaluation criterion BIC is obtained by ignoring terms with order less than $O(1)$ with respect to the sample size n

$$\text{BIC} = -2\log f(\boldsymbol{y}|\hat{\boldsymbol{\theta}}) + p\log n. \tag{5.89}$$

We see that BIC was obtained by approximating the marginal likelihood associated with the posterior probability of the model by Laplace's method for integrals, and that it is not an information-theoretic criterion, leading to an estimator of the Kullback-Leibler information. It can also be seen that BIC is an evaluation criterion for models estimated by the methods of maximum likelihood. Konishi et al. (2004) extended the BIC in such a way that it can be applied to the evaluation of models estimated by the regularization methods discussed in Section 3.4.

The Laplace approximation for integrals is, in general, given as follows. Let $q(\boldsymbol{\theta})$ be a real-valued function of a p-dimensional parameter vector $\boldsymbol{\theta}$, and let $\hat{\boldsymbol{\theta}}$ be the mode of $q(\boldsymbol{\theta})$. Then the Laplace approximation of the integral is given by

$$\int \exp\{nq(\boldsymbol{\theta})\}d\boldsymbol{\theta} \approx \frac{(2\pi)^{p/2}}{n^{p/2}|J_q(\hat{\boldsymbol{\theta}})|^{1/2}} \exp\left\{nq(\hat{\boldsymbol{\theta}})\right\}, \tag{5.90}$$

where

$$J_q(\hat{\boldsymbol{\theta}}) = -\left.\frac{\partial^2 q(\boldsymbol{\theta})}{\partial\boldsymbol{\theta}\partial\boldsymbol{\theta}^T}\right|_{\boldsymbol{\theta}=\hat{\boldsymbol{\theta}}}. \tag{5.91}$$

The use of Laplace's method for integrals has been extensively investigated as a useful tool for approximating Bayesian predictive distributions, Bayes factors, and Bayesian model selection criteria (Davison, 1986; Tierney and Kadane, 1986; Kass and Wasserman, 1995; Kass and Raftery, 1995; O'Hagan, 1995; Konishi and Kitagawa, 1996; Neath and Cavanaugh, 1997; Pauler, 1998; Lanterman, 2001; Konishi et al., 2004).

5.3.3 *Bayesian Inference and Model Averaging*

In statistical modeling, as we have seen, the BIC is focused on selection of a single optimum approximation model for prediction of future phenomena. In multimodel inference, in contrast, the focus has shifted to multiple models as a basis for prediction of phenomena (Burnham and Anderson, 2002). In this section, we consider multimodel inference through the construction of a predictive distribution by *model averaging* based on the Bayesian approach (Hoeting et al., 1999; Wasserman, 2000).

As in Section 5.3.1, we designate the r candidate models as $M_1, M_2, \cdots, M_r$ and characterize each model M_i by the probability distribution $f_i(y|\boldsymbol{\theta}_i)$ ($\boldsymbol{\theta}_i \in \Theta_i \subset R^{p_i}$) and the prior distribution $\pi_i(\boldsymbol{\theta}_i)$ of the parameter vector $\boldsymbol{\theta}_i$. The predictive distribution is the model used in inference of future data $Z = z$ and thus in the predictive perspective, and is defined as

$$h_i(z|\boldsymbol{y}) = \int f_i(z|\boldsymbol{\theta}_i)\pi_i(\boldsymbol{\theta}_i|\boldsymbol{y})d\boldsymbol{\theta}_i, \qquad i = 1, 2, \cdots, r, \tag{5.92}$$

where $\pi_i(\boldsymbol{\theta}_i|\boldsymbol{y})$ is the posterior distribution defined by Bayes' theorem as

$$\pi_i(\boldsymbol{\theta}_i|\boldsymbol{y}) = \frac{f_i(\boldsymbol{y}|\boldsymbol{\theta}_i)\pi_i(\boldsymbol{\theta}_i)}{\int f_i(\boldsymbol{y}|\boldsymbol{\theta}_i)\pi_i(\boldsymbol{\theta}_i)d\boldsymbol{\theta}_i}. \tag{5.93}$$

The basic concept of model averaging in the Bayesian approach comprises model construction incorporating some form of weighting as in

$$h(z|\boldsymbol{y}) = \sum_{i=1}^{r} w_i h_i(z|\boldsymbol{y}), \tag{5.94}$$

rather than selection of a single optimum model from among the r predictive distribution models. Posterior probability is used for this weighting in predictive distribution modeling by model averaging in the Bayesian inference and is defined essentially as follows.

By Bayes' theorem, if $P(M_i)$ is given as the prior probability of the i-th model occurrence, the posterior probability of the i-th model is then

$$P(M_i|\boldsymbol{y}) = \frac{p_i(\boldsymbol{y})P(M_i)}{\sum_{j=1}^{r} p_j(\boldsymbol{y})P(M_j)}, \qquad i = 1, 2, \cdots, r, \tag{5.95}$$

where $p_i(\boldsymbol{y})$ is the marginal distribution in (5.75) defined as the integral for the prior distribution $\pi_i(\boldsymbol{\theta}_i)$ of the parameter vector $\boldsymbol{\theta}_i$. The predictive distribution obtained by model averaging in Bayesian inference using the posterior probability is then given by

$$h(z|\boldsymbol{y}) = \sum_{i=1}^{r} P(M_i|\boldsymbol{y})h_i(z|\boldsymbol{y}). \tag{5.96}$$

The posterior probability represents the probability that data, when observed, will have originated in the i-th model. The relative certainty of each model in the model set is assessed in terms of its posterior probability and the predictive distribution is thus obtained using the weighted average of the model weights represented by their posterior probabilities.

Exercises

5.1 Verify that the Kullback-Leibler information defined by (5.21) has the property

$$I\{g(z); f(z|\hat{\theta})\} \geq 0,$$

and that equality holds only when $g(z) = f(z|\hat{\theta})$.

5.2 Suppose that the true distribution $g(y)$ $(G(y))$ generating data and the specified model $f(y)$ have normal distributions $N(m, \tau^2)$ and $N(\mu, \sigma^2)$, respectively.

(a) Show that

$$E_G[\log g(Y)] = -\frac{1}{2}\log(2\pi\tau^2) - \frac{1}{2},$$

where $E_G[\cdot]$ is an expectation with respect to the true distribution $N(m, \tau^2)$.

(b) Show that

$$E_G[\log f(Y)] = -\frac{1}{2}\log(2\pi\sigma^2) - \frac{\tau^2 + (m-\mu)^2}{2\sigma^2}.$$

(c) Show that the Kullback-Leibler information of $f(y)$ with respect to $g(y)$ is given by

$$\begin{aligned} I\{g(y), f(y)\} &= E_G[\log g(Y)] - E_G[\log f(Y)] \\ &= \frac{1}{2}\left\{\log\frac{\sigma^2}{\tau^2} + \frac{\tau^2 + (m-\mu)^2}{\sigma^2} - 1\right\}. \end{aligned}$$

5.3 Suppose that the true distribution is a double exponential (Laplace) distribution $g(y) = \frac{1}{2}\exp(-|y|)$ $(-\infty < y < \infty)$ and that the specified model $f(y)$ is $N(\mu, \sigma^2)$.

(a) Show that

$$E_G[\log g(Y)] = -\log 2 - 1,$$

where $E_G[\cdot]$ is an expectation with respect to the double exponential distribution.

(b) Show that

$$E_G[\log f(Y)] = -\frac{1}{2}\log(2\pi\sigma^2) - \frac{1}{4\sigma^2}(4 + 2\mu^2).$$

(c) Show that the Kullback-Leibler information of $f(y)$ with respect to $g(y)$ is given by

$$\begin{aligned} I\{g(y), f(y)\} &= E_G[\log g(Y)] - E_G[\log f(Y)] \\ &= \frac{1}{2}\log(2\pi\sigma^2) + \frac{2+\mu^2}{2\sigma^2} - \log 2 - 1. \end{aligned}$$

(d) Find the values of σ^2 and μ that minimize the Kullback-Leibler information.

5.4 Assume that there are two dice that have the following probabilities for rolling the numbers one to six:

$$\begin{aligned} f_a &= \{0.20,\ 0.12,\ 0.18,\ 0.12,\ 0.20,\ 0.18\}, \\ f_b &= \{0.18,\ 0.12,\ 0.14,\ 0.19,\ 0.22,\ 0.15\}. \end{aligned}$$

In terms of the Kullback-Leibler information, which is the fairer dice?

5.5 Suppose that two sets of data $G_1 = \{y_1, y_2, \cdots, y_n\}$ and $G_2 = \{y_{n+1}, y_{n+2}, \cdots, y_{n+m}\}$ are given. To check the homogeneity of the two data sets in question, we assume the following models:

$$G_1 : y_1, y_2, \cdots, y_n \sim N(\mu_1, \sigma_1^2),$$

$$G_2 : y_{n+1}, y_{n+2}, \cdots, y_{n+m} \sim N(\mu_2, \sigma_2^2).$$

Derive the AIC under the following three restricted cases:

(a) $\mu_1 = \mu_2 = \mu$ and $\sigma_1^2 = \sigma_2^2 = \sigma^2$.

(b) $\sigma_1^2 = \sigma_2^2 = \sigma^2$.

(c) $\mu_1 = \mu_2 = \mu$.

5.6 Suppose that there exist k possible outcomes $E_1, \cdots, E_k$ in a trial. Let $\mathrm{P}(E_i) = p_i$, where $\sum_{i=1}^k p_i = 1$, and let Y_i $(i = 1, \cdots, k)$ denote the number of times outcome E_i occurs in n trials, where $\sum_{i=1}^k Y_i = n$. If the trials are repeated independently, then a multinomial distribution with parameters n, p_1, $\cdots$, p_k is defined as a discrete distribution having

$$f(y_1, y_2, \cdots, y_k | p_1, p_2, \cdots, p_k) = \frac{n!}{y_1! y_2! \cdots y_k!} p_i^{y_1} p_2^{y_2} \cdots p_k^{y_k},$$

where $y_i = 0, 1, 2, \cdots, n$ $(\sum_{i=1}^{k} y_i = n)$.
Assume that we have a set of data $Y_1 = n_1$, $Y_2 = n_2$, $\cdots$ $Y_k = n_k$ having k categories. Then, show that the AIC is given by

$$\text{AIC} = -2\left\{\log n! - \sum_{i=1}^{k} \log n_i! + \sum_{i=1}^{k} n_i \log\left(\frac{n_i}{n}\right)\right\} + 2(k-1).$$

This equation can be used to determine the optimal bin size of a histogram (Sakamoto et al., 1986).

Chapter 6

Discriminant Analysis

The objective of *discriminant analysis* is to construct an effective rule for classifying previously unassigned individuals to two or more predetermined classes or groups based on several measurements. It is currently used in fields ranging from medicine, agriculture, and economics to life science, earth and environmental science, biology, and other sciences, and investigations are in progress for its application to new fields and problems.

The basic concept of statistical discriminant analysis was introduced in the 1930s by R. A. Fisher. It has taken its present form as a result of subsequent research by P. C. Mahalanobis, C. R. Rao, and others, centering on linear discriminant analysis expressed chiefly by linear combination of variables (see, e.g., Seber, 1984; Siotani et al., 1985; Anderson, 2003; McLachlan, 2004). In this chapter, we begin with discussion and examples of the basic concept of statistical discriminant analysis, centering chiefly on the linear and quadratic discrimination of Fisher and Mahalanobis. We then proceed to discussion and examples of the basic concepts of multiclass classification, and canonical discriminant analysis, which enables visualization of data dispersed in multiple classes in higher-dimensional spaces.

6.1 Fisher's Linear Discriminant Analysis

In this section, the focus is two-class classification. We discuss statistical discriminant analysis formulated by linear combination of variables with compression of the information in the observed data into sample mean vectors and sample variance-covariance matrices.

6.1.1 Basic Concept

Let us assume that certain evergreen trees can be broadly divided into two varieties on the basis of their leaf shape, and consider the use of width and length as two properties (variables) in constructing a formula

Table 6.1 *The 23 two-dimensional observed data from the varieties A and B.*

		1	2	3	4	5	6	7	8	9	10	11	12
A:	L(x_1)	5	7	6	8	5	9	6	9	7	6	7	9
	W(x_2)	5	7	7	6	6	8	6	7	5	5	8	4
B:	L(x_1)	6	8	7	9	7	10	8	10	9	8	7	
	W(x_2)	2	4	4	5	3	5	5	6	6	3	6	

for classification of newly obtained data known to be from one of the two varieties A and B but not from which one. We first take sample leaves for which the variety is known and measure their length (x_1) and width (x_2), resulting in the 23 two dimensional observed data (Table 6.1).

This dataset for the known varieties is used to construct the classification formula, and is thus called training data. Here, let us consider a linear equation

$$y = w_1 x_1 + w_2 x_2. \tag{6.1}$$

If we were to perform the classification based solely on length x_1, we would take $w_1 = 1$ and $w_2 = 0$ as the variable coefficients and thus perform the classification by projecting the two-dimensional data onto the axis $y = x_1$, as shown in Figure 6.1. Similarly, if we were to perform the classification based solely on width x_2, we would take the coefficients as $w_1 = 0$ and $w_2 = 1$ and thus project the two-dimensional data onto the axis $y = x_2$, as also shown in Figure 6.1. To perform the classification based on the information presented by both variables, the question is then what kind of axis to use for the projection. This question can be reduced to variable weighting based on a criterion, by projecting the two-dimensional data onto $y = w_1x_1 + w_2x_2$ in the figure and selecting the weighting that yields the best separation of the two classes.

Suppose that for random variables $\boldsymbol{x} = (x_1, x_2)^T$ we have n_1 two-dimensional data $\boldsymbol{x}_i^{(1)} = (x_{i1}^{(1)}, x_{i2}^{(1)})^T$ $(i = 1, 2, \cdots, n_1)$ from class G_1 and n_2 two-dimensional data $\boldsymbol{x}_i^{(2)} = (x_{i1}^{(2)}, x_{i2}^{(2)})^T$ $(i = 1, 2, \cdots, n_2)$ from class G_2. In Figure 6.2 the total $n = (n_1 + n_2)$ training data are plotted in the x_1–x_2 plane, and we tentatively assume three projection axes and express the distribution of the data when projected on each one. In this figure we take the values μ_{y_1} and μ_{y_2} on axis y in (b) as representing the class G_1 and class G_2 means, respectively. Similarly, we take the values μ_{z_1} and μ_{z_2} on axis z in (c) as representing the class G_1 and class G_2 means, respectively.

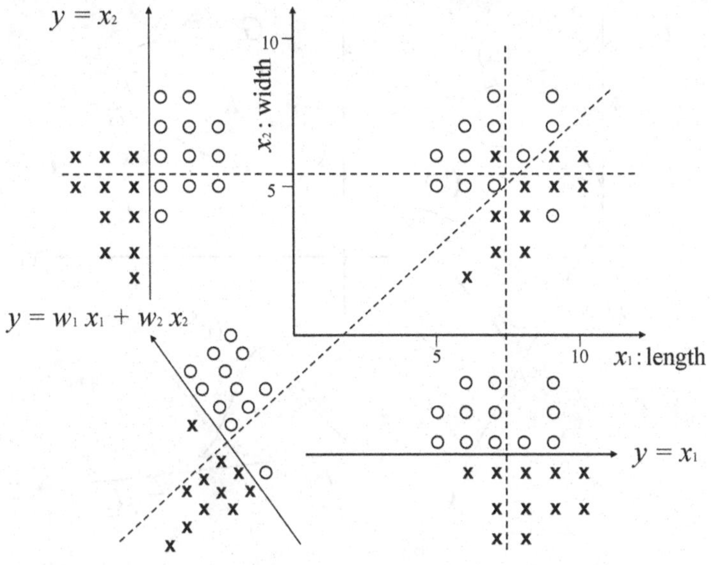

Figure 6.1 *Projecting the two-dimensional data in Table 6.1 onto the axes* $y = x_1$, $y = x_2$ *and* $y = w_1x_1 + w_2x_2$.

Figure 6.2 then shows the following.

(1) Greater separation between the two class means is obtained by projection onto axis z rather than onto axis y. That is,

$$(\mu_{y_1} - \mu_{y_2})^2 < (\mu_{z_1} - \mu_{z_2})^2. \tag{6.2}$$

In the projection of the data onto axis y, $(\mu_{y_1} - \mu_{y_2})^2$ can be regarded as a measure of the degree of separation of the two classes, referred to as the *between-class variance* or between-class variation on axis y.

(2) The variance of class G_1 is smaller on axis y than on axis z. The same holds true for G_2. To determine the degree of data dispersion within each class with projection onto axis y, we consider the sum weighted for the number of data

$$\frac{(n_1 - 1)(\text{var. of } G_1 \text{ on } y) + (n_2 - 1)(\text{var. of } G_2 \text{ on } y)}{(n_1 + n_2 - 2)} \tag{6.3}$$

referred to as the *within-class variance* or within-class variation on

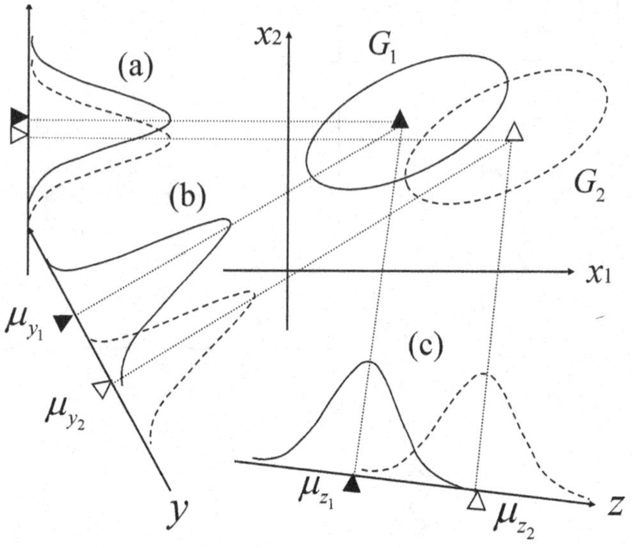

Figure 6.2 *Three projection axes (a), (b), and (c) and the distributions of the class G_1 and class G_2 data when projected on each one.*

axis y. The within-class variance around the mean within each class serves as a measure of the degree of data concentration.

A high degree of separation between the two class means generally facilitates classification. If it is attributable to a large dispersion, however, the region of overlap between the two classes will also tend to be large, with an adverse effect on the performance of the classification. The question then becomes how to determine the optimum axis with respect to these advantageous and adverse effects. One approach is to set coefficients w_1 and w_2 so as to obtain a high ratio of between-class variance to within-class variance in the projection onto axis $y = w_1x_1 + w_2x_2$ as follows:

$$\lambda = \frac{\text{between-class variance}}{\text{within-class variance}}. \tag{6.4}$$

In this approach, the projection axis is selected to obtain the largest possible between-class variance in the numerator together with the smallest possible within-class variance in the denominator. For this purpose, the ratio of between-class variance to within-class variance is expressed on the basis of the training data and the optimum projection axis is determined by the maximum ratio as next described.

6.1.2 Linear Discriminant Function

In the above approach, we first determine the projection axis that best separates the observed data. By projecting the i-th two-dimensional data $\boldsymbol{x}_i^{(1)} = (x_{i1}^{(1)}, x_{i2}^{(1)})^T$ of class G_1 onto $y = w_1 x_1 + w_2 x_2$, we have

$$y_i^{(1)} = w_1 x_{i1}^{(1)} + w_2 x_{i2}^{(1)}, \qquad i = 1, 2, \cdots, n_1. \tag{6.5}$$

Similarly the projection of the i-th data $\boldsymbol{x}_i^{(2)} = (x_{i1}^{(2)}, x_{i2}^{(2)})^T$ of class G_2 onto axis y is given by

$$y_i^{(2)} = w_1 x_{i1}^{(2)} + w_2 x_{i2}^{(2)}, \qquad i = 1, 2, \cdots, n_2. \tag{6.6}$$

In this way, by projection onto y, the two-dimensional data are reduced to one-dimensional data.

The sample means of classes G_1 and G_2, as obtained from the one-dimensional data on y, may be given by

$$\begin{aligned} \overline{y}^{(1)} &= \frac{1}{n_1}\sum_{i=1}^{n_1} y_i^{(1)} = \frac{1}{n_1}\sum_{i=1}^{n_1}(w_1 x_{i1}^{(1)} + w_2 x_{i2}^{(1)}) = w_1 \overline{x}_1^{(1)} + w_2 \overline{x}_2^{(1)}, \\ \overline{y}^{(2)} &= \frac{1}{n_2}\sum_{i=1}^{n_2} y_i^{(2)} = \frac{1}{n_2}\sum_{i=1}^{n_2}(w_1 x_{i1}^{(2)} + w_2 x_{i2}^{(2)}) = w_1 \overline{x}_1^{(2)} + w_2 \overline{x}_2^{(2)}. \end{aligned} \tag{6.7}$$

Accordingly, the between-class variance defined by the formula (6.2) can be expressed as

$$\begin{aligned} (\overline{y}^{(1)} - \overline{y}^{(2)})^2 &= \left\{ w_1(\overline{x}_1^{(1)} - \overline{x}_1^{(2)}) + w_2(\overline{x}_2^{(1)} - \overline{x}_2^{(2)}) \right\}^2 \\ &= \left\{ \boldsymbol{w}^T(\overline{\boldsymbol{x}}_1 - \overline{\boldsymbol{x}}_2) \right\}^2, \end{aligned} \tag{6.8}$$

where

$$\boldsymbol{w} = \begin{pmatrix} w_1 \\ w_2 \end{pmatrix}, \qquad \overline{\boldsymbol{x}}_1 = \begin{pmatrix} \overline{x}_1^{(1)} \\ \overline{x}_2^{(1)} \end{pmatrix}, \qquad \overline{\boldsymbol{x}}_2 = \begin{pmatrix} \overline{x}_1^{(2)} \\ \overline{x}_2^{(2)} \end{pmatrix}. \tag{6.9}$$

Also, when we project the data of class G_1 onto y, the sample variance on y is given by

$$\frac{1}{n_1 - 1}\sum_{i=1}^{n_1}\left(y_i^{(1)} - \overline{y}^{(1)}\right)^2$$

$$= \frac{1}{n_1 - 1} \sum_{i=1}^{n_1} \left\{ w_1^2 \left(x_{i1}^{(1)} - \overline{x}_1^{(1)} \right)^2 + 2w_1 w_2 \left(x_{i1}^{(1)} - \overline{x}_1^{(1)} \right) \left(x_{i2}^{(1)} - \overline{x}_2^{(1)} \right) \right.$$
$$\left. + w_2^2 \left(x_{i2}^{(1)} - \overline{x}_2^{(1)} \right)^2 \right\}$$
$$= w_1^2 s_{11}^{(1)} + 2w_1 w_2 s_{12}^{(1)} + w_2^2 s_{22}^{(1)} = \boldsymbol{w}^T S_1 \boldsymbol{w}, \tag{6.10}$$

and similarly we have the sample variance on y for the data of class G_2

$$\frac{1}{n_2 - 1} \sum_{i=1}^{n_2} \left(y_i^{(2)} - \overline{y}^{(2)} \right)^2 = w_1^2 s_{11}^{(2)} + 2w_1 w_2 s_{12}^{(2)} + w_2^2 s_{22}^{(2)} = \boldsymbol{w}^T S_2 \boldsymbol{w},$$

where S_1 and S_2 are, respectively, the sample variance-covariance matrices of G_1 and G_2 given by

$$S_1 = \begin{pmatrix} s_{11}^{(1)} & s_{12}^{(1)} \\ s_{21}^{(1)} & s_{22}^{(1)} \end{pmatrix}, \qquad S_2 = \begin{pmatrix} s_{11}^{(2)} & s_{12}^{(2)} \\ s_{21}^{(2)} & s_{22}^{(2)} \end{pmatrix} \tag{6.11}$$

Hence the within-class variance defined by the formula (6.3) can be written as

$$\frac{1}{n_1 + n_2 - 2} \left\{ (n_1 - 1) \boldsymbol{w}^T S_1 \boldsymbol{w} + (n_2 - 1) \boldsymbol{w}^T S_2 \boldsymbol{w} \right\} = \boldsymbol{w}^T S \boldsymbol{w}, \tag{6.12}$$

where

$$S = \frac{1}{n_1 + n_2 - 2} \{ (n_1 - 1) S_1 + (n_2 - 1) S_2 \}. \tag{6.13}$$

The matrix S is called the *pooled sample variance-covariance matrix*.

Therefore it follows from (6.8) and (6.12) that the ratio of between-class variance to within-class variance in the projection onto axis $y = w_1 x_1 + w_2 x_2$ can be expressed as

$$\lambda = \frac{\left\{ \boldsymbol{w}^T (\overline{\boldsymbol{x}}_1 - \overline{\boldsymbol{x}}_2) \right\}^2}{\boldsymbol{w}^T S \boldsymbol{w}}. \tag{6.14}$$

It can be shown from the result (6.90) in Section 6.4 that the coefficient vector $\boldsymbol{w}$ which maximizes the ratio λ is

$$\hat{\boldsymbol{w}} = S^{-1} (\overline{\boldsymbol{x}}_1 - \overline{\boldsymbol{x}}_2). \tag{6.15}$$

Thus, we obtain the optimum projection axis

$$y = \hat{w}_1 x_1 + \hat{w}_2 x_2 = \hat{\boldsymbol{w}}^T \boldsymbol{x} = (\overline{\boldsymbol{x}}_1 - \overline{\boldsymbol{x}}_2)^T S^{-1} \boldsymbol{x} \tag{6.16}$$

for maximum separation of the two classes. This linear function is called *Fisher's linear discriminant function.*

An observation $\boldsymbol{x}$ can be classified to one of the two classes G_1 and G_2 based on the projected point $\hat{\boldsymbol{w}}^T\boldsymbol{x}$. Projecting the sample mean vectors $\overline{\boldsymbol{x}}_1$ and $\overline{\boldsymbol{x}}_2$ onto axis y gives

$$G_1 : (\overline{\boldsymbol{x}}_1 - \overline{\boldsymbol{x}}_2)^T S^{-1}\overline{\boldsymbol{x}}_1, \qquad G_2 : (\overline{\boldsymbol{x}}_1 - \overline{\boldsymbol{x}}_2)^T S^{-1}\overline{\boldsymbol{x}}_2. \tag{6.17}$$

Then $\boldsymbol{x}$ is classified to the class whose projected sample mean $(\hat{\boldsymbol{w}}^T\overline{\boldsymbol{x}}_i)$ is closer to $\hat{\boldsymbol{w}}^T\boldsymbol{x}$. This is equivalent to comparing $\hat{\boldsymbol{w}}^T\boldsymbol{x}$ with the midpoint

$$\frac{1}{2}\left\{(\overline{\boldsymbol{x}}_1 - \overline{\boldsymbol{x}}_2)^T S^{-1}\overline{\boldsymbol{x}}_1 + (\overline{\boldsymbol{x}}_1 - \overline{\boldsymbol{x}}_2)^T S^{-1}\overline{\boldsymbol{x}}_2\right\} = \frac{1}{2}(\overline{\boldsymbol{x}}_1 - \overline{\boldsymbol{x}}_2)^T S^{-1}(\overline{\boldsymbol{x}}_1 + \overline{\boldsymbol{x}}_2), \tag{6.18}$$

and consequently we have the classification rule

$$h(\boldsymbol{x}) = (\overline{\boldsymbol{x}}_1 - \overline{\boldsymbol{x}}_2)^T S^{-1}\boldsymbol{x} - \frac{1}{2}(\overline{\boldsymbol{x}}_1 - \overline{\boldsymbol{x}}_2)^T S^{-1}(\overline{\boldsymbol{x}}_1 + \overline{\boldsymbol{x}}_2)\begin{cases} \geq 0 & \Rightarrow \quad G_1 \\ < 0 & \Rightarrow \quad G_2. \end{cases} \tag{6.19}$$

(See Figure 6.3.)

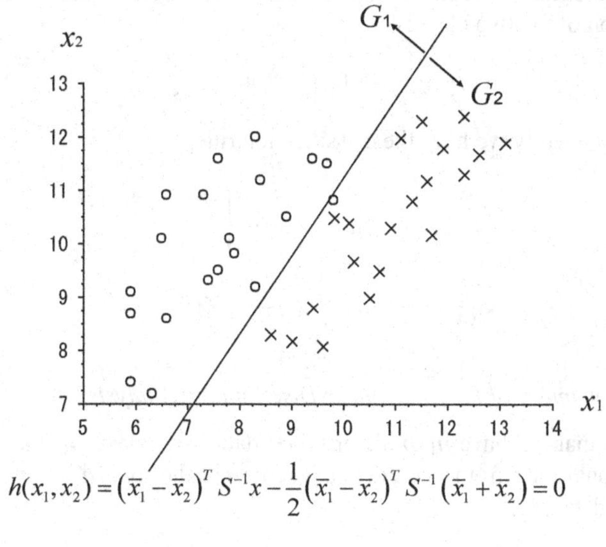

Figure 6.3 *Fisher's linear discriminant function.*

Example 6.1 (Linear discriminant function) Consider the 23 two dimensional leaf shape data in Table 6.1 in which the measurements are taken on the length (x_1) and width (x_2) for the classes A and B. The sample mean vectors and the sample variance-covariance matrices are

$$A: \overline{\boldsymbol{x}}_1 = \begin{pmatrix} 7.0 \\ 6.17 \end{pmatrix}, \qquad B: \overline{\boldsymbol{x}}_2 = \begin{pmatrix} 8.09 \\ 4.46 \end{pmatrix}, \tag{6.20}$$

$$A: S_1 = \begin{pmatrix} 2.18 & 0.36 \\ 0.36 & 1.61 \end{pmatrix}, \quad B: S_2 = \begin{pmatrix} 1.69 & 1.16 \\ 1.16 & 1.87 \end{pmatrix}. \tag{6.21}$$

The pooled sample variance-covariance matrix in (6.13) is then

$$S = \begin{pmatrix} 1.95 & 0.74 \\ 0.74 & 1.73 \end{pmatrix}. \tag{6.22}$$

Therefore, the coefficient vector in (6.16) is

$$\hat{\boldsymbol{w}} = S^{-1}(\overline{\boldsymbol{x}}_1 - \overline{\boldsymbol{x}}_2) = \begin{pmatrix} 0.61 & -0.26 \\ -0.26 & 0.69 \end{pmatrix} \begin{pmatrix} -1.09 \\ 1.71 \end{pmatrix} = \begin{pmatrix} -1.12 \\ 1.46 \end{pmatrix}.$$

Thus, we obtain the following optimum projection axis for maximum separation of the two classes

$$y = -1.12x_1 + 1.46x_2, \tag{6.23}$$

and consequently we have the classification rule

$$h(\boldsymbol{x}) = -1.12x_1 + 1.46x_2 + 0.59 \begin{cases} \geq 0 & \Rightarrow \quad A \\ < 0 & \Rightarrow \quad B, \end{cases} \tag{6.24}$$

where $(\overline{\boldsymbol{x}}_1 - \overline{\boldsymbol{x}}_2)^T S^{-1}(\overline{\boldsymbol{x}}_1 + \overline{\boldsymbol{x}}_2)/2 = -0.59$.

6.1.3 *Summary of Fisher's Linear Discriminant Analysis*

Suppose that we have n_1 p-dimensional data from class G_1 and n_2 p-dimensional data from class G_2, and represent the total $n = (n_1 + n_2)$ training data as

$$G_1: \boldsymbol{x}_1^{(1)}, \boldsymbol{x}_2^{(1)}, \cdots, \boldsymbol{x}_{n_1}^{(1)}, \qquad G_2: \boldsymbol{x}_1^{(2)}, \boldsymbol{x}_2^{(2)}, \cdots, \boldsymbol{x}_{n_2}^{(2)}. \tag{6.25}$$

Then the sample mean vectors and variance-covariance matrices for each class are given by

$$G_1: \ \overline{x}_1 = \frac{1}{n_1}\sum_{i=1}^{n_1} x_i^{(1)}, \qquad S_1 = \frac{1}{n_1 - 1}\sum_{i=1}^{n_1}(x_i^{(1)} - \overline{x}_1)(x_i^{(1)} - \overline{x}_1)^T, \tag{6.26}$$

$$G_2: \ \overline{x}_2 = \frac{1}{n_2}\sum_{i=1}^{n_2} x_i^{(2)}, \qquad S_2 = \frac{1}{n_2 - 1}\sum_{i=1}^{n_2}(x_i^{(2)} - \overline{x}_2)(x_i^{(2)} - \overline{x}_2)^T.$$

To determine the projection axis that best separates the observed data, we project the n p-dimensional data in (6.25) onto axis

$$y = w_1 x_1 + w_2 x_2 + \cdots + w_p x_p = \boldsymbol{w}^T \boldsymbol{x}. \tag{6.27}$$

The ratio of between-class variance to within-class variance in the projection onto axis $y = \boldsymbol{w}^T \boldsymbol{x}$ can be expressed as

$$\lambda = \frac{\left\{\boldsymbol{w}^T(\overline{x}_1 - \overline{x}_2)\right\}^2}{\boldsymbol{w}^T S \boldsymbol{w}} \tag{6.28}$$

with S pooled variance-covariance matrix given by $S = (n_1 + n_2 - 2)^{-1}\{(n_1 - 1)S_1 + (n_2 - 1)S_2\}$. We find the p-dimensional coefficient vector such that the between-class variance is maximized relative to the within-class variance. It follows from the result (6.90) in Section 6.4 that the solution is

$$\hat{\boldsymbol{w}} = S^{-1}(\overline{x}_1 - \overline{x}_2). \tag{6.29}$$

Thus we obtain *Fisher's linear discriminant function*

$$y = \hat{\boldsymbol{w}}^T \boldsymbol{x} = (\overline{x}_1 - \overline{x}_2)^T S^{-1} \boldsymbol{x}, \tag{6.30}$$

the optimum projection axis for maximum separation of the two classes. A future observation $\boldsymbol{x}$ is classified to the class whose projected sample mean $(\hat{\boldsymbol{w}}^T \overline{x}_i)$ is closer to $\hat{\boldsymbol{w}}^T \boldsymbol{x}$.

Noting that the midpoint between the projected sample means $(\overline{x}_1 - \overline{x}_2)^T S^{-1}\overline{x}_1$ and $(\overline{x}_1 - \overline{x}_2)^T S^{-1}\overline{x}_2$ is

$$\frac{1}{2}\left\{(\overline{x}_1 - \overline{x}_2)^T S^{-1}\overline{x}_1 + (\overline{x}_1 - \overline{x}_2)^T S^{-1}\overline{x}_2\right\} = \frac{1}{2}(\overline{x}_1 - \overline{x}_2)^T S^{-1}(\overline{x}_1 + \overline{x}_2),$$

we have the classification rule based on Fisher linear discriminant function in the form

$$h(\boldsymbol{x}) = (\overline{x}_1 - \overline{x}_2)^T S^{-1}\boldsymbol{x} - \frac{1}{2}(\overline{x}_1 - \overline{x}_2)^T S^{-1}(\overline{x}_1 + \overline{x}_2) \begin{cases} \geq 0 & \Rightarrow \ G_1 \\ < 0 & \Rightarrow \ G_2. \end{cases} \tag{6.31}$$

It is also possible, as next described, to adjust the decision boundary using the concept of prior probability and loss in cases in which the classification depends on whether the value of the linear discriminant function is positive or negative.

The maximum of the ratio in (6.28) attained at $\hat{\boldsymbol{w}} = S^{-1}(\overline{\boldsymbol{x}}_1 - \overline{\boldsymbol{x}}_2)$ is given by

$$(\overline{\boldsymbol{x}}_1 - \overline{\boldsymbol{x}}_2)^T S^{-1} (\overline{\boldsymbol{x}}_1 - \overline{\boldsymbol{x}}_2) \equiv D^2, \tag{6.32}$$

which is the *Mahalanobis distance* between the sample mean vectors $\overline{\boldsymbol{x}}_1$ and $\overline{\boldsymbol{x}}_2$. This distance measure is described in detail in Section 6.2.

6.1.4 Prior Probability and Loss

Up to this point, we have not considered the cost of the loss involved in classification performed with incorrect data incidence, or frequency of occurrence, in the two classes. Let us consider the case of stomach ulcers and stomach cancer. We first organize the medical test data in relation to several properties and divide the patients into class (G_1) for those with stomach ulcers and class (G_2) for those with stomach cancer, and construct a linear discriminant function on this basis. Using this function, we then attempt to assign new patients to one or the other class on the basis of their medical test data. It is safe to presume that the number of stomach ulcer patients is inherently quite different from the number of stomach cancer patients. In other words, a large difference in incidence naturally exists between the two diseases. In one approach, the incidences represent a form of information acquired in advance that is incorporated into the construction of the determinant function as a *prior probability*.

Let the relative incidence of stomach ulcers and stomach cancer be represented by π_1 and π_2 $(\pi_1 + \pi_2 = 1)$, respectively. In ordinary linear classification, as discussed above, the assignment for future data is based on the value of the linear discriminant function with 0 as the classification point. In linear determination incorporating prior probability, in contrast, assignment to class G_1 is performed when

$$h(\boldsymbol{x}) = (\overline{\boldsymbol{x}}_1 - \overline{\boldsymbol{x}}_2)^T S^{-1} \boldsymbol{x} - \frac{1}{2}(\overline{\boldsymbol{x}}_1 - \overline{\boldsymbol{x}}_2)^T S^{-1} (\overline{\boldsymbol{x}}_1 + \overline{\boldsymbol{x}}_2) > \log\left(\frac{\pi_2}{\pi_1}\right).$$

The stomach ulcer incidence π_1 is higher than the stomach cancer incidence π_2, and accordingly the value of $\log(\pi_2/\pi_1)$ is negative. Shifting the classification point from 0 toward the class with the lower incidence (in this case, the stomach cancer class) will presumably facilitate classification-based assignment to the higher-incidence class G_1 (the stomach ulcer patient class). This type of classification point operation may be considered in cases such as plant variety classification in which one variety is extremely rare and the relative proportion of observation data acquired for it is inherently quite small. In such a case, incorporating this difference can be presumed to be meaningful.

The question remains, however, whether this would also be true in cases such as classification between the stomach ulcer patient class (G_1) and the stomach cancer patient class (G_2). When incidence is incorporated, it effectively raises the bar for assignment of patients to the stomach cancer class. In this regard, however, it must be noted that if a patient who actually has stomach cancer is judged to have a stomach ulcer, the loss may be irreparable and very large. If on the other hand, a stomach ulcer is mistakenly judged to be stomach cancer, the loss will presumably be substantially smaller. Some method is therefore necessary to incorporate the concept of the *cost of loss* due to mistaken classification and thereby further adjust the classification point.

For this purpose, let the cost of the loss be $c(2 \mid 1)$ if a determination results in assignment to the stomach cancer class G_2 for a patient who actually belongs in the stomach ulcer class G_1, and $c(1 \mid 2)$ if the reverse occurs and assignment is made to the stomach ulcer class G_1 when the patient actually belongs in the stomach cancer class G_2. It is of course necessary to make $c(1 \mid 2)$ large relative to $c(2 \mid 1)$. On this basis, if the value of the linear discriminant function for medical test data from a new patient is found to be

$$h(\boldsymbol{x}) = (\overline{\boldsymbol{x}}_1 - \overline{\boldsymbol{x}}_2)^T S^{-1}\boldsymbol{x} - \frac{1}{2}(\overline{\boldsymbol{x}}_1 - \overline{\boldsymbol{x}}_2)^T S^{-1}(\overline{\boldsymbol{x}}_1 + \overline{\boldsymbol{x}}_2) > \log\left[\frac{\pi_2 c(1 \mid 2)}{\pi_1 c(2 \mid 1)}\right],$$

then the patient will be assigned to G_1. In cases such as this example of stomach ulcer and stomach cancer, the value of $\log\,[\pi_2 c(1 \mid 2)/\pi_1 c(2 \mid 1)]$ becomes positive by making $c(1 \mid 2)$ relatively large and effectively lowers the bar for assignment to the class having the lower incidence.

It is actually possible to estimate the incidence from the number of observed data: if the numbers of observations in classes G_1 and G_2 are taken as n_1 and n_2, respectively, the estimated incidence is then $\pi_1 = n_1/(n_1 + n_2)$ and $\pi_2 = n_2/(n_1 + n_2)$. This estimation, however, can be applied only if the data acquisition is random for both classes. If it is not random, then the estimate will not correctly represent the true incidence, and considerable care is required in this regard. The appropriate method for determination of the classification point varies with the problem, and it is always necessary to make this determination carefully on the basis of an appropriate variety of information obtained by effective information gathering.

6.2 Classification Based on Mahalanobis Distance

With Fisher's linear discriminant function, the classification consists of projecting p-dimensional observed data from both classes onto axes obtained by linear combination of variables and selecting as the optimum projection axis the one that maximizes the ratio of between-class variance to within-class variance. In this section, we discuss a method in which the distance of an observation to the class sample mean vector is defined and the observation is assigned to the class being the closest distance. The sample mean vector representing each class can be considered as its center of mass.

6.2.1 *Two-Class Classification*

Suppose that we have n_1 p-dimensional data from class G_1 and n_2 p-dimensional data from class G_2, and represent the sample mean vector and variance-covariance matrix for each class as

$$
\begin{aligned}
G_1: \quad \overline{\boldsymbol{x}}_1 &= \frac{1}{n_1}\sum_{i=1}^{n_1} \boldsymbol{x}_i^{(1)}, \qquad S_1 = \frac{1}{n_1 - 1}\sum_{i=1}^{n_1} (\boldsymbol{x}_i^{(1)} - \overline{\boldsymbol{x}}_1)(\boldsymbol{x}_i^{(1)} - \overline{\boldsymbol{x}}_1)^T, \\
G_2: \quad \overline{\boldsymbol{x}}_2 &= \frac{1}{n_2}\sum_{i=1}^{n_2} \boldsymbol{x}_i^{(2)}, \qquad S_2 = \frac{1}{n_2 - 1}\sum_{i=1}^{n_2} (\boldsymbol{x}_i^{(2)} - \overline{\boldsymbol{x}}_2)(\boldsymbol{x}_i^{(2)} - \overline{\boldsymbol{x}}_2)^T.
\end{aligned} \tag{6.33}
$$

The (squared) *Mahalanobis distance* of $\boldsymbol{x}$ to the i-th class sample mean vector $\overline{\boldsymbol{x}}_i$ is then defined by

$$
D_i^2 = (\boldsymbol{x} - \overline{\boldsymbol{x}}_i)^T S_i^{-1} (\boldsymbol{x} - \overline{\boldsymbol{x}}_i), \qquad i = 1, 2. \tag{6.34}
$$

A new observed data $\boldsymbol{x}$ is assigned to the class whose sample mean vector is closest to $\boldsymbol{x}$ in Mahalanobis distance. The classification rule assigns $\boldsymbol{x}$ to G_1 if

$$
\begin{aligned}
Q(\boldsymbol{x}) &= D_2^2 - D_1^2 \qquad\qquad (6.35) \\
&= (\boldsymbol{x} - \overline{\boldsymbol{x}}_2)^T S_2^{-1} (\boldsymbol{x} - \overline{\boldsymbol{x}}_2) - (\boldsymbol{x} - \overline{\boldsymbol{x}}_1)^T S_1^{-1} (\boldsymbol{x} - \overline{\boldsymbol{x}}_1) \geq 0,
\end{aligned}
$$

and to G_2 otherwise, which gives a *quadratic discriminant function* with quadratic decision boundaries.

The sample variance-covariance matrices S_1 and S_2 are combined to obtain an unbiased estimator of the common population variance-covariance matrix Σ in the following:

$$S = \frac{1}{n_1 + n_2 - 2}\{(n_1 - 1)S_1 + (n_2 - 1)S_2\}. \qquad (6.36)$$

Then the quadratic function $Q(\boldsymbol{x}) = D_2^2 - D_1^2$ in (6.35) can be rewritten as

$$h(\boldsymbol{x}) = (\overline{\boldsymbol{x}}_1 - \overline{\boldsymbol{x}}_2)^T S^{-1}\boldsymbol{x} - \frac{1}{2}\left(\overline{\boldsymbol{x}}_1^T S^{-1}\overline{\boldsymbol{x}}_1 - \overline{\boldsymbol{x}}_2^T S^{-1}\overline{\boldsymbol{x}}_2\right). \qquad (6.37)$$

Noting that $\left(\overline{\boldsymbol{x}}_1^T S^{-1}\overline{\boldsymbol{x}}_1 - \overline{\boldsymbol{x}}_2^T S^{-1}\overline{\boldsymbol{x}}_2\right) = (\overline{\boldsymbol{x}}_1 - \overline{\boldsymbol{x}}_2)^T S^{-1}(\overline{\boldsymbol{x}}_1 + \overline{\boldsymbol{x}}_2)$, the discriminant function based on Mahalanobis distance with the pooled sample variance-covariance matrix corresponds to Fisher's linear discriminant function in (6.31).

In quadratic discriminant analysis, in contrast to linear discriminant analysis, it is difficult to assess the relative importance of the variables from their estimated coefficients. To investigate variable x_1, for example, it is necessary to consider the estimated values of the coefficients of both x_1^2 and x_1, x_2 in the quadratic equation at the same time, making it quite difficult to determine whether it is effective for separation of the two classes. If the sample variance-covariance matrices of the two classes differ substantially and the number of samples from each class is quite large, on the other hand, quadratic discriminant analysis is more effective than linear discriminant analysis for good separation of the two classes.

6.2.2 *Multiclass Classification*

When training data from three or more classes have been observed, the construction of a classification procedure based on Mahalanobis distance is straightforward. The basic concept is similar to that of two-class classification: when an observation $\boldsymbol{x}$ has been obtained, the distance to the center of mass of each class (the sample mean vector) is computed and $\boldsymbol{x}$ is assigned to the class achieving the shortest distance.

Suppose that we have n_j p-dimensional data from class G_j ($j = 1, 2, \cdots, g$)

$$G_j: \ \boldsymbol{x}_1^{(j)}, \boldsymbol{x}_2^{(j)}, \cdots, \boldsymbol{x}_{n_j}^{(j)}, \qquad j = 1, 2, \cdots, g. \qquad (6.38)$$

For each class G_j the sample mean vector and the sample variance-covariance matrix are given by

$$G_j : \overline{\boldsymbol{x}}_j = \frac{1}{n_j}\sum_{i=1}^{n_j} \boldsymbol{x}_i^{(j)}, \quad S_j = \frac{1}{n_j - 1}\sum_{i=1}^{n_j} (\boldsymbol{x}_i^{(j)} - \overline{\boldsymbol{x}}_j)(\boldsymbol{x}_i^{(j)} - \overline{\boldsymbol{x}}_j)^T. \quad (6.39)$$

Then the classification rule is constructed by assigning $\boldsymbol{x}$ to class G_j, $j = 1, 2, \cdots, g$, for which

$$D_j^2 = (\boldsymbol{x} - \overline{\boldsymbol{x}}_j)^T S_j^{-1} (\boldsymbol{x} - \overline{\boldsymbol{x}}_j), \qquad j = 1, 2, \cdots, g \qquad (6.40)$$

is smallest, which gives quadratic discriminant functions with quadratic decision boundaries

$$Q_{jk}(\boldsymbol{x}) = D_j^2 - D_k^2 \qquad j \neq k;\ j, k = 1, 2, \cdots, g. \qquad (6.41)$$

Taking the pooled sample variance-covariance matrix as

$$S = \frac{1}{n - g}\left\{(n_1 - 1)S_1 + (n_2 - 1)S_2 + \cdots + (n_g - 1)S_g\right\}, \quad (6.42)$$

where $n = n_1 + n_2 + \cdots + n_g$, we have linear discriminant functions with linear decision boundaries. The g predicted classes are separated by linear functions

$$h_{jk}(\boldsymbol{x}) = D_j^2 - D_k^2 \qquad j \neq k;\ j, k = 1, 2, \cdots, g. \qquad (6.43)$$

What is Mahalanobis distance? Figure 6.4 shows the data dispersion in the form of an ellipse. This probability ellipse is analogous to the representation of elevation in a mountain contour map and shows that the data concentration increases as the center of mass is approached. In the left panel the sample mean vector and variance-covariance matrix are

$$\overline{\boldsymbol{x}} = \begin{bmatrix} 7.6 \\ 10.0 \end{bmatrix}, \qquad S = \begin{bmatrix} 1.56 & 1.17 \\ 1.17 & 1.89 \end{bmatrix}. \qquad (6.44)$$

The squared Euclidean distances from the points $\boldsymbol{x}_P = (9.5, 11.8)^T$ and $\boldsymbol{x}_Q = (9.5, 8.2)^T$ to $\overline{\boldsymbol{x}}$ are $(\boldsymbol{x}_P - \overline{\boldsymbol{x}})^T(\boldsymbol{x}_P - \overline{\boldsymbol{x}}) = (\boldsymbol{x}_Q - \overline{\boldsymbol{x}})^T(\boldsymbol{x}_Q - \overline{\boldsymbol{x}}) = 6.85$. In contrast the squared Mahalanobis distances are $D_P^2 = (\boldsymbol{x}_P - \overline{\boldsymbol{x}})^T S^{-1}(\boldsymbol{x}_P - \overline{\boldsymbol{x}}) = 2.48$ and $D_Q^2 = (\boldsymbol{x}_Q - \overline{\boldsymbol{x}})^T S^{-1}(\boldsymbol{x}_Q - \overline{\boldsymbol{x}}) = 12.61$. The Mahalanobis distances of points P and Q from $\overline{\boldsymbol{x}}$ are clearly different; point P is substantially closer. The inclusion of point P in the class is stronger than that of Q, and this is effectively captured by the Mahalanobis distance. In the mountain analogy, the Mahalanobis distance is an explicit expression of the relatively gentle gradient (slope) of a climb from point P to the center of mass, and the relatively steep gradient from point Q.

$$S = \begin{pmatrix} s_{11} & s_{12} \\ s_{21} & s_{22} \end{pmatrix} \qquad I = \begin{pmatrix} 1 & 0 \\ 0 & 1 \end{pmatrix}$$

$$(x-\overline{x})^T S^{-1}(x-\overline{x}) \qquad (x-\overline{x})^T (x-\overline{x})$$

Squared Mahalanobis distance   Squared Euclidean distance

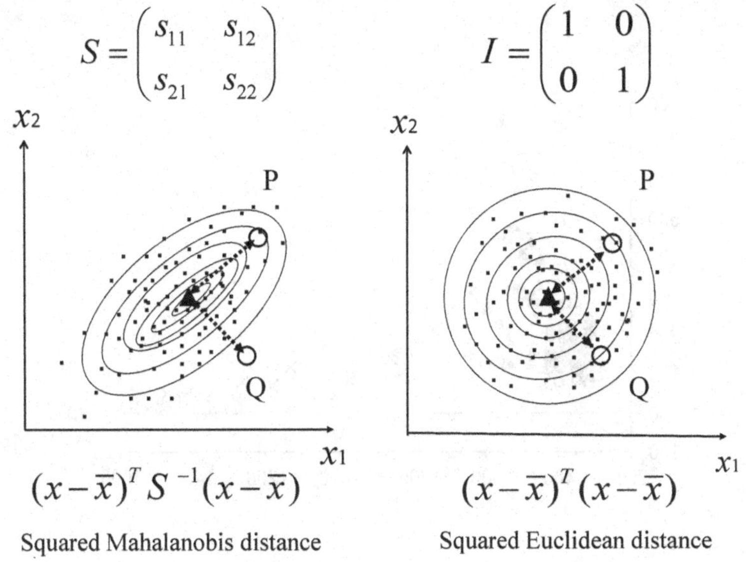

Figure 6.4 *Mahalanobis distance and Euclidean distance.*

6.2.3   Example: Diagnosis of Diabetes

Reaven and Miller (1979) have described a classification for 145 non-obese adult subjects by three variables, with assignment to three classes: a normal class (G_1), a chemical diabetes class (G_2) with no finding of clinical abnormality but with a glucose tolerance test finding of abnormality, and a clinical diabetes class (G_3) showing clinical symptoms characteristic of diabetes. Here, we first configure the three-class classification based on two of the test variables. The two diabetes test variables are as follows.

x_1 : Glucose is ingested after fasting by each subject, the plasma glucose level is then measured every 30 min for 3 h, and the area under the resulting curve is taken as a test value.

x_2 : In the same way, the plasma insulin level is measured and the area under the resulting curve is taken as a test value.

Figure 6.5 shows the plot of 145 training data for a normal class G_1 (∘), a chemical diabetes class G_2 (▲) and clinical diabetes class G_3 (×), in which G_1, G_2, and G_3 include $n_1 = 76$, $n_2 = 36$, and $n_3 = 33$

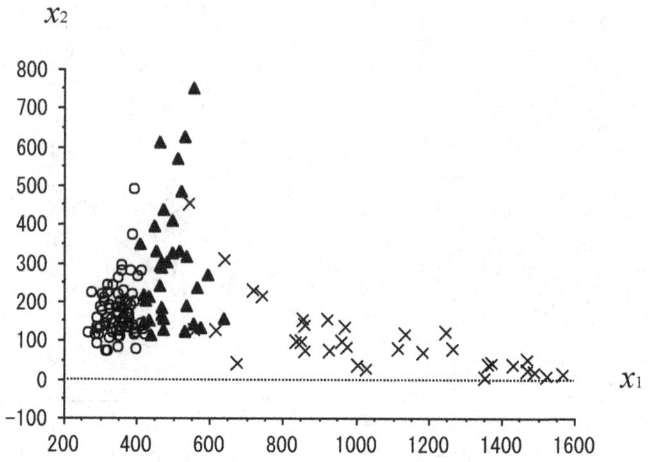

Figure 6.5 *Plot of 145 training data for a normal class G_1 (○), a chemical diabetes class G_2 (▲), and clinical diabetes class G_3 (×).*

observations, respectively. The objective is to formulate a method for effective separation of these three classes.

The sample mean vectors and sample variance-covariance matrices are

$$\begin{aligned}
G_1 &: \ \bar{\boldsymbol{x}}_1 = \begin{bmatrix} 350.0 \\ 172.6 \end{bmatrix}, & S_1 &= \begin{bmatrix} 1359.4 & 574.7 \\ 574.7 & 4740.9 \end{bmatrix}, \\
G_2 &: \ \bar{\boldsymbol{x}}_2 = \begin{bmatrix} 493.9 \\ 288.0 \end{bmatrix}, & S_2 &= \begin{bmatrix} 3070.5 & 1082.5 \\ 1082.5 & 24910.9 \end{bmatrix}, \qquad (6.45) \\
G_3 &: \ \bar{\boldsymbol{x}}_3 = \begin{bmatrix} 1043.8 \\ 106.0 \end{bmatrix}, & S_3 &= \begin{bmatrix} 95725.4 & -19843.2 \\ -19843.2 & 8728.3 \end{bmatrix}.
\end{aligned}$$

Let us consider an individual who newly undergoes the tests with resulting test values of $\boldsymbol{x} = (x_1, x_2)^T = (820, 150)^T$. The question is then which of the three classes, normal, chemical diabetes, or clinical diabetes, this person belongs to. For this diagnosis of diabetes, we use linear discriminant analysis based on a pooled sample variance-covariance matrix for the three classes.

It follows from (6.45) and (6.42) with $g = 3$ that the pooled sample variance-covariance matrix is

$$S = \begin{bmatrix} 23046.8 & -3901.4 \\ -3901.4 & 10610.9 \end{bmatrix}.$$

The squared Mahalanobis distances of $\boldsymbol{x} = (x_1, x_2)^T = (820, 150)^T$ to $\overline{\boldsymbol{x}}_i$, the sample mean vector of G_i, are given by

$$\begin{aligned} G_1 &: \ D_1^2 = (\boldsymbol{x} - \overline{\boldsymbol{x}}_1)^T S^{-1} (\boldsymbol{x} - \overline{\boldsymbol{x}}_1) = 9.91, \\ G_2 &: \ D_2^2 = (\boldsymbol{x} - \overline{\boldsymbol{x}}_2)^T S^{-1} (\boldsymbol{x} - \overline{\boldsymbol{x}}_2) = 5.3, \\ G_3 &: \ D_3^2 = (\boldsymbol{x} - \overline{\boldsymbol{x}}_3)^T S^{-1} (\boldsymbol{x} - \overline{\boldsymbol{x}}_3) = 2.18. \end{aligned} \tag{6.46}$$

The sample mean vector $\overline{\boldsymbol{x}}_3$ in G_3 is closest to $\boldsymbol{x} = (820, 150)^T$ in Mahalanobis distance, and thus we assign this data $\boldsymbol{x}$ as belonging to the clinical diabetes class.

Figure 6.6 shows the lines separating the three classes based on Mahalanobis distance. With line L_{12}, the Mahalanobis distances D_1^2 to the center of mass $\overline{\boldsymbol{x}}_1 = (350.0, 172.6)^T$ of class G_1 and D_2^2 to that $\overline{\boldsymbol{x}}_2 = (493.9, 288.0)^T$ of class G_2 each represents the set (x_1, x_2) such that $D_1^2 = D_2^2$ ($D_2^2 - D_1^2 = 0$), and the locus is clearly a straight line. The L_{12} represents a linear decision boundary that separates the normal class G_1 and the chemical diabetes class G_2. Similarly, L_{23} separates the chemical diabetes class G_2 and the clinical diabetes class G_3, and L_{31} the clinical diabetes class G_3 and the normal class G_1. These linear discriminant functions are given by

$$\begin{aligned} L_{12} &= -0.0086x_1 - 0.014x_2 + 6.873, \\ L_{23} &= -0.0223x_1 + 0.0089x_2 + 15.418, \\ L_{31} &= 0.031x_1 + 0.0051x_2 - 22.291. \end{aligned}$$

As shown in Figure 6.6, the separation of the regions in the input space by a straight line clearly demonstrates that we are in this case essentially performing linear discriminant analysis of multiple class combinations. If such predicted class regions are formed in classification based on two variables, then the class membership of new test values can be readily determined by plotting them in the figure. The course of the symptoms can also be ascertained to some degree from the positional relationship of each class. Finally, by narrowing the normal class region, it is possible to reduce the probability of diagnostic failure by assigning an individual actually having an abnormality to the normal class. For data dispersed in a higher-dimensional space we consider a method of projection onto a two-dimensional plane or a three-dimensional space for visualization in Section 6.4.

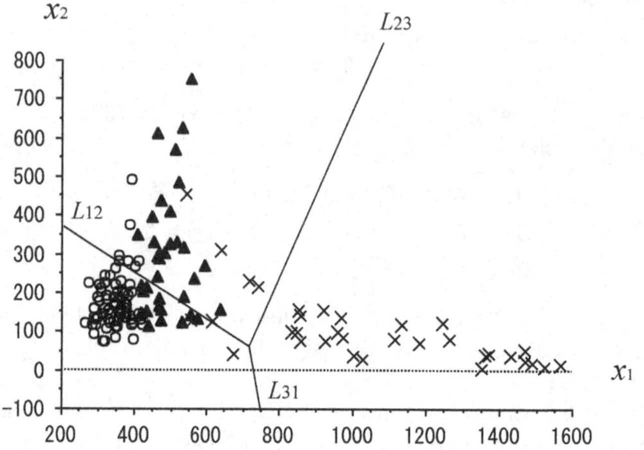

Figure 6.6 *Linear decision boundaries that separate the normal class* G_1, *the chemical diabetes class* G_2, *and the clinical diabetes class* G_3.

6.3 Variable Selection

We have discussed the basic concept of constructing classification rules based on training data with clearly known class membership. With careful investigation of the formulated discriminant function, it may become possible to effectively separate two classes through the use of certain of the variables in combination, rather than all of the variables. If we can find a set of fewer variables that is effective for classification, then we may well be able to formulate a more stable discriminant function. This brings us to the problem of variable selection in discriminant analysis. Particularly in clinical differential diagnosis, identification of a variables combination that is effective for classification of many variables (i.e., the results of many different tests) can substantially reduce the burden and cost of patient testing. It is also important in regard to analysis of disease factors. In this section, we consider this problem.

6.3.1 Prediction Errors

We consider the problem of finding the optimum set of variables for effective classification. One such variable selection criterion is the error rate (the *prediction error*) of a formulated classification rule in predicting future data. Classification rules based on linear, quadratic, nonlinear, and various other discriminant procedures are generally denoted as $r(\boldsymbol{x}|\boldsymbol{Z}_n)$, where $\boldsymbol{Z}_n$ is the training data having known class labels given by

$$\boldsymbol{Z}_n = \{z_1, z_2, \cdots, z_n\} \quad \text{with } z_i = (\boldsymbol{x}_i, y_i). \tag{6.47}$$

In classifying data $\boldsymbol{x}$, the class label y outputs 1 or 0 in two-class classification or a class number in multiclass classification.

To assess the goodness of a classification rule with regard to observed data $(\boldsymbol{x}_0, y_0)$ from one of the target classes, we use the indicator

$$Q[y_0, r(\boldsymbol{x}_0|\boldsymbol{Z}_n)] = \begin{cases} 0 & \text{when } y_0 = r(\boldsymbol{x}_0|\boldsymbol{Z}_n) \\ 1 & \text{when } y_0 \neq r(\boldsymbol{x}_0|\boldsymbol{Z}_n). \end{cases} \tag{6.48}$$

This function returns 0 if the data membership is correctly determined and 1 if it is not, since y_0 serves as a class label representing the class membership of the observed data $\boldsymbol{x}_0$.

Once we have constructed a classification rule based on the training data $\boldsymbol{Z}_n$, it is then necessary to assess its capability for correct classification of future data. Suppose that $(\boldsymbol{X}_0, Y_0) = (\boldsymbol{x}_0, y_0)$ is a future data randomly drawn from the probability distribution F that generated the training data. Then the *prediction error* is defined by

$$\text{PRE}(\boldsymbol{Z}_n, F) = \text{E}_F\left[Q[Y_0, r(\boldsymbol{X}_0|\boldsymbol{Z}_n)]\right]. \tag{6.49}$$

Finding an effective method of estimating its value is the essential problem of prediction error estimation.

One method of estimating the prediction error is to reuse the training data $\boldsymbol{Z}_n$ that were originally used to construct the classification rule, in place of future data. This yields the *apparent error rate* given by

$$\text{APE}(\boldsymbol{Z}_n, \hat{F}) = \frac{1}{n}\sum_{i=1}^{n} Q[y_i, r(\boldsymbol{x}_i|\boldsymbol{Z}_n)]. \tag{6.50}$$

This function gives the proportion of input data $\boldsymbol{x}_i$ that do not yield an output of $y_i = r(\boldsymbol{x}_i|\boldsymbol{Z}_n)$, and thus the proportion of data that have been misclassified. In this estimation, the unknown probability distribution F is taken to be an *empirical distribution function*, with equal probability of $1/n$ for each of the training data. We may note that taking the expectation in (6.49) with respect to this empirical distribution function gives

$$\text{PRE}(\boldsymbol{Z}_n, \hat{F}) = \text{E}_{\hat{F}}\left[Q[Y_0, r(\boldsymbol{X}_0|\boldsymbol{Z}_n)]\right] = \frac{1}{n}\sum_{i=1}^{n} Q[y_i, r(\boldsymbol{x}_i|\boldsymbol{Z}_n)]. \tag{6.51}$$

The apparent error rate tends to underestimate the prediction error because of its use of the same data in constructing and evaluating the classification rule. *Cross-validation* (Stone, 1974) provides a method of reducing this tendency, since the data used to estimate the prediction error in (6.49) are separate from the training data used to construct the classification rule. In cross-validation, the classification rule is first constructed from the $n-1$ data $\boldsymbol{Z}_n^{(-i)}$ that remain after removing the i-th data $(\boldsymbol{x}_i, y_i)$ from the n observed data, and denoted by $r(\boldsymbol{x}|\boldsymbol{Z}_n^{(-i)})$. Then the removed data $(\boldsymbol{x}_i, y_i)$ is classified, and the indicator $Q[y_i, r(\boldsymbol{x}_i|\boldsymbol{Z}_n^{(-i)})]$ is applied to assess whether the classification is correct. This process is repeated for all of the data, and thus provides an estimate of the prediction error (6.49) as

$$\mathrm{CV} = \frac{1}{n}\sum_{i=1}^{n} Q[y_i, r(\boldsymbol{x}_i|\boldsymbol{Z}_n^{(-i)})]. \tag{6.52}$$

In general, the n observed data are divided into K approximately equal-sized data sets $\{\boldsymbol{Z}_1, \boldsymbol{Z}_2, \cdots, \boldsymbol{Z}_K\}$, the model is estimated from the $K-1$ data sets that remain after removal of the i-th data set $\boldsymbol{Z}_i$, and the resulting model is then evaluated using $\boldsymbol{Z}_i$, which comprises the previously removed n/K data. This process is applied in turn to $i = 1, 2, \cdots, K$, and the mean value of the results is taken as the estimate of the prediction error rate. This is accordingly known as K-fold cross-validation. In the case where $K = n$ mentioned previously, it is known as leave-one-out cross-validation.

6.3.2 Bootstrap Estimates of Prediction Errors

The apparent error rate in (6.50) generally underestimates the prediction error (6.49), because the same data are used in constructing and evaluating a classification rule. The bootstrap methods may be used to assess the degree of underestimation, that is, the bias of the apparent error rate in estimating the prediction error defined by

$$\mathrm{bias}(F) = \mathrm{E}_{F(\boldsymbol{Z})}\left[\mathrm{E}_F\left[Q[Y_0, r(\boldsymbol{X}_0|\boldsymbol{Z}_n)]\right] - \frac{1}{n}\sum_{i=1}^{n} Q[Y_i, r(\boldsymbol{X}_i|\boldsymbol{Z}_n)]\right]. \tag{6.53}$$

By correcting the bias of the apparent error rate, we obtain a more refined result for the estimation of the prediction error. This can be performed using the bootstrap, essentially as follows.

The basic idea of the bootstrap is to use the empirical distribution function $\hat{F}$ to estimate the probability distribution F that has generated the data (see Appendix A). We perform repeated extractions of size n bootstrap samples from the empirical distribution function. The training data $\boldsymbol{Z}_n = \{(\boldsymbol{x}_i, y_i);\ i = 1, 2, \cdots, n\}$ are sampled with replacement, thus forming the B bootstrap samples

$$\boldsymbol{Z}_n^*(b) = \{(\boldsymbol{x}_i^*(b), y_i^*(b));\ i = 1, \cdots, n\}, \quad b = 1, 2, \cdots, B. \quad (6.54)$$

We then take a bootstrap sample $\boldsymbol{Z}_n^*(b)$ as training data and use it to construct the classification rule, which we denote as $r(\boldsymbol{x}|\boldsymbol{Z}_n^*(b))$. By replacing the unknown probability distribution function F in (6.49) with the empirical distribution function $\hat{F}$, we obtain the prediction error

$$\begin{aligned} \mathrm{PRE}(\boldsymbol{Z}_n^*(b), \hat{F}) &= \mathrm{E}_{\hat{F}}\left[Q[Y, r(\boldsymbol{X}|\boldsymbol{Z}_n^*(b))]\right] \\ &= \frac{1}{n}\sum_{i=1}^{n} Q[y_i, r(\boldsymbol{x}_i|\boldsymbol{Z}_n^*(b))], \end{aligned} \quad (6.55)$$

which may be viewed as an expression of the use of the training data to evaluate the classification rule constructed with the bootstrap sample.

Because the data that have been used to construct the classification rule are reused to estimate the prediction error, the apparent error rate is given by

$$\mathrm{APE}(\boldsymbol{Z}_n^*(b), \hat{F}^*) = \frac{1}{n}\sum_{i=1}^{n} Q[y_i^*(b), r(\boldsymbol{x}_i^*(b)|\boldsymbol{Z}_n^*(b))], \quad (6.56)$$

where $\hat{F}^*$ is the empirical distribution function with an equal probability $1/n$ for each point on $\boldsymbol{Z}_n^*(b) = \{(\boldsymbol{x}_i^*(b), y_i^*(b));\ i = 1, \cdots, n\}$. The bootstrap estimate of the bias in (6.53) is therefore given by

$$\begin{aligned} &\mathrm{bias}(\hat{F}) \quad (6.57) \\ &= \mathrm{E}_{\hat{F}}\left[\frac{1}{n}\sum_{i=1}^{n} Q[y_i, r(\boldsymbol{x}_i|\boldsymbol{Z}_n^*(b))] - \frac{1}{n}\sum_{i=1}^{n} Q[y_i^*(b), r(\boldsymbol{x}_i^*(b)|\boldsymbol{Z}_n^*(b))]\right]. \end{aligned}$$

The expected value for the empirical distribution function is numerically approximated by using the B bootstrap samples as follows:

$$\begin{aligned} &\mathrm{bias}(\hat{F})^{(BS)} \quad (6.58) \\ &= \frac{1}{B}\sum_{b=1}^{B}\left\{\frac{1}{n}\sum_{i=1}^{n} Q[y_i, r(\boldsymbol{x}_i|\boldsymbol{Z}_n^*(b))] - \frac{1}{n}\sum_{i=1}^{n} Q[y_i^*(b), r(\boldsymbol{x}_i^*(b)|\boldsymbol{Z}_n^*(b))]\right\}. \end{aligned}$$

This estimate of the bias is added to the apparent error rate to correct for its underestimation, as

$$\mathrm{APE}^{(BS)}(\boldsymbol{Z}_n, \hat{F}) = \frac{1}{n}\sum_{i=1}^{n} Q[y_i, r(\boldsymbol{x}_i|\boldsymbol{Z}_n)] + \mathrm{bias}(\hat{F})^{(BS)}, \quad (6.59)$$

which is then used as the prediction error estimate.

In cross-validation, wc estimated the prediction error by taking part of the data as the test data. In bootstrapping, in contrast, we have estimated the prediction error by constructing the classification rule from bootstrap samples obtained from the training data by repeated sampling with replacement and then using the training data themselves as the test data.

6.3.3 The .632 Estimator

Even with the above bootstrap estimate of the bias in the apparent error rate and the addition of this bias estimate to correct for underestimation of the prediction error, as shown in (6.59), the tendency for underestimation is diminished but not eliminated. This indicates that the bias obtained by (6.58) is itself an underestimate. The bias underestimation can be attributed to replacement of the prediction error as performed in the bootstrap.

It may be recalled that the prediction error was defined as $\mathrm{PRE}(\boldsymbol{Z}_n, F) = \mathrm{E}_F[Q[Y_0, r(\boldsymbol{X}_0|\boldsymbol{Z}_n)]]$, which applies to random sampling of future data $(\boldsymbol{X}_0, Y_0) = (\boldsymbol{x}_0, y_0)$ by the same probability distribution F independently of the training data. An error thus arises from the random sampling of data $((\boldsymbol{x}_0, y_0) \notin \boldsymbol{Z}_n)$ (different from the training data) that are at some distance from the training data. In the numerical approximation within the bootstrap framework, as may be seen from the first term on the right-hand side of (6.58), this is expressed as

$$\frac{1}{B}\sum_{b=1}^{B}\frac{1}{n}\sum_{i=1}^{n} Q\left[y_i, r(\boldsymbol{x}_i|\boldsymbol{Z}_n^*(b))\right]. \quad (6.60)$$

The problem with this formula is the extremely high probability that the data $(\boldsymbol{x}_i, y_i)$ used in evaluating the classification rule $r(\boldsymbol{x}_i|\boldsymbol{Z}_n^*(b))$ constructed from the bootstrap samples $\boldsymbol{Z}_n^*(b)$ have been included in the bootstrap samples themselves. This probability is given by

$$\Pr\{(\boldsymbol{x}_i, y_i) \in \boldsymbol{Z}_n^*(b)\} = 1 - \left(1 - \frac{1}{n}\right)^n$$

$$\approx 1 - e^{-1} \qquad (n \to +\infty) \tag{6.61}$$
$$= 0.632.$$

It should then be possible to reduce the degree of bias underestimation by ensuring that the bias is computed only for cases in which the i-th data are not included in the bootstrap samples $\boldsymbol{Z}_n^*(b)$. In this light, the expected value of cross-validation estimated by the bootstrap method may be expressed as

$$\mathrm{CV}_{BS} = \frac{1}{n}\sum_{b=1}^{B}\sum_{i=1}^{n}\left\{\frac{1}{B_i}\sum_{b\in C_i} Q\left[y_i, r(\boldsymbol{x}_i|\boldsymbol{Z}_n^*(b))\right]\right\}, \tag{6.62}$$

where C_i represents the set of bootstrap sample numbers, among bootstrap samples $\boldsymbol{Z}_n^*(b)$, that do not include data $(\boldsymbol{x}_i, y_i)$, and B_i is the number of such bootstrap samples. If, for example, we have 10 bootstrap samples $\boldsymbol{Z}_n^*(1), \cdots, \boldsymbol{Z}_n^*(10)$ and data $(\boldsymbol{x}_i, y_i)$ are not included in $\boldsymbol{Z}_n^*(2)$, $\boldsymbol{Z}_n^*(5)$ and $\boldsymbol{Z}_n^*(8)$, then $C_i = \{2, 5, 8\}$ and $B_i = 3$.

Efron (1983) assessed the degree of underestimation of the prediction error by the apparent error rate in (6.58) as

$$\mathrm{bias}(\hat{F})^{.632} = 0.632\left\{\mathrm{CV}_{BS} - \mathrm{APE}(\boldsymbol{Z}_n, \hat{F})\right\}. \tag{6.63}$$

The estimate of the prediction error, known as the *.632 estimateor*, is then obtained by adding the correction for this bias to the apparent error rate, and is given by

$$0.632\ \mathrm{EST} = \mathrm{APE}(\boldsymbol{Z}_n, \hat{F}) + 0.632\left\{\mathrm{CV}_{BS} - \mathrm{APE}(\boldsymbol{Z}_n, \hat{F})\right\}$$
$$= 0.368\ \mathrm{APE}(\boldsymbol{Z}_n, \hat{F}) + 0.632\ \mathrm{CV}_{BS}. \tag{6.64}$$

The 0.632 estimator may thus be viewed as a new prediction error estimator, obtained with weighting assigned to the apparent error rate $\mathrm{APE}(\boldsymbol{Z}_n, \hat{F})$ and to the form of bootstrap cross-validation CV_{BS}. Efron (1983) further noted that the CV_{BS} of (6.62) is nearly equivalent to the cross-validation (CV_{HF}) in which half of the observed data is removed at once, and showed that the 0.632 estimate can therefore be approximated as

$$0.632\ \mathrm{EST}_{AP} = 0.368\ \mathrm{APE}(\boldsymbol{Z}_n, \hat{F}) + 0.632\ \mathrm{CV}_{HF}. \tag{6.65}$$

For further information on the 0.632 estimator, see Efron (1983) and Efron and Tibshirani (1997).

6.3.4 Example: Calcium Oxalate Crystals

Calcium oxalate crystals, if present in the body, may lead to the formation of kidney or ureteral stones, but thorough medical testing is required for their detection. Here, we consider the formulation of an effective classification procedure based on six properties of urine that are believed to be involved in the crystal formation and are readily detected in medical testing for construction of a relatively simple diagnostic method. We begin with the results of a series of medical tests that showed the presence of calcium oxalate crystals in the urine of 33 individuals constituting a patient class (G_1) but not in the urine of 44 normal individuals (G_2), and the resulting six-dimensional 77 data per test parameter available for the two classes. The six test variables are as follows (Andrews and Herzberg, 1985, p. 249).

x_1 : specific gravity (SG), x_2 : pH,
x_3 : osmolality (mOsm), x_4 : conductivity (CD),
x_5 : urea concentration (UREA),
x_6 : calcium concentration (CALC).

The classification rule constructed by the linear discriminant function is

$$h(\boldsymbol{x}) = (\overline{\boldsymbol{x}}_1 - \overline{\boldsymbol{x}}_2)^T S^{-1}\boldsymbol{x} - \frac{1}{2}(\overline{\boldsymbol{x}}_1 - \overline{\boldsymbol{x}}_2)^T S^{-1}(\overline{\boldsymbol{x}}_1 + \overline{\boldsymbol{x}}_2)\begin{cases} \geq 0 & \Rightarrow \quad G_1 \\ < 0 & \Rightarrow \quad G_2, \end{cases}$$

where $\overline{\boldsymbol{x}}_1$ and $\overline{\boldsymbol{x}}_2$ are respectively the sample mean vectors of the patient class G_1 and the normal class G_2, and S is the pooled sample variance-covariance matrix.

Comparison of prediction error estimation

We first estimate the error rates of the linear classification rule based on all six variables, using the apparent error rate (6.50), the cross-validation (6.52), the bias-corrected apparent error rate (6.59), and the .632 estimator (6.64). Table 6.2 compares the results for the 77 six-dimensional observed data, in which bootstrap samples are extracted repeatedly 100 times.

From Table 6.2 we see that the apparent error rate underestimates the prediction error substantially, because the same data are used in constructing and evaluating the classification rule. The bootstrap bias-corrected apparent error rate provides a method of reducing this ten-

Table 6.2 *Comparison of prediction error estimates for the classification rule constructed by the linear discriminant function.*

Apparent error rate	0.169
Cross-validation	0.221
Bias-corrected apparent error rate	0.202
.632 estimator	0.221

Table 6.3 *Variable selection via the apparent error rates* (APE).

No. of variables	APE	Selected variables
1	0.260	CALC
2	0.234	SG, CALC
3	0.182	SG, mOsm, CALC
4	0.182	SG, mOsm, CD, CALC
5	0.169	SG, pH, CD, UREA, CALC
6	0.169	SG, pH, mOsm, CD, UREA, CALC

dency, but it still underestimates. The cross-validation and .632 estimator give more refined results for the estimation of the prediction error.

Variable selection

In the apparent error rate the training data used to derive the discriminant function is also used to estimate the error rate. Care in its application is necessary, particularly because apparent error rates inherently tend to decrease as the number of variables increases. Table 6.3 shows the results of the repetitive process of first formulating a classification rule for each of the variables to find the variable that minimizes the apparent error rate and next formulating a rule for each pair of variables to find the two variables that minimize the apparent error rate. As shown in the table, the apparent error rate is smallest when all six variables are used to obtain the linear discriminant function.

We then apply cross-validation to these results to estimate the prediction error for each variable combination and find the combination that yields the smallest error estimate, which in this case leads to the selection of three variables (*specific gravity, osmolality, and calcium concen-*

tration). The prediction error of the classification rule formulated with these three variables is 0.208. In contrast, the prediction error using all six variables is 0.221.

Bootstrap selection probability

We performed repetitive bootstrapping to extract bootstrap samples from the 77 six-dimensional observed data. The variable selection is made for each bootstrap sample and the optimum set of variables is obtained on the basis of prediction error, using the cross-validation and .632 estimator. We calculate the number of times selected per 100 repetitions for each variable combination. The selection rates, called the *bootstrap selection probability*, in these cases are highest for the three variables (*specific gravity, osmolality, and calcium concentration*), indicating that these three are key variables for assessment of the presence of calcium oxalate crystals in urine by the linear discriminant procedure. For the bootstrap methods, see Appendix A.

6.3.5 Stepwise Procedures

If the number of variables is large, the computation for variable selection using all the variable combinations tends to present a problem. With twenty variables ($p = 20$), for example, the number of combinations for all of the variables is $2^p - 1 = 2^{20} - 1 = 1{,}048{,}575$. In cases where the computer processing capability is insufficient, *stepwise procedures* may be preferable. Statistical packages usually include stepwise selection procedures for linear discriminant analysis, referred to as step-up, step-down, or forward selection. They perform stepwise addition or elimination of variables to enable formulation of more effective classifiers. Stepwise procedures are based on the concept that separation between two classes improves with increasing Mahalanobis distance between their means (centers of mass), thus enabling more effective classification rules. In this light, we next discuss the methods of verification as to whether the addition of a single variable x_{r+1} to r variables is actually effective for classification.

Suppose that we have n_1 r-dimensional data from class G_1 and n_2 r-dimensional data from class G_2. The sample mean vector, the variance-covariance matrix for each class, and the pooled variance-covariance matrix are given by

$$G_1 : \ \overline{\boldsymbol{x}}_1, \ \ S_1, \qquad G_2 : \ \overline{\boldsymbol{x}}_2, \ \ S_2,$$

$$S = \frac{1}{(n_1 - 1) + (n_2 - 1)}\{(n_1 - 1)S_1 + (n_2 - 1)S_2\}.$$

Then the Mahalanobis squared distance between the centers of mass of the two classes is

$$D_r^2 = (\overline{x}_1 - \overline{x}_2)^T S^{-1}(\overline{x}_1 - \overline{x}_2). \qquad (6.66)$$

We take D_{r+1}^2 as the Mahalanobis squared distance between the two classes based on the $(r + 1)$-dimensional data observed for $(r + 1)$ variables. The D_{r+1}^2 is generally larger than D_r^2, and thus it seems to be preferable to include the additional variable if it increases the Mahalanobis squared distance. A question that must be investigated, however, is whether the added variable x_{r+1} actually contributes to more effective classification or the observed increase is simply the result of an error due to sample variation. We treat this as an inferential problem and perform a hypothesis test using the result that the test statistic

$$F = \frac{(n - r)(D_{r+1}^2 - D_r^2)}{\dfrac{n(n+2)}{n_1 n_2} + D_r^2}; \qquad n = n_1 + n_2 - 2 \qquad (6.67)$$

follows F-distribution with the degree of freedom $(1, n - r)$. If the value of F is larger than 2.0, for example, then variable x_{r+1} is judged effective for classification.

Example 6.2 (Stepwise variable selection for calcium oxalate crystals data) We apply a step-up selection procedure to data relating to the formation of calcium oxalate crystals described in Section 6.3.4. Step-up procedure enters the variables one at a time, and in the first step calcium concentration (CALC) was selected as the variable with the largest F-value of 30.859. In the 2nd step CALC and specific gravity were entered into the model with F-value of 2.815. In combination with these variables already entered into the model, osmolarity (mOsm) was selected as the variable with the largest F-value of 11.149. We stop adding variables, since none of the remaining variables are significant in combination with the selected three variables, and finally {CALC, specific gravity, osmolality} were selected.

When we employ a linear discriminant analysis using the variables (x_1, x_3, x_6) obtained by a stepwise selection procedure, a question may arise as to how large the error rate is and whether some other variable combination might exist that would give a smaller error rate. To address this question in the present case, we obtained the linear discriminant functions for all combinations of the six variables and estimated the error rate by cross-validation. The results show that the variables (x_1, x_3, x_6) yielded the smallest error rate of 0.208, which is the same as the results obtained by the stepwise selection procedure. It is necessary to note here that, in general, the model selected by a stepwise selection procedure will not always match that obtained by an all possible subset selection.

6.4 Canonical Discriminant Analysis

Figure 6.5 shows a plot of the values obtained in two tests for individuals belonging to a normal, chemical diabetes, or clinical diabetes class. In that figure, it was possible to visually ascertain the features of the test values and the degree and direction of data dispersion in each of the three classes. With four or more variables, the data for each individual becomes four-dimensional or higher and direct graphical representation in a visually recognizable form is therefore difficult. In such a case, we may consider a method of projection onto a two-dimensional plane or a three-dimensional space for visualization of the positional relationships and data features of multiple classes dispersed in higher-dimensional space. This can be done by the technique known as *canonical discriminant analysis*.

6.4.1 Dimension Reduction by Canonical Discriminant Analysis

The projection of data dispersed in a higher-dimensional space to a two-dimensional plane or a three-dimensional space so that we can intuitively comprehend the data dispersion is generally known as *dimension reduction* or *dimensional compression*. One representative technique used for this purpose is principal component analysis, which is described in Chapter 9. It should be noted that a primary purpose of discriminant analysis is class separation, and it is therefore desirable to separate the classes to the degree possible in the related projection.

This may be likened to the observation of three stationary balls held in midair. When viewed from one direction, two of the balls may appear to overlap or even be seen as just one ball. From a certain direction, however, it may be possible to observe them as three quite separate balls and thus most readily ascertain their relative positions. Canonical discriminant analysis is a method of multivariate analysis used to select coordinate axes to provide that direction.

Canonical discriminant analysis may be regarded as an extension to multiple classes, of the basic concept described in Section 6.1 for the derivation of Fisher's linear discriminant function for two classes. We first project p-dimensional observed data from multiple classes onto axes that can be expressed by linear combination of variables to find an optimal projection axis that maximizes the ratio of between-class variation (which represents the class separation) to within-class variation (which represents the degree of data dispersion within each class). We next find a projection axis that maximizes the ratio of between-class variation to within-class variation under certain conditions. This process proceeds stepwise, to compose plane and spatial coordinates that most effectively separate the multiple classes. The determination of these projection axes is thus reduced to a generalized eigenvalue problem.

Suppose that we have n_j p-dimensional data from class G_j ($j = 1, 2, \cdots, g$)

$$G_j : \boldsymbol{x}_1^{(j)}, \boldsymbol{x}_2^{(j)}, \cdots, \boldsymbol{x}_{n_j}^{(j)}, \qquad j = 1, 2, \cdots, g. \tag{6.68}$$

We first project the total n ($\sum_{j=1}^{g} n_j$) p-dimensional data onto

$$y = w_1 x_1 + w_2 x_2 + \cdots + w_p x_p = \boldsymbol{w}^T \boldsymbol{x}, \tag{6.69}$$

and then the p-dimensional data are reduced to one-dimensional data

$$y_i^{(j)} = \boldsymbol{w}^T \boldsymbol{x}_i^{(j)}, \qquad i = 1, 2, \cdots, n_j, \quad j = 1, 2, \cdots, g. \tag{6.70}$$

The sample mean and the sample variance for each class G_j, as obtained from the one-dimensional data on axis y, are given by

$$\overline{y}^{(j)} = \frac{1}{n_j} \sum_{i=1}^{n_j} y_i^{(j)} = \frac{1}{n_j} \sum_{i=1}^{n_j} \boldsymbol{w}^T \boldsymbol{x}_i^{(j)} = \boldsymbol{w}^T \overline{\boldsymbol{x}}_j, \tag{6.71}$$

and

$$\begin{aligned} \frac{1}{n_j - 1} \sum_{i=1}^{n_j} (y_i^{(j)} - \overline{y}^{(j)})^2 &= \boldsymbol{w}^T \left\{ \frac{1}{n_j - 1} \sum_{i=1}^{n_j} (\boldsymbol{x}_i^{(j)} - \overline{\boldsymbol{x}}_j)(\boldsymbol{x}_i^{(j)} - \overline{\boldsymbol{x}}_j)^T \right\} \boldsymbol{w} \\ &= \boldsymbol{w}^T S_j \boldsymbol{w}. \end{aligned} \tag{6.72}$$

Also, the sample mean on axis y based on n observations is

$$\overline{y} = \frac{1}{n} \sum_{j=1}^{g} \sum_{i=1}^{n_j} y_i^{(j)} = \frac{1}{n} \sum_{j=1}^{g} \sum_{i=1}^{n_j} \boldsymbol{w}^T \boldsymbol{x}_i^{(j)} = \boldsymbol{w}^T \overline{\boldsymbol{x}}. \tag{6.73}$$

The between-class variance can then be expressed as

$$\sum_{j=1}^{g} n_j (\overline{y}^{(j)} - \overline{y})^2 = \boldsymbol{w}^T \left\{ \sum_{j=1}^{g} n_j (\overline{\boldsymbol{x}}_j - \overline{\boldsymbol{x}})(\overline{\boldsymbol{x}}_j - \overline{\boldsymbol{x}})^T \right\} \boldsymbol{w} = \boldsymbol{w}^T B \boldsymbol{w}, \tag{6.74}$$

where

$$B = \sum_{j=1}^{g} n_j(\overline{\boldsymbol{x}}_j - \overline{\boldsymbol{x}})(\overline{\boldsymbol{x}}_j - \overline{\boldsymbol{x}})^T. \tag{6.75}$$

The $p \times p$ matrix B is called the *between-class matrix*. When projecting the data onto axis y, the within-class variance can be written as

$$\sum_{j=1}^{g}\sum_{i=1}^{n_j}(y_i^{(j)} - \overline{y}^{(j)})^2 = \boldsymbol{w}^T\left\{\sum_{j=1}^{g}(n_j - 1)S_j\right\}\boldsymbol{w} = \boldsymbol{w}^T W\boldsymbol{w}, \tag{6.76}$$

where

$$W = \sum_{j=1}^{g}(n_j - 1)S_j. \tag{6.77}$$

The $p \times p$ matrix W is called the *within-class matrix*. It can be shown that

$$B + W = \sum_{i=1}^{n}(\boldsymbol{x}_i - \overline{\boldsymbol{x}})(\boldsymbol{x}_i - \overline{\boldsymbol{x}})^T \equiv T, \tag{6.78}$$

where T is the *total variance-covariance matrix*.

By an argument similar to that described in Section 6.1 for Fisher's linear discriminant function for two classes, the problem can be reduced to obtain the coefficient vector $\boldsymbol{w}$, which maximizes the ratio

$$\lambda = \frac{\boldsymbol{w}^T B\boldsymbol{w}}{\boldsymbol{w}^T W\boldsymbol{w}}. \tag{6.79}$$

The coefficient vector $\boldsymbol{w}_1$ is determined by solving the generalized eigenvalue problem described below.

The *generalized eigenvalue problem* is to find the solutions of the equations

$$B\boldsymbol{w}_j = \lambda_j W\boldsymbol{w}_j, \qquad \boldsymbol{w}_j^T S\boldsymbol{w}_k = \delta_{jk}, \tag{6.80}$$

where $S = W/(n-g)$. We denote the eigenvalues of the matrix $W^{-1}B$ in descending order of magnitude, and the corresponding eigenvectors as

$$\begin{array}{ccccccccc} \lambda_1 & \geq & \lambda_2 & \geq & \cdots & \geq & \lambda_d & > & 0, \\ \boldsymbol{w}_1, & & \boldsymbol{w}_2, & & \cdots, & & \boldsymbol{w}_d & & \end{array} \tag{6.81}$$

where $d = \min\{g-1, p\}$ and the p-dimensional eigenvectors are normalized to $\boldsymbol{w}_j^T S \boldsymbol{w}_k = \delta_{jk}$.

As will be shown in the result (2) given below, the generalized eigenvalue problem gives the discriminant coordinates as the eigenvectors corresponding to the ordered eigenvalues of $W^{-1}B$. The first discriminant variable or canonical variable is the linear combination $y_1 = \boldsymbol{w}_1^T \boldsymbol{x}$ with the normalized coefficient vector $\boldsymbol{w}_1$ corresponding to the largest eigenvalue λ_1 of $W^{-1}B$. The second discriminant variable is given by $y_2 = \boldsymbol{w}_2^T \boldsymbol{x}$ with $\boldsymbol{w}_2$ that maximizes the ratio λ in (6.79) subject to the conditions $\boldsymbol{w}_2^T S \boldsymbol{w}_2 = 1$ and $\boldsymbol{w}_2^T S \boldsymbol{w}_1 = 0$. This process of finding discriminant variables can be repeated until the number of discriminant variables equals $d = \min\{g-1, p\}$, and we have $y_j = \boldsymbol{w}_j^T \boldsymbol{x}$ $(j = 1, 2, \cdots, d)$. The discriminant variables are thus uncorrelated in the sense that $\boldsymbol{w}_j^T S \boldsymbol{w}_k = 0$ $(j \neq k)$. By projecting p-dimensional observed data from multiple classes onto axes (y_1, y_2), we may observe the mutual relationship of several classes and their data dispersion.

Generalized eigenvalue and maximization problems

(1) Let A be a $p \times p$ symmetric matrix with ordered eigenvalues $\lambda_1 \geq \lambda_2 \geq \cdots \geq \lambda_p$, and corresponding orthonormal eigenvectors $\boldsymbol{h}_1, \boldsymbol{h}_2, \cdots, \boldsymbol{h}_p$. Then we have the following results (a) and (b):

$$\text{(a)} \qquad \max_{\boldsymbol{x}} \frac{\boldsymbol{x}^T A \boldsymbol{x}}{\boldsymbol{x}^T \boldsymbol{x}} = \lambda_1, \tag{6.82}$$

and the maximum is attained when $\boldsymbol{x} = \boldsymbol{h}_1$, the eigenvector corresponding to the largest eigenvalue λ_1 of A.

Under the conditions that $\boldsymbol{h}_j^T \boldsymbol{x} = 0$ for $j = 1, 2, \cdots, k$ $(k < p)$, it follows that

$$\text{(b)} \qquad \max_{H_k^T \boldsymbol{x} = 0} \frac{\boldsymbol{x}^T A \boldsymbol{x}}{\boldsymbol{x}^T \boldsymbol{x}} = \lambda_{k+1}, \tag{6.83}$$

where $H_k = (\boldsymbol{h}_1, \boldsymbol{h}_2, \cdots, \boldsymbol{h}_k)$. The maximum is attained when $\boldsymbol{x} = \boldsymbol{h}_{k+1}$, the eigenvector corresponding to the $(k+1)$-th eigenvalue λ_{k+1} of A.

Proof Define $H = (\boldsymbol{h}_1, \boldsymbol{h}_2, \cdots, \boldsymbol{h}_p)$. Let $\boldsymbol{x} = H\boldsymbol{y} = y_1 \boldsymbol{h}_1 + y_2 \boldsymbol{h}_2 + \cdots + y_p \boldsymbol{h}_p$. Then we see that

$$\frac{\boldsymbol{x}^T A \boldsymbol{x}}{\boldsymbol{x}^T \boldsymbol{x}} = \frac{\boldsymbol{y}^T H^T A H \boldsymbol{y}}{\boldsymbol{y}^T \boldsymbol{y}} = \frac{\sum_{i=1}^{p} \lambda_i y_i^2}{\boldsymbol{y}^T \boldsymbol{y}} \leq \frac{\lambda_1 \boldsymbol{y}^T \boldsymbol{y}}{\boldsymbol{y}^T \boldsymbol{y}} = \lambda_1, \tag{6.84}$$

and equality holds when $y_1 = 1$ and $y_2 = y_3 = \cdots = y_p = 0$, that is, $\boldsymbol{x} = \boldsymbol{h}_1$.

If $\boldsymbol{h}_j^T \boldsymbol{x} = 0$ for $j = 1, 2, \cdots, k$, then $y_1 = y_2 = \cdots = y_k = 0$. The same argument can be used to prove the result (b).

(2) Let B be a $p \times p$ symmetric matrix and W be a $p \times p$ positive definite matrix. Let $\lambda_1 \geq \lambda_2 \geq \cdots \geq \lambda_p$ be the ordered eigenvalues of $W^{-1}B$ with corresponding eigenvectors $\boldsymbol{w}_1, \boldsymbol{w}_2, \cdots, \boldsymbol{w}_p$. Then we have the following results:

$$\text{(c)} \qquad \max_{\boldsymbol{x}} \frac{\boldsymbol{x}^T B \boldsymbol{x}}{\boldsymbol{x}^T W \boldsymbol{x}} = \lambda_1 \quad \text{and} \quad \min_{\boldsymbol{x}} \frac{\boldsymbol{x}^T B \boldsymbol{x}}{\boldsymbol{x}^T W \boldsymbol{x}} = \lambda_p, \tag{6.85}$$

and the maximum and minimum are attained when $\boldsymbol{x} = \boldsymbol{w}_1$ and $\boldsymbol{x} = \boldsymbol{w}_p$, respectively.

Proof Since W is positive definite, there exists a nonsingular matrix Q such that $W = Q^T Q$, the Cholesky decomposition of W. Setting $\boldsymbol{y} = Q\boldsymbol{x}$ ($\boldsymbol{x} = Q^{-1}\boldsymbol{y}$) and using the result (a) in (6.82) yields

$$\max_{\boldsymbol{x}} \frac{\boldsymbol{x}^T B \boldsymbol{x}}{\boldsymbol{x}^T W \boldsymbol{x}} = \max_{\boldsymbol{y}} \frac{\boldsymbol{y}^T (Q^{-1})^T B Q^{-1} \boldsymbol{y}}{\boldsymbol{y}^T \boldsymbol{y}} = \lambda_1, \tag{6.86}$$

where λ_1 is the largest eigenvalue of $(Q^{-1})^T B Q^{-1}$, and the maximum is attained when $\boldsymbol{x} = \boldsymbol{h}_1$, the unit eigenvector corresponding to the largest eigenvalue of $(Q^{-1})^T B Q^{-1}$. It is easily checked that $W^{-1}B$ and $(Q^{-1})^T B Q^{-1}$ have the same eigenvalues.

We note that

$$(Q^{-1})^T B Q^{-1} \boldsymbol{h}_1 = \lambda_1 \boldsymbol{h}_1, \qquad \boldsymbol{h}_1^T \boldsymbol{h}_1 = 1. \tag{6.87}$$

By multiplying on the left by $(n-g)^{1/2} Q^{-1}$, we have

$$\begin{aligned} (n-g)^{1/2} (Q^T Q)^{-1} B Q^{-1} \boldsymbol{h}_1 &= W^{-1} B (n-g)^{1/2} Q^{-1} \boldsymbol{h}_1 \\ &= \lambda_1 (n-g)^{1/2} Q^{-1} \boldsymbol{h}_1. \end{aligned} \tag{6.88}$$

We therefore find that $(n-g)^{1/2} Q^{-1} \boldsymbol{h}_1 = \boldsymbol{w}_1$ is the eigenvector corresponding to the largest eigenvalue λ_1 of $W^{-1}B$. We further notice that

$$\boldsymbol{h}_1^T \boldsymbol{h}_1 = \frac{1}{n-g} \boldsymbol{w}_1^T Q^T Q \boldsymbol{w}_1 = \frac{1}{n-g} \boldsymbol{w}_1^T W \boldsymbol{w}_1 = \boldsymbol{w}_1^T S \boldsymbol{w}_1, \tag{6.89}$$

where $S = W/(n-g)$ defined in (6.81). The same argument can be applied to prove the minimization result in (c).

(3) Let S be a $p \times p$ positive definite matrix. Then for any p-dimensional vector $\boldsymbol{d}$, it follows that

$$\text{(d)} \qquad \max_{\boldsymbol{x}} \frac{(\boldsymbol{x}^T \boldsymbol{d})^2}{\boldsymbol{x}^T S \boldsymbol{x}} = \boldsymbol{d}^T S^{-1} \boldsymbol{d}. \tag{6.90}$$

The maximum occurs when $\boldsymbol{x}$ is proportional to $S^{-1}\boldsymbol{d}$.

Proof The result can be obtained by applying (c) with $\boldsymbol{x}^T B\boldsymbol{x} = \boldsymbol{x}^T(\boldsymbol{d}\boldsymbol{d}^T)\boldsymbol{x}$. Since $\boldsymbol{d}\boldsymbol{d}^T$ has rank one, $S^{-1}\boldsymbol{d}\boldsymbol{d}^T$ has only one non-zero eigenvalue $\boldsymbol{d}^T S^{-1}\boldsymbol{d}$ and the corresponding eigenvector $\boldsymbol{x} = S^{-1}\boldsymbol{d}$. This can be seen by noting that $S^{-1}\boldsymbol{d}\boldsymbol{d}^T\boldsymbol{x} = \lambda\boldsymbol{x}$ is expressible in the form $\boldsymbol{d}^T S^{-1}\boldsymbol{d}y = \lambda y$, where $y = \boldsymbol{d}^T\boldsymbol{x}$ and $\boldsymbol{d}^T S^{-1}\boldsymbol{d}$ is a scalar. Taking S to be a pooled sample variance-covariance matrix and setting $\boldsymbol{d} = \overline{\boldsymbol{x}}_1 - \overline{\boldsymbol{x}}_2$ yields the coefficient vector of Fisher's linear discriminant function given by (6.30).

For multiclass classification involving extremely high-dimensional data, other techniques that may be considered include dimension reduction for all of the data followed by execution of the classification and the concurrent use of the clustering technique discussed in Chapter 10 to extract information and patterns from the high-dimensional data.

Example 6.3 (Dimension reduction by canonical discriminant analysis) We illustrate the dimension reduction technique by canonical discriminant analysis, using three of the test parameters in the diagnosis of diabetes described in Section 6.2.3. The three variables will thus be x_1, and x_2, and in addition x_3, the blood sugar level at equilibrium after intravenous injection of given quantity of insulin and glucose.

In this application, the 145 non-obese subjects have been divided into three classes based on the results of the three tests: a normal class (G_1; 76 subjects), a chemical diabetes class (G_2; 36 subjects), and a clinical diabetes class (G_3; 33 subjects). We have a space containing a dispersion of 145 three-dimensional data for the three variables x_1, x_2, and x_3. To examine the mutual relationship of these three classes, we project the data onto a two-dimensional plane. The objective is to find the coefficients on axes y_1, y_2 that best separate the three classes based on the 145 data points.

The between-class and within-class matrices are

$$B = 76(\overline{\boldsymbol{x}}_1 - \overline{\boldsymbol{x}})(\overline{\boldsymbol{x}}_1 - \overline{\boldsymbol{x}})^T + 36(\overline{\boldsymbol{x}}_2 - \overline{\boldsymbol{x}})(\overline{\boldsymbol{x}}_2 - \overline{\boldsymbol{x}})^T + 33(\overline{\boldsymbol{x}}_3 - \overline{\boldsymbol{x}})(\overline{\boldsymbol{x}}_3 - \overline{\boldsymbol{x}})^T$$

$$= \begin{bmatrix} 1.12 \times 10^7 & -1.31 \times 10^6 & 3.21 \times 10^6 \\ -1.31 \times 10^6 & 5.99 \times 10^5 & -1.93 \times 10^5 \\ 3.21 \times 10^6 & -1.93 \times 10^5 & 9.95 \times 10^5 \end{bmatrix},$$

and

$$W = (76-1)S_1 + (36-1)S_2 + (33-1)S_3$$
$$= \begin{bmatrix} 3.27\times 10^6 & -5.54\times 10^5 & 5.19\times 10^5 \\ -5.54\times 10^5 & 1.51\times 10^6 & 2.08\times 10^5 \\ 5.19\times 10^5 & 2.08\times 10^5 & 6.24\times 10^5 \end{bmatrix}. \quad (6.91)$$

The discriminant coordinates are given as the normalized eigenvectors corresponding to the ordered eigenvalues of $W^{-1}B$. The first two discriminant variables are

$$y_1 = 0.00543x_1 - 0.00014x_2 + 0.00511x_3,$$
$$y_2 = -0.00046x_1 + 0.00851x_2 + 0.00465x_3. \quad (6.92)$$

Figure 6.7 shows a plot of the values obtained by projecting the 145 observed data from three classes onto axes (y_1, y_2) in (6.92). We may observe the mutual relationship of a normal class (G_1), a chemical diabetes class (G_2), and a clinical diabetes class (G_3) through the projected test scores.

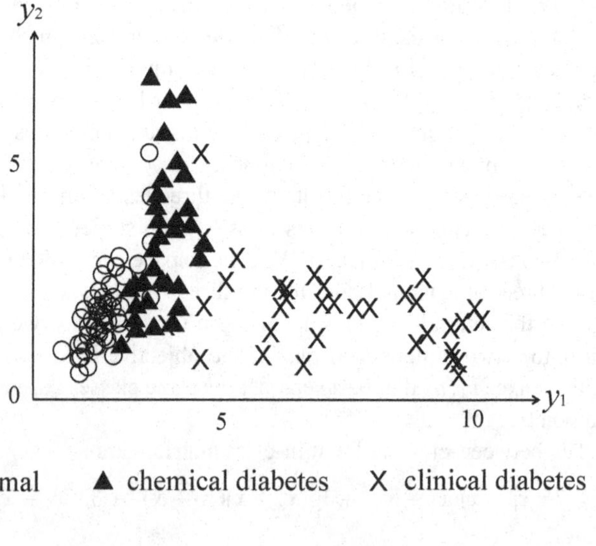

Figure 6.7 *Plot of the values obtained by projecting the 145 observed data from three classes onto the first two discriminant variables* (y_1, y_2) *in (6.92).*

This technique is inherently useful for projecting non-intuitive high-dimensional data in four or more dimensions to a two-dimensional plane or a three-dimensional space and thereby enabling observation from an appropriate direction for visual comprehension of the positional relationship of several classes and their data dispersion. In the case of data projection into a three-dimensional space, we can rotate the coordinate axes in various directions while using graphical software to obtain a visual presentation of the spatial dispersion, which may enable extraction of more effective information from the high-dimensional data.

Exercises

6.1 Suppose that we project the n p-dimensional data in (6.25) onto axis $y = w_1x_1 + w_2x_2 + \cdots + w_px_p = \boldsymbol{w}^T\boldsymbol{x}$.

(a) Show that the between-class variance in the projection onto axis $y = \boldsymbol{w}^T\boldsymbol{x}$ is given by $\{\boldsymbol{w}^T(\overline{\boldsymbol{x}}_1 - \overline{\boldsymbol{x}}_2)\}^2$, where $\overline{\boldsymbol{x}}_i$ are the sample mean vectors.

(b) Show that the within-class variance in the projection onto axis $y = \boldsymbol{w}^T\boldsymbol{x}$ is given by $\boldsymbol{w}^TS\boldsymbol{w}$, where S is a pooled sample variance-covariance matrix.

6.2 Show that the p-dimensional coefficient vector $\boldsymbol{w}$ that maximizes the ratio

$$\lambda = \frac{\left\{\boldsymbol{w}^T(\overline{\boldsymbol{x}}_1 - \overline{\boldsymbol{x}}_2)\right\}^2}{\boldsymbol{w}^TS\boldsymbol{w}}$$

is given by $\hat{\boldsymbol{w}} = S^{-1}(\overline{\boldsymbol{x}}_1 - \overline{\boldsymbol{x}}_2)$, and that the maximum of the ratio is $(\overline{\boldsymbol{x}}_1 - \overline{\boldsymbol{x}}_2)^T S^{-1} (\overline{\boldsymbol{x}}_1 - \overline{\boldsymbol{x}}_2)$, the Mahalanobis distance between the sample mean vectors $\overline{\boldsymbol{x}}_1$ and $\overline{\boldsymbol{x}}_2$ (see (6.90)).

6.3 Suppose that the sample mean vectors of the classes G_1 and G_2 and the pooled sample variance-covariance matrix are given by

$$G_1 : \overline{\boldsymbol{x}}_1 = \begin{pmatrix} 0 \\ 1 \end{pmatrix}, \quad G_2 : \overline{\boldsymbol{x}}_2 = \begin{pmatrix} 1 \\ 0 \end{pmatrix}, \quad S = \begin{pmatrix} 0.4 & -0.2 \\ -0.2 & 0.6 \end{pmatrix}.$$

(a) Find the Mahalanobis distances D_i^2 $(i = 1, 2)$ between the sample mean vectors $\overline{\boldsymbol{x}}_i$ and $\boldsymbol{x} = (x_1, x_2)^T$.

(b) Show that $h(x_1, x_2) = D_2^2 - D_1^2$ can be expressed as a linear combination of x_1 and x_2.

(c) Which class does the observed data $(x_1, x_2) = (0.5, 0.8)$ belong to?

6.4 We project the total n $(\sum_{j=1}^{g} n_j)$ p-dimensional data in (6.68) onto $y = w_1x_1 + w_2x_2 + \cdots + w_px_p = \boldsymbol{w}^T\boldsymbol{x}$. When projecting the data onto axis y, show that the between-class variance and the within-class variance are, respectively, given by (6.74) and (6.76).

Chapter 7

Bayesian Classification

In this chapter, we discuss the formulation of linear and quadratic discriminant analysis based on Bayes' theorem and probability distributions, with a focus on two-class classification problems. We then proceed to the technique, which is based on Bayesian classification and is related to the logistic regression modeling. In addition, we consider the nonlinear discriminant analysis by non-linearization of the logistic model.

7.1 Bayes' Theorem

In classification by the Bayesian approach, the basic concept is embodied in Bayes' theorem. As one example, let us assume that we have observed the occurrence of a symptom in a certain patient in the form of a fever and need to assess whether it has been caused by a cold or by influenza. If the probability of a cold as the cause of a fever is higher than that of influenza, we may then at least tentatively attribute the fever in this patient to a cold. This is the underlying concept of Bayes classification. Bayes' theorem gives the relation between the conditional probability of an event based on acquired information, which in this case may be described as follows.

First, we denote the probability of a fever (D) as a symptom of a cold (G_1) and that of influenza (G_2) by

$$\mathrm{P}(D|G_1) = \frac{\mathrm{P}(D \cap G_1)}{\mathrm{P}(G_1)}, \qquad \mathrm{P}(D|G_2) = \frac{\mathrm{P}(D \cap G_2)}{\mathrm{P}(G_2)}, \tag{7.1}$$

respectively. The $\mathrm{P}(D|G_1)$ and $\mathrm{P}(D|G_2)$ represent the probabilities of a fever being the result of a cold and of influenza, respectively, and are called *conditional probabilities*. Here, $\mathrm{P}(G_1)$ and $\mathrm{P}(G_2)$ ($\mathrm{P}(G_1)+\mathrm{P}(G_2) = 1$) are the relative incidences of colds and influenza, and are called *prior probabilities*. It is assumed that these conditional and prior probabilities can be estimated on the basis of observations and accumulated information. Then the probability $\mathrm{P}(D)$ is given by

$$\mathrm{P}(D) = \mathrm{P}(G_1)\mathrm{P}(D|G_1) + \mathrm{P}(G_2)\mathrm{P}(D|G_2), \tag{7.2}$$

the probability of a fever being the result of a cold or of influenza, called the *law of total probability*.

In our example, we want to know the probabilities that the fever which has occurred has been caused by a cold or by influenza, respectively, as represented by conditional probabilities $\mathrm{P}(G_1|D)$ and $\mathrm{P}(G_2|D)$. Bayes' theorem provides these probabilities on the basis of known priors $\mathrm{P}(G_i)$ and conditional probabilities $\mathrm{P}(D|G_i)$. That is, the conditional probabilities $\mathrm{P}(G_i|D)$ are given by

$$\begin{aligned}\mathrm{P}(G_i|D) &= \frac{\mathrm{P}(G_i \cap D)}{\mathrm{P}(D)} \\ &= \frac{\mathrm{P}(G_i)\mathrm{P}(D|G_i)}{\mathrm{P}(G_1)\mathrm{P}(D|G_1) + \mathrm{P}(G_2)\mathrm{P}(D|G_2)}, \qquad i = 1, 2, \qquad (7.3)\end{aligned}$$

where $\mathrm{P}(D)$ is replaced by (7.2). After the occurrence of result D, the conditional probabilities $\mathrm{P}(G_i|D)$ then become *posterior probabilities*. In general *Bayes' theorem* is formulated as follows:

Bayes' theorem Suppose that the sample space Ω is divided into r mutually disjoint events G_j as $\Omega = G_1 \cup G_2 \cup \cdots \cup G_r$ $(G_i \cap G_j = \emptyset)$. Then, for any event D, the conditional probability $\mathrm{P}(G_i|D)$ is given by

$$\mathrm{P}(G_i|D) = \frac{\mathrm{P}(G_i \cap D)}{\mathrm{P}(D)} = \frac{\mathrm{P}(G_i)\mathrm{P}(D|G_i)}{\sum_{j=1}^{r} \mathrm{P}(G_j)\mathrm{P}(D|G_j)}, \qquad i = 1, 2, \cdots, r, \quad (7.4)$$

where $\sum_{j=1}^{r} \mathrm{P}(G_j) = 1$.

Example 7.1 (Bayesian diagnosis) Let us consider a case of stomach pain (D), with the three possible causes being gastritis (G_1), stomach ulcer (G_2), and stomach cancer (G_3). The relative incidence of each disease may be taken as the prior probability $\mathrm{P}(G_i)$ $(i = 1, 2, 3)$ of the existence of each of these three possible causes. Assume further that the doctor has learned from cumulative data on these relative frequencies and on the frequency of occurrence $\mathrm{P}(D|G_i)$ $(i = 1, 2, 3)$ of this symptom (stomach pain) in cases in which the three diseases have been identified as the cause.

By application of Bayes' theorem, if a certain symptom is taken as prior, then in reverse, it is possible to assess which of the causes has given rise that symptom (i.e., to assess $\mathrm{P}(G_i|D)$, $i = 1, 2, 3$) by its relative probability (e.g., $\mathrm{P}(G_2|D)$, the probability that the cause of the pain is a stomach ulcer). It is possible to formalize this mode of thought as posterior probability by Bayes' theorem and apply it to many different recognition and classification problems. For cases in which the data (the symptom) for the individual cannot be obtained and only its probability is known in advance, it is only possible to make a judgment based on that known probability (on prior probability of occurrence). If the data can be obtained, however, then a diagnosis can be made through assessment of the relative probability of its cause, using the posterior probability derived from Bayes' theorem.

7.2 Classification with Gaussian Distributions

In this section, the purpose is to perform classification for class assignment of newly observed p-dimensional data, based on the posterior probability of its membership in each of the classes. We discuss the application of Bayes' theorem and expression of the posterior probability using a probability distribution model, and the method of formulating linear and quadratic discriminant analyses, for the class assignment.

7.2.1 Probability Distributions and Likelihood

Suppose that we have n_1 p-dimensional data from class G_1 and n_2 p-dimensional data from class G_2, and represent the total $n = (n_1 + n_2)$ training data as

$$G_1 : \boldsymbol{x}_1^{(1)}, \boldsymbol{x}_2^{(1)}, \cdots, \boldsymbol{x}_{n_1}^{(1)}, \qquad G_2 : \boldsymbol{x}_1^{(2)}, \boldsymbol{x}_2^{(2)}, \cdots, \boldsymbol{x}_{n_2}^{(2)}. \quad (7.5)$$

Let us assume that the training data for classes G_i $(i = 1, 2)$ have been observed according to p-dimensional normal distributions $N_p(\boldsymbol{\mu}_i, \Sigma_i)$ with mean vectors $\boldsymbol{\mu}_i$ and variance-covariance matrices Σ_i as follows:

$$\begin{aligned} G_1 :&\quad N_p(\boldsymbol{\mu}_1, \Sigma_1) \sim \boldsymbol{x}_1^{(1)}, \boldsymbol{x}_2^{(1)}, \cdots, \boldsymbol{x}_{n_1}^{(1)}, \\ G_2 :&\quad N_p(\boldsymbol{\mu}_2, \Sigma_2) \sim \boldsymbol{x}_1^{(2)}, \boldsymbol{x}_2^{(2)}, \cdots, \boldsymbol{x}_{n_2}^{(2)}. \end{aligned} \quad (7.6)$$

Given this type of probability distribution model, then if we tentatively assume that certain data $\boldsymbol{x}_0$ belongs to class G_1 or G_2, the relative level of occurrence of that data in each class (the "plausibility" or degree of certainty) can be quantified as $f(\boldsymbol{x}_0|\boldsymbol{\mu}_i, \Sigma_i)$, using the p-dimensional normal distribution. This corresponds to the conditional probability $\mathrm{P}(D|G_i)$ described in the previous paragraph and may be termed the *likelihood* of data $\boldsymbol{x}_0$. As one example, we may consider an observation on male height

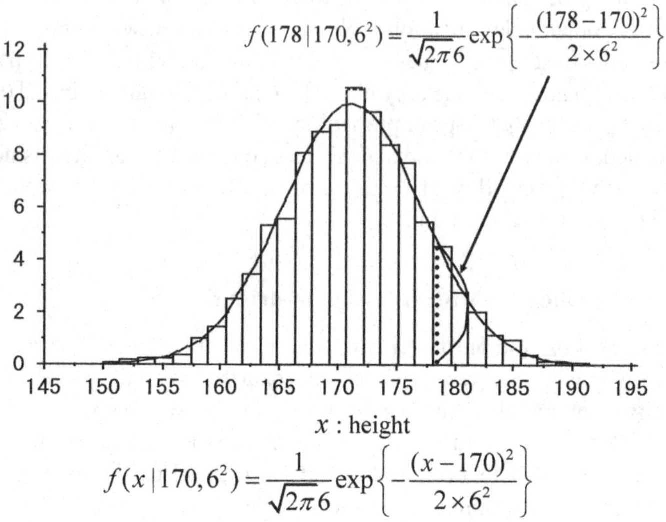

Figure 7.1 *Likelihood of the data: The relative level of occurrence of males 178 cm in height can be determined as* $f(178|170, 6^2)$.

drawn from the normal distribution $N(170, 6^2)$. Then by using the probability density function, the relative level of occurrence of males 178 cm in height can be determined as $f(178|170, 6^2)$ (Figure 7.1).

In order to obtain the likelihood of data, we replace the unknown parameters $\boldsymbol{\mu}_i$ and Σ_i in (7.6) with their maximum likelihood estimates

$$\bar{\boldsymbol{x}}_i = \frac{1}{n_i}\sum_{j=1}^{n_i}\boldsymbol{x}_j^{(i)}, \quad S_i = \frac{1}{n_i}\sum_{j=1}^{n_i}(\boldsymbol{x}_j^{(i)} - \bar{\boldsymbol{x}}_i)(\boldsymbol{x}_j^{(i)} - \bar{\boldsymbol{x}}_i)^T, \quad i = 1,2, \qquad (7.7)$$

respectively. By applying Bayes' theorem and using the posterior probability expressed as a probability distribution, we formulate the Bayesian classification and derive linear and quadratic discriminant functions for the class assignment.

7.2.2 *Discriminant Functions*

The essential purpose of discriminant analysis is to construct a classification rule based on the training data and predict the class membership of future data $\boldsymbol{x}$ in two or more predetermined classes. Let us now put this in a Bayesian framework by considering the two classes G_1 and G_2. Our goal is to obtain the posterior probability $P(G_i|D) = P(G_i|\boldsymbol{x})$ when data $D = \{\boldsymbol{x}\}$ is observed. For this purpose, we apply Bayes' theorem to obtain the posterior probability, and assign future data $\boldsymbol{x}$ to the class with the higher probability. We thus perform *Bayesian classification* based on the ratio of posterior probabilities

$$\frac{\mathrm{P}(G_1|\boldsymbol{x})}{\mathrm{P}(G_2|\boldsymbol{x})} \begin{cases} \geq 1 & \Rightarrow \quad \boldsymbol{x} \in G_1 \\ < 1 & \Rightarrow \quad \boldsymbol{x} \in G_2. \end{cases} \tag{7.8}$$

Taking logarithms of both sides yields

$$\log \frac{\mathrm{P}(G_1|\boldsymbol{x})}{\mathrm{P}(G_2|\boldsymbol{x})} \begin{cases} \geq 0 & \Rightarrow \quad \boldsymbol{x} \in G_1 \\ < 0 & \Rightarrow \quad \boldsymbol{x} \in G_2. \end{cases} \tag{7.9}$$

From Bayes' theorem in (7.3), the posterior probabilities are given by

$$\mathrm{P}(G_i|\boldsymbol{x}) = \frac{\mathrm{P}(G_i)\mathrm{P}(\boldsymbol{x}|G_i)}{\mathrm{P}(G_1)\mathrm{P}(\boldsymbol{x}|G_1) + \mathrm{P}(G_2)\mathrm{P}(\boldsymbol{x}|G_2)}, \qquad i = 1, 2. \tag{7.10}$$

Using the estimated p-dimensional normal distributions $f(\boldsymbol{x}|\overline{\boldsymbol{x}}_i, S_i)$ $(i = 1, 2)$, the conditional probabilities can be written as

$$\mathrm{P}(\boldsymbol{x}|G_i) = \frac{f(\boldsymbol{x}|\overline{\boldsymbol{x}}_i, S_i)}{f(\boldsymbol{x}|\overline{\boldsymbol{x}}_1, S_1) + f(\boldsymbol{x}|\overline{\boldsymbol{x}}_2, S_2)}, \qquad i = 1, 2, \tag{7.11}$$

(see Figure 7.2). By substituting these equations into (7.10), the ratio of posterior probabilities is expressed as

$$\frac{\mathrm{P}(G_1|\boldsymbol{x})}{\mathrm{P}(G_2|\boldsymbol{x})} = \frac{\mathrm{P}(G_1)\mathrm{P}(\boldsymbol{x}|G_1)}{\mathrm{P}(G_2)\mathrm{P}(\boldsymbol{x}|G_2)} = \frac{\mathrm{P}(G_1) f(\boldsymbol{x}|\overline{\boldsymbol{x}}_1, S_1)}{\mathrm{P}(G_2) f(\boldsymbol{x}|\overline{\boldsymbol{x}}_2, S_2)}. \tag{7.12}$$

Taking the logarithm of this expression under the assumption that the prior probabilities are equal, we have the Bayesian classification based on the probability distributions

$$h(\boldsymbol{x}) = \log \frac{f(\boldsymbol{x}|\overline{\boldsymbol{x}}_1, S_1)}{f(\boldsymbol{x}|\overline{\boldsymbol{x}}_2, S_2)} \begin{cases} \geq 0 & \Rightarrow \quad \boldsymbol{x} \in G_1 \\ < 0 & \Rightarrow \quad \boldsymbol{x} \in G_2. \end{cases} \tag{7.13}$$

$$\mathrm{P}(x \mid G_i) = \frac{f(x \mid \overline{x}_i, S_i)}{f(x \mid \overline{x}_1, S_1) + f(x \mid \overline{x}_2, S_2)} \qquad (i = 1, 2)$$

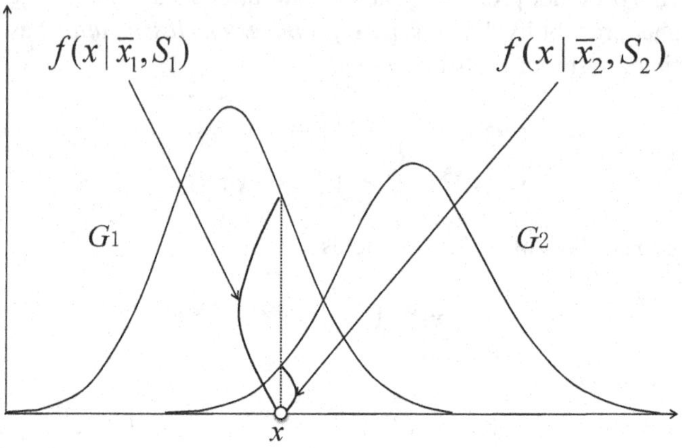

Figure 7.2 *The conditional probability* P(x|G_i) *that gives the relative level of occurrence of data* **x** *in each class.*

The discriminant function $h(\boldsymbol{x})$ with the estimated p-dimensional normal distributions $N_p(\overline{\boldsymbol{x}}_i, S_i)$ $(i = 1, 2)$ is given by

$$h(\boldsymbol{x}) = \log f(\boldsymbol{x}|\overline{\boldsymbol{x}}_1, S_1) - \log f(\boldsymbol{x}|\overline{\boldsymbol{x}}_2, S_2) \qquad (7.14)$$

$$= \frac{1}{2}\left\{(\boldsymbol{x} - \overline{\boldsymbol{x}}_2)^T S_2^{-1}(\boldsymbol{x} - \overline{\boldsymbol{x}}_2) - (\boldsymbol{x} - \overline{\boldsymbol{x}}_1)^T S_1^{-1}(\boldsymbol{x} - \overline{\boldsymbol{x}}_1) - \log\left(\frac{|S_1|}{|S_2|}\right)\right\},$$

called the *quadratic discriminant function*. By replacing S_i with the pooled sample variance-covariance matrix $S = (n_1 S_1 + n_2 S_2)/(n_1 + n_2)$, the discriminant function is further reduced to the *linear discriminant function*

$$h(\boldsymbol{x}) = (\boldsymbol{x} - \overline{\boldsymbol{x}}_2)^T S^{-1}(\boldsymbol{x} - \overline{\boldsymbol{x}}_2) - (\boldsymbol{x} - \overline{\boldsymbol{x}}_1)^T S^{-1}(\boldsymbol{x} - \overline{\boldsymbol{x}}_1)$$

$$= (\overline{\boldsymbol{x}}_1 - \overline{\boldsymbol{x}}_2)^T S^{-1}\boldsymbol{x} - \frac{1}{2}\left(\overline{\boldsymbol{x}}_1^T S^{-1}\overline{\boldsymbol{x}}_1 - \overline{\boldsymbol{x}}_2^T S^{-1}\overline{\boldsymbol{x}}_2\right). \qquad (7.15)$$

We obtain the Bayes classification rule (7.13) based on the sign of the logarithm of the ratio between the estimated probability distributions that characterize the classes. The functions $h(\boldsymbol{x})$ expressed by the

p-dimensional normal distributions give the linear and quadratic discriminant functions. Furthermore, the Bayesian classification based on the multivariate normal distributions is equivalent to the method based on the Mahalanobis distance described in Section 6.2 in which the distance of the data $\boldsymbol{x}$ from the center of mass $\overline{\boldsymbol{x}}$ of each class was measured in terms of $(\boldsymbol{x}-\overline{\boldsymbol{x}})^T S^{-1}(\boldsymbol{x}-\overline{\boldsymbol{x}})$. It may also be seen, finally, that the linear discriminant function (7.15) corresponds to the Fisher linear discriminant function discussed in Section 6.1.3 in which observed data are projected onto the axis with linear combination of variables and the ratio of between-class variance to within-class variance is maximized.

7.3 Logistic Regression for Classification

In the preceding section, we assumed the probability distribution of the random vectors characterizing the individuals to be a multivariate normal distribution and constructed the classification rule on the basis of the posterior probability derived from Bayes' theorem. We now turn to the question of how to formulate the classification when assumption of the multivariate normal distribution is inappropriate. For this purpose, let us discuss the method of formulating linear and nonlinear classifications by estimating posterior probability in Bayesian classification through the use of a logistic regression model.

7.3.1 Linear Logistic Regression Classifier

Suppose that we have the following $n = (n_1 + n_2)$ p-dimensional training data from two-classes G_1 and G_2:

$$\begin{aligned} G_1: \quad f(\boldsymbol{x}|\boldsymbol{\theta}_1) &\sim \boldsymbol{x}_1^{(1)}, \boldsymbol{x}_2^{(1)}, \cdots, \boldsymbol{x}_{n_1}^{(1)}, \\ G_2: \quad f(\boldsymbol{x}|\boldsymbol{\theta}_2) &\sim \boldsymbol{x}_1^{(2)}, \boldsymbol{x}_2^{(2)}, \cdots, \boldsymbol{x}_{n_2}^{(2)}. \end{aligned} \tag{7.16}$$

Assume that the training data for classes G_i $(i = 1, 2)$ have been observed according to p-dimensional probability distributions $f(\boldsymbol{x}|\boldsymbol{\theta}_i)$. Then the estimated probability distributions $f(\boldsymbol{x}|\hat{\boldsymbol{\theta}}_i)$ are obtained by replacing the unknown parameters $\boldsymbol{\theta}_i$ by their sample estimates $\hat{\boldsymbol{\theta}}_i$.

When data $\boldsymbol{x}$ having unknown class membership is obtained, if we tentatively assume the data to be a member of class G_i, then the relative plausibility of its observation in class G_i can be given as the likelihood $f(\boldsymbol{x}|\hat{\boldsymbol{\theta}}_i)$ $(i = 1, 2)$. Accordingly, under the assumption that the prior probabilities of the two classes are the same, then with data $\boldsymbol{x}$ as result D and its cause as G_1 or G_2, we again apply Bayes' theorem. Relating this to the formulation derived by tentatively assuming a multivariate normal distribution, the Bayesian classification based on the posterior probabilities can be formulated as

$$\log \frac{\mathrm{P}(G_1|\boldsymbol{x})}{\mathrm{P}(G_2|\boldsymbol{x})} = \log \frac{f(\boldsymbol{x}|\hat{\boldsymbol{\theta}}_1)}{f(\boldsymbol{x}|\hat{\boldsymbol{\theta}}_2)}. \tag{7.17}$$

The essential point in constructing a classification rule is to consider the logarithm of the ratio of the posterior probabilities given by the linear combination of p random variables $x_1, x_2, \cdots, x_p$ as follows:

$$\log \frac{\mathrm{P}(G_1|\boldsymbol{x})}{\mathrm{P}(G_2|\boldsymbol{x})} = \beta_0 + \beta_1 x_1 + \beta_2 x_2 + \cdots + \beta_p x_p = \beta_0 + \boldsymbol{\beta}^T \boldsymbol{x}, \tag{7.18}$$

where $\boldsymbol{\beta} = (\beta_1, \beta_2, \cdots, \beta_p)^T$. That is, the log-odds of the posterior probabilities is assumed to be the linear combination of p variables. Noting that $\mathrm{P}(G_1|\boldsymbol{x}) + \mathrm{P}(G_2|\boldsymbol{x}) = 1$, the log-odds can be rewritten as

$$\mathrm{P}(G_1|\boldsymbol{x}) = \frac{\exp(\beta_0 + \boldsymbol{\beta}^T \boldsymbol{x})}{1 + \exp(\beta_0 + \boldsymbol{\beta}^T \boldsymbol{x})}. \tag{7.19}$$

We have been able to derive the equation corresponding to (4.11) for the logistic regression model in Section 4.2 of Chapter 4. There, we discussed the prediction of risk probability by considering each variable as a causal risk factor. Here, in contrast, we take the p-dimensional data $\boldsymbol{x}$ as multiple stimulus levels of combined risk factors and the corresponding response probabilities as posterior probabilities, and in this way reduce the estimation of posterior probability in (7.19) to the logistic regression model estimation in Section 4.2. For this purpose, the response variable representing the occurrence or non-occurrence is replaced by the label variable Y representing the class membership of the $n = n_1 + n_2$ p-dimensional data from classes G_1 and G_2.

Suppose we have n independent observations

$$(\boldsymbol{x}_1, y_1), (\boldsymbol{x}_2, y_2), \cdots, (\boldsymbol{x}_n, y_n), \quad y_i = \begin{cases} 1 & \boldsymbol{x}_i \in G_1 \\ 0 & \boldsymbol{x}_i \in G_2. \end{cases} \tag{7.20}$$

By introducing variables representing class membership, the training data correspond to the data recorded as occurrence or non-occurrence of a response to the risk factor $\boldsymbol{x}$ described in Section 4.2, and we see the relationship between the posterior probabilities and the response probabilities

$$\mathrm{P}(G_1|\boldsymbol{x}) \to \mathrm{P}(Y = 1|\boldsymbol{x}) = \pi(\boldsymbol{x}), \quad \mathrm{P}(G_2|\boldsymbol{x}) \to \mathrm{P}(Y = 0|\boldsymbol{x}) = 1 - \pi(\boldsymbol{x}).$$

The class-indicator variable Y is distributed according to the Bernoulli distribution

$$f(y|\pi(\boldsymbol{x})) = \pi(\boldsymbol{x})^y\{1 - \pi(\boldsymbol{x})\}^{1-y}, \qquad y = 0, 1, \tag{7.21}$$

conditional on $\boldsymbol{x}$, where

$$\pi(\boldsymbol{x}) = \frac{\exp(\beta_0 + \boldsymbol{\beta}^T \boldsymbol{x})}{1 + \exp(\beta_0 + \boldsymbol{\beta}^T \boldsymbol{x})}. \tag{7.22}$$

The likelihood function based on the training data in (7.20) is thus given by

$$L(\beta_0, \boldsymbol{\beta}) = \prod_{i=1}^{n} \pi_i^{y_i}(1 - \pi_i)^{1-y_i}, \qquad y_i = 0, 1, \tag{7.23}$$

where

$$\pi_i = \pi(\boldsymbol{x}_i) = \frac{\exp(\beta_0 + \boldsymbol{\beta}^T \boldsymbol{x}_i)}{1 + \exp(\beta_0 + \boldsymbol{\beta}^T \boldsymbol{x}_i)}, \qquad i = 1, 2, \cdots, n. \tag{7.24}$$

By substituting π_i into the likelihood function (7.23) and taking the logarithm, we obtain the log-likelihood function of the parameters

$$\ell(\beta_0, \boldsymbol{\beta}) = \sum_{i=1}^{n} y_i(\beta_0 + \boldsymbol{\beta}^T \boldsymbol{x}_i) - \sum_{i=1}^{n} \log\left\{1 + \exp(\beta_0 + \boldsymbol{\beta}^T \boldsymbol{x}_i)\right\}. \tag{7.25}$$

The unknown parameters β_0 and $\boldsymbol{\beta}$ are estimated by the maximum likelihood methods within the framework of the logistic regression model described in Section 4.2.

The estimated posterior probabilities of class membership for the future observation are given by

$$\mathrm{P}(G_1|\boldsymbol{x}) = \frac{\exp\left(\hat{\beta}_0 + \hat{\boldsymbol{\beta}}^T \boldsymbol{x}\right)}{1 + \exp\left(\hat{\beta}_0 + \hat{\boldsymbol{\beta}}^T \boldsymbol{x}\right)}. \tag{7.26}$$

Allocation is then carried out by evaluating the posterior probabilities, and the future observation is assigned according to the following classification rule:

$$\mathrm{P}(G_1|\boldsymbol{x}) = \frac{\exp\left(\hat{\beta}_0 + \hat{\boldsymbol{\beta}}^T \boldsymbol{x}\right)}{1 + \exp\left(\hat{\beta}_0 + \hat{\boldsymbol{\beta}}^T \boldsymbol{x}\right)} \quad \begin{cases} \geq \dfrac{1}{2} & \Rightarrow \quad G_1 \\ < \dfrac{1}{2} & \Rightarrow \quad G_2. \end{cases} \tag{7.27}$$

By taking the logit transformation, we see that the classification rule is equivalent to the rule

$$\log \frac{P(G_1|\boldsymbol{x})}{1 - P(G_1|\boldsymbol{x})} = \hat{\beta}_0 + \hat{\boldsymbol{\beta}}^T \boldsymbol{x} \quad \begin{cases} \geq 0 & \Rightarrow & G_1 \\ < 0 & \Rightarrow & G_2. \end{cases} \tag{7.28}$$

Example 7.2 (Presence of calcium oxalate crystals) Let us now apply the linear logistic regression classifier to data relating to the formation of calcium oxalate crystals, which are a known cause of kidney and ureteral stones, as described in Section 6.3.4. The data were obtained for the purpose of constructing methods for assessing the presence of calcium oxalate, which causes calculus formation in individuals, from six physical properties of a person's urine that are thought to contribute to the formation of calcium oxalate crystals and can readily be measured in clinical testing (Andrews and Herzberg, 1985, p. 249).

The data are from two classes of individuals in whom the presence or absence of calcium oxalate crystals had been determined by thorough medical examination. The abnormal class (G_1), consisting of 33 individuals in whom calcium oxalate crystals were found, is assigned the label $y = 1$. The normal class (G_2) consists of individuals found to be free of the crystals and is assigned the label $y = 0$. The clinical test data obtained from urine samples received from the individuals in these two classes comprise a total of 77 data for six physical properties: specific gravity, x_1; pH, x_2; osmolality, x_3; conductivity, x_4; urea concentration, x_5; and calcium concentration, x_6. The objective is to construct a classification method (and thus a differential diagnosis method) based on these 77 training data that could then be applied in the future to classify individuals as belonging to G_1 or G_2 according to their test results regarding these properties.

Let $P(G_1|\boldsymbol{x})$ and $P(G_2|\boldsymbol{x})$ denote the posterior probabilities obtained under the assumption that the observed six-dimensional data $\boldsymbol{x}$ belong in abnormal class G_1 and normal class G_2, respectively. In the Bayesian classification model, the logarithm of the ratio of these two posterior probabilities is assumed to be expressed by a linear combination of the six physical properties

$$\log \frac{P(G_1|\boldsymbol{x})}{P(G_2|\boldsymbol{x})} = \beta_0 + \beta_1 x_1 + \beta_2 x_2 + \beta_3 x_3 + \beta_4 x_4 + \beta_5 x_5 + \beta_6 x_6. \tag{7.29}$$

Estimation of the parameters of the model, based on the training data in this example, yields the linear discriminant function $h(\boldsymbol{x}) = -355.34 +$

$355.94x_1 - 0.5x_2 + 0.02x_3 - 0.43x_4 + 0.03x_5 + 0.78x_6$. An individual with test results $\boldsymbol{x}$ is then assigned to abnormal class G_1 if $h(\boldsymbol{x}) \geq 0$ and to normal class G_2 if $h(\boldsymbol{x}) < 0$. For example, the value of the discriminant function for a particular individual's $\boldsymbol{x} = (1.017, 5.74, 577, 20, 296, 4.49)$ is 1.342, and then that individual is classified as belonging to G_1. From (7.27), moreover, the estimated posterior probability of that individual's membership in G_1 is $\mathrm{P}(G_1|\boldsymbol{x}) = \exp(1.342)/\{1 + \exp(1.342)\} = 0.794$, which indicates a fairly high probability that the classification as G_1 is correct. In performing logistic classification, we can in this way obtain classification probabilities (or risk probabilities) based on the posterior probabilities of individuals' data.

Let us next consider the question of which of the six variables are most closely involved in the formation of calcium oxalate crystals. As their units differ, the size of their estimated discriminant coefficients in the discriminant function $h(\boldsymbol{x})$ cannot be directly compared for this purpose. Therefore, each coefficient must be standardized by multiplying by the standard deviation of the corresponding variable. For example, in the 77 training data, specific gravity (x_1) has a standard deviation of 0.00667. Then the coefficient of the specific gravity is standardized as

$$\hat{\beta}_1 x_1 = 355.94 \times 0.00667 \times \frac{x_1}{0.00667} = 2.374 \times \frac{x_1}{0.00667},$$

which gives the standardized coefficient 2.374. Calculating all coefficients in the same manner, the standardized coefficients of specific gravity, pH, osmolality, conductivity, urea concentration, and calcium concentration are $2.3, -0.3, 4.7, -3.4, -3.9$, and 2.5, respectively. Comparison of these values clearly indicates that individuals with higher values in specific gravity, osmolality, and calcium concentration will tend to output higher posterior probabilities, and thus are more likely to be classified as members of the abnormal class. The results thus suggest that these three variables in particular are closely involved in the formation of calcium oxalate crystals.

7.3.2 Nonlinear Logistic Regression Classifier

Figure 7.3 is a plot of the data obtained from class G_1 ($\circ$) and class G_2 ($\triangle$) individuals, together with the decision boundary generated by the linear function. It clearly shows that, because of the complex structure of the two classes, the performance of this linear classification with respect to future data will not be high. In general, classification of phenomena exhibiting complex class structures requires a nonlinear function able to

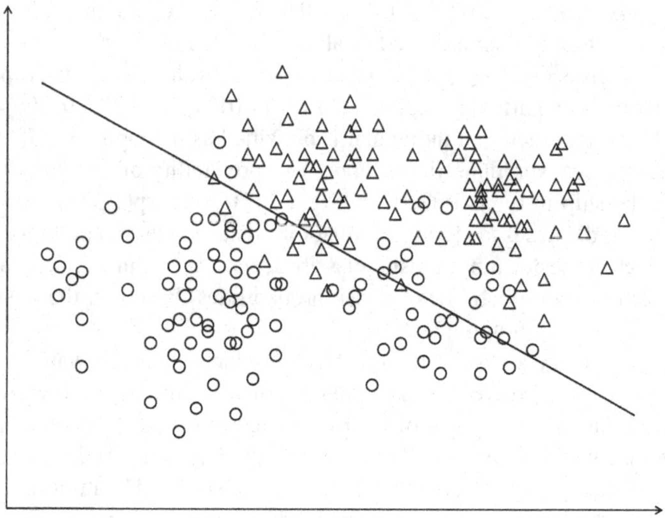

Figure 7.3 *Decision boundary generated by the linear function.*

generate the complexities of the original data, such as that giving the decision boundary shown in Figure 7.4 rather than a linear function. To meet this need, we can perform nonlinearization of the linear logistic regression for classification by a process conceptually similar to that applied to the logistic regression model in Section 4.3.

Let us consider a case in which p-dimensional data $\boldsymbol{x}$ from class G_1 or class G_2 has been observed. In linear logistic classification, it is assumed that the logarithm of the ratio between the posterior probabilities $\mathrm{P}(G_1|\boldsymbol{x})$ and $\mathrm{P}(G_2|\boldsymbol{x})$ for data is expressed as a linear combination of the variables. Nonlinearization of the logistic classification, in contrast, is generally obtained by expressing the logarithm of the ratio between the posterior probabilities as a nonlinear function of the variables. Here, we represent the nonlinear function as a linear combination of basis functions $b_0(\boldsymbol{x}) \equiv 1$, $b_1(\boldsymbol{x})$, $b_2(\boldsymbol{x})$, $\cdots$, $b_m(\boldsymbol{x})$, and then the logarithm of the ratio between the posterior probabilities is expressed as

$$\log \frac{\mathrm{P}(G_1|\boldsymbol{x})}{\mathrm{P}(G_2|\boldsymbol{x})} = \sum_{j=0}^{m} w_j b_j(\boldsymbol{x}) = \boldsymbol{w}^T \boldsymbol{b}(\boldsymbol{x}), \tag{7.30}$$

where $\boldsymbol{w} = (w_0, w_1, w_2, \cdots, w_m)^T$ and $\boldsymbol{b}(\boldsymbol{x}) = (b_0(\boldsymbol{x}),\ b_1(\boldsymbol{x}),\ b_2(\boldsymbol{x}),\ \cdots,$

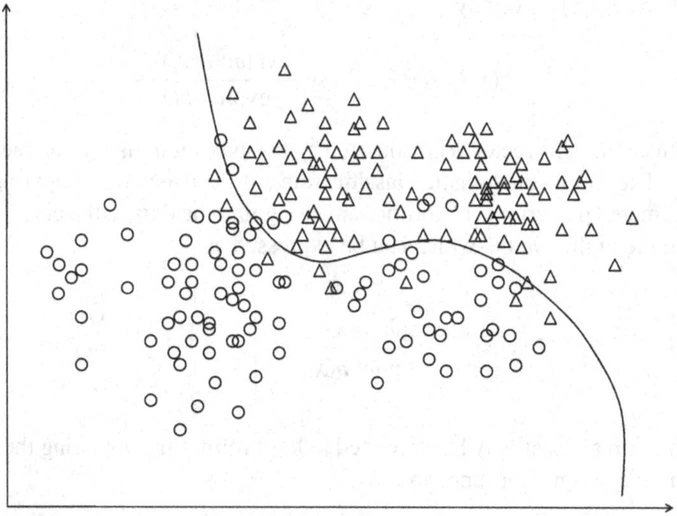

Figure 7.4 *Classification of phenomena exhibiting complex class structures requires a nonlinear discriminant function.*

$b_m(\boldsymbol{x}))^T$ is a vector of basis functions. For a detailed description of basis function nonlinearization, refer to Section 3.3.

From the relationship $\mathrm{P}(G_2|\boldsymbol{x}) = 1 - \mathrm{P}(G_1|\boldsymbol{x})$ between the posterior probabilities, (7.30) can be expressed as

$$\mathrm{P}(G_1|\boldsymbol{x}) = \frac{\exp\{\boldsymbol{w}^T \boldsymbol{b}(\boldsymbol{x})\}}{1 + \exp\{\boldsymbol{w}^T \boldsymbol{b}(\boldsymbol{x})\}}, \tag{7.31}$$

and by taking the posterior probability as a response probability, we find that it corresponds to the nonlinear logistic model in (4.27). We can therefore estimate the coefficient vector $\boldsymbol{w}$ for the n $(= n_1 + n_2)$ p-dimensional observed data from classes G_1 and G_2 on the basis of training data $\{(\boldsymbol{x}_i, y_i);\ i = 1, 2, \cdots, n\}$ incorporating a class-indicator variable Y. Thus, as in linear logistic regression for classification, we construct a likelihood function based on the Bernoulli distribution of the variables Y_i representing the class membership as follows:

$$f(y_i|\pi(\boldsymbol{x}_i)) = \pi(\boldsymbol{x}_i)^{y_i}\{1 - \pi(\boldsymbol{x}_i)\}^{1-y_i}, \quad y_i = 0, 1, \quad i = 1, \cdots, n, \tag{7.32}$$

where $\pi(\boldsymbol{x}_i)$ is given by

$$\pi(\boldsymbol{x}_i) = \mathrm{P}(G_1|\boldsymbol{x}_i) = \frac{\exp\{\boldsymbol{w}^T \boldsymbol{b}(\boldsymbol{x}_i)\}}{1 + \exp\{\boldsymbol{w}^T \boldsymbol{b}(\boldsymbol{x}_i)\}}. \tag{7.33}$$

The coefficient vector $\boldsymbol{w}$ is estimated by the maximum likelihood method.

The nonlinear logistic classification rule consists of comparing the estimated posterior probabilities and assigning the data to the class yielding the higher value, which can be expressed as

$$\mathrm{P}(G_1|\boldsymbol{x}) = \frac{\exp\{\hat{\boldsymbol{w}}^T \boldsymbol{b}(\boldsymbol{x})\}}{1 + \exp\{\hat{\boldsymbol{w}}^T \boldsymbol{b}(\boldsymbol{x})\}} \begin{cases} \geq \dfrac{1}{2} & \Rightarrow \quad G_1 \\ < \dfrac{1}{2} & \Rightarrow \quad G_2. \end{cases} \tag{7.34}$$

This can alternatively be converted to logit form, thus obtaining the nonlinear discriminant function

$$\log \frac{\mathrm{P}(G_1|\boldsymbol{x})}{1 - \mathrm{P}(G_1|\boldsymbol{x})} = \hat{\boldsymbol{w}}^T \boldsymbol{b}(\boldsymbol{x}) \begin{cases} \geq 0 & \Rightarrow \quad G_1 \\ < 0 & \Rightarrow \quad G_2. \end{cases} \tag{7.35}$$

Figure 7.5 shows the decision boundary that separates the two classes in the nonlinear logistic regression model based on the Gaussian basis functions of (3.20) in Section 3.2.3. Here we used

$$b_j(\boldsymbol{x}) = \phi_j(x_1, x_2) = \exp\left\{ -\frac{(x_1 - \mu_{j1})^2 + (x_2 - \mu_{j2})^2}{2h_j^2} \right\},$$

where the center vector $\boldsymbol{\mu}_j = (\mu_{j1}, \mu_{j2})^T$ determines the position of the basis functions and the quantity h_j^2 represents their spread. The nonlinear discriminant function that separates the two classes in the figure comprises 15 basis functions with the coefficient vector estimated by maximum likelihood as the weighting, which is expressed as

$$h(x_1, x_2) = \hat{w}_0 + \sum_{j=1}^{15} \hat{w}_j \phi_j(x_1, x_2) = 0. \tag{7.36}$$

Nonlinear logistic regression for classification is effective as an analytical technique for character and voice recognition, image analysis, and other pattern recognition applications. It must be noted, however, that the maximum likelihood method might not function effectively for model estimation with high-dimensional data and in such cases can therefore be

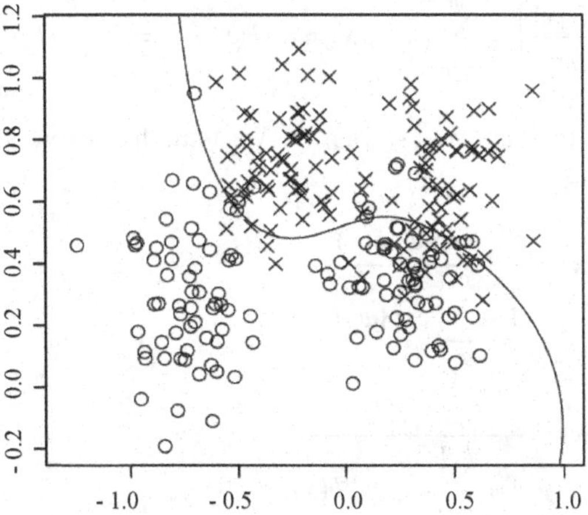

Figure 7.5 *Decision boundary that separates the two classes in the nonlinear logistic regression model based on the Gaussian basis functions.*

replaced by the regularized maximum likelihood technique described in Section 3.4. With appropriate choice of the regularization parameters, it is then possible to construct a nonlinear discriminant function that is effective for prediction.

7.3.3 *Multiclass Nonlinear Logistic Regression Classifier*

Let us now turn from classification into two classes to classification into multiple (i.e., three or more) classes and the related construction of nonlinear discriminants from the training data obtained from multiple classes. Suppose that n p-dimensional data $\{\boldsymbol{x}_1, \boldsymbol{x}_2, \cdots, \boldsymbol{x}_n\}$ from L classes $G_1, G_2, \cdots, G_L$ is observed and used as training data. When a future observation $\boldsymbol{x}$ from one of these classes is observed, the posterior probability that the individual is from the k-th class is denoted as $\mathrm{P}(G_k|\boldsymbol{x})$ $(k = 1, 2, \cdots, L)$. Our objective is now to construct a nonlinear classification by extending (7.30) for just two classes to multiple classes. In this extension, the multiclass classification is formalized from the logarithms of the ratios of the posterior probabilities of each of the first $L-1$ classes to the L-th class G_L, which are expressed as

$$\log\left\{\frac{\mathrm{P}(G_k|\boldsymbol{x})}{\mathrm{P}(G_L|\boldsymbol{x})}\right\} = \sum_{j=0}^{m} w_{kj}b_j(\boldsymbol{x}) = \boldsymbol{w}_k^T\boldsymbol{b}(\boldsymbol{x}), \quad k = 1, \cdots, L-1, \tag{7.37}$$

where $\boldsymbol{w}_k = (w_{k0}, w_{k1}, w_{k2}, \cdots, w_{km})^T$. We may then express posterior probabilities as

$$\begin{aligned}
\mathrm{P}(G_k|\boldsymbol{x}) &= \frac{\exp\{\boldsymbol{w}_k^T\boldsymbol{b}(\boldsymbol{x})\}}{1 + \sum_{g=1}^{L-1}\exp\{\boldsymbol{w}_g^T\boldsymbol{b}(\boldsymbol{x})\}}, \qquad k = 1, 2, \cdots, L-1, \\
\mathrm{P}(G_L|\boldsymbol{x}) &= \frac{1}{1 + \sum_{g=1}^{L-1}\exp\{\boldsymbol{w}_g^T\boldsymbol{b}(\boldsymbol{x})\}}.
\end{aligned} \tag{7.38}$$

We also note that because the observed data $\boldsymbol{x}$ belongs to one of the L classes, the sum of all L posterior probabilities must be exactly 1, i.e., $\sum_{g=1}^{L}\mathrm{P}(G_g|\boldsymbol{x}) = 1$.

Because each of the training data belongs to one of the L classes, for each data $\boldsymbol{x}_i$, we define a corresponding variable vector $\boldsymbol{y}_i$ representing its class membership, which is defined as

$$\begin{aligned}
\boldsymbol{y}_i &= (y_{i1}, y_{i2}, \cdots, y_{i(L-1)})^T \\
&= \begin{cases} (0, \cdots, 0, \overset{(k)}{1}, 0, \cdots, 0)^T & \text{if } \boldsymbol{x}_i \in G_k, \ k = 1, 2, \cdots, L-1, \\ (0, \cdots, 0, \cdots, 0)^T & \text{if } \boldsymbol{x}_i \in G_L. \end{cases}
\end{aligned} \tag{7.39}$$

That is, if data $\boldsymbol{x}_i$ is from the k-th class ($k \neq L$), this is indicated by an $(L-1)$-dimensional vector in which the k-th component is 1 and all other components are 0. In the case that the data belong to the L-th class, this is indicated by an $(L-1)$-dimensional vector having 0 as all components. In the case $L = 2$, we have $\boldsymbol{y}_i = y_{i1}$, and if $\boldsymbol{x}_i \in G_1$ then $y_{i1} = 1$, whereas if $\boldsymbol{x}_i \in G_2$ then $y_{i1} = 0$, which is just the two-class classification using a class-indicator variable defined in the previous section.

As can be seen from (7.38), the posterior probabilities depend on the coefficient vectors of the basis functions. We therefore denote as $\boldsymbol{w} = \{\boldsymbol{w}_1, \boldsymbol{w}_2, \cdots, \boldsymbol{w}_{L-1}\}$ the parameters estimated from the training data and express the posterior probabilities by $\mathrm{P}(G_k|\boldsymbol{x}_i) = \pi_k(\boldsymbol{x}_i; \boldsymbol{w})$. While the class-indicator variable of the two classes follows a Bernoulli distribution, the multiclass $(L-1)$-dimensional variable vector $\boldsymbol{y}_i$ follows the multinomial distribution

$$f(\boldsymbol{y}_i|\boldsymbol{x}_i;\boldsymbol{w}) = \prod_{k=1}^{L-1} \pi_k(\boldsymbol{x}_i;\boldsymbol{w})^{y_{ik}} \pi_L(\boldsymbol{x}_i;\boldsymbol{w})^{1-\sum_{g=1}^{L-1} y_{ig}}. \tag{7.40}$$

The log-likelihood function based on the n training data is then given by

$$\ell(\boldsymbol{w}) = \sum_{i=1}^{n} \left[\sum_{k=1}^{L-1} y_{ik} \log \pi_k(\boldsymbol{x}_i;\boldsymbol{w}) + \left(1 - \sum_{g=1}^{L-1} y_{ig}\right) \log \pi_L(\boldsymbol{x}_i;\boldsymbol{w}) \right], \tag{7.41}$$

where

$$\begin{aligned} \pi_k(\boldsymbol{x}_i;\boldsymbol{w}) &= \frac{\exp\left\{\boldsymbol{w}_k^T \boldsymbol{b}(\boldsymbol{x}_i)\right\}}{1 + \sum_{g=1}^{L-1} \exp\left\{\boldsymbol{w}_g^T \boldsymbol{b}(\boldsymbol{x}_i)\right\}}, \qquad k = 1, 2, \cdots, L-1, \\ \pi_L(\boldsymbol{x}_i;\boldsymbol{w}) &= \frac{1}{1 + \sum_{g=1}^{L-1} \exp\left\{\boldsymbol{w}_g^T \boldsymbol{b}(\boldsymbol{x}_i)\right\}}. \end{aligned} \tag{7.42}$$

The maximum likelihood estimates of the coefficient vectors $\boldsymbol{w} = \{\boldsymbol{w}_1, \boldsymbol{w}_2, \cdots, \boldsymbol{w}_{L-1}\}$ of the basis functions are given by solution of the likelihood equation

$$\frac{\partial \ell(\boldsymbol{w})}{\partial \boldsymbol{w}_k} = \sum_{i=1}^{n} \{y_{ik} - \pi_k(\boldsymbol{x}_i;\boldsymbol{w})\} \boldsymbol{b}(\boldsymbol{x}_i) = \boldsymbol{0}, \quad k = 1, 2, \cdots, L-1. \tag{7.43}$$

The solution is obtained by the Newton-Raphson method or other numerical optimization. The required second derivative of the likelihood function is given by

$$\frac{\partial^2 \ell(\boldsymbol{w})}{\partial \boldsymbol{w}_j \partial \boldsymbol{w}_k^T} = \begin{cases} -\sum_{i=1}^{n} \pi_j(\boldsymbol{x}_i;\boldsymbol{w})\{1 - \pi_j(\boldsymbol{x}_i;\boldsymbol{w})\} \boldsymbol{b}(\boldsymbol{x}_i)\boldsymbol{b}(\boldsymbol{x}_i)^T, & k = j, \\ \sum_{i=1}^{n} \pi_j(\boldsymbol{x}_i;\boldsymbol{w})\pi_k(\boldsymbol{x}_i;\boldsymbol{w}) \boldsymbol{b}(\boldsymbol{x}_i)\boldsymbol{b}(\boldsymbol{x}_i)^T, & k \neq j. \end{cases} \tag{7.44}$$

By substituting the estimates $\hat{\boldsymbol{w}} = \{\hat{\boldsymbol{w}}_1, \hat{\boldsymbol{w}}_2, \cdots, \hat{\boldsymbol{w}}_{L-1}\}$ obtained by numerical optimization for the unknown coefficient vectors in (7.38), we can estimate the posterior probabilities of class membership of data $\boldsymbol{x}$. When new data are observed, we can then compare their L posterior probabilities and classify them as to class membership on the basis of the maximum posterior probability. This, in summary, is the technique referred to as *multiclass nonlinear logistic regression for classification*, which is widely used in remote-sensing data analysis, voice recognition, image analysis, and other fields.

Exercises

7.1 Suppose that the sample space Ω is divided into r mutually disjoint events G_j as $\Omega = G_1 \cup G_2 \cup \cdots \cup G_r$ $(G_i \cap G_j = \emptyset)$. Show that for any event D,

$$\mathrm{P}(D) = \mathrm{P}(G_1)\mathrm{P}(D|G_1) + \mathrm{P}(G_2)\mathrm{P}(D|G_2) + \cdots + \mathrm{P}(G_r)\mathrm{P}(D|G_r).$$

7.2 Prove Bayes's theorem given by (7.4).

7.3 Suppose that the two classes G_1 and G_2 are, respectively, characterized by the p-dimensional normal distributions

$$G_1 : \ f(\boldsymbol{x}|\overline{\boldsymbol{x}}_1, S) = \frac{1}{(2\pi)^{p/2}|S|^{1/2}} \exp\left\{-\frac{1}{2}(\boldsymbol{x}-\overline{\boldsymbol{x}}_1)^T S^{-1}(\boldsymbol{x}-\overline{\boldsymbol{x}}_1)\right\},$$

$$G_2 : \ f(\boldsymbol{x}|\overline{\boldsymbol{x}}_2, S) = \frac{1}{(2\pi)^{p/2}|S|^{1/2}} \exp\left\{-\frac{1}{2}(\boldsymbol{x}-\overline{\boldsymbol{x}}_2)^T S^{-1}(\boldsymbol{x}-\overline{\boldsymbol{x}}_2)\right\},$$

where $\overline{\boldsymbol{x}}_i$ are the sample mean vectors and S is the pooled sample variance-covariance matrix.

(a) Under the assumption that the prior probabilities are equal, construct the Bayes classification rule based on the probability distributions $f(\boldsymbol{x}|\overline{\boldsymbol{x}}_i, S)$ $(i = 1, 2)$.

(b) Let $h(\boldsymbol{x}) = \log f(\boldsymbol{x}|\overline{\boldsymbol{x}}_1, S) - \log f(\boldsymbol{x}|\overline{\boldsymbol{x}}_2, S) = \boldsymbol{w}^T\boldsymbol{x} + c$. Find the weight vector $\boldsymbol{w}$ and the constant c.

7.4 Show that the Bayesian classification based on the multivariate normal distributions is equivalent to the method based on the Mahalanobis distance described in Section 6.2.

7.5 Let $\mathrm{P}(G_i|\boldsymbol{x})$ $(i = 1, 2)$ be the posterior probabilities when p-dimensional data $D = \{\boldsymbol{x}\}$ is observed. Assume that the logarithm of the ratio of these posterior probabilities is expressed as

$$\log \frac{\mathrm{P}(G_1|\boldsymbol{x})}{\mathrm{P}(G_2|\boldsymbol{x})} = \beta_0 + \boldsymbol{\beta}^T \boldsymbol{x}.$$

(a) Show that the posterior probabilities are given by

$$\mathrm{P}(G_1|\boldsymbol{x}) = \frac{\exp(\beta_0 + \boldsymbol{\beta}^T \boldsymbol{x})}{1 + \exp(\beta_0 + \boldsymbol{\beta}^T \boldsymbol{x})}, \quad \mathrm{P}(G_2|\boldsymbol{x}) = \frac{1}{1 + \exp(\beta_0 + \boldsymbol{\beta}^T \boldsymbol{x})}.$$

(b) Derive the log-likelihood function of the parameters $(\beta_0, \boldsymbol{\beta})$ when the class-indicator variable Y is distributed according to the Bernoulli distribution.

Chapter 8

Support Vector Machines

In statistical classification such as linear and quadratic discriminant analyses, discussed in Chapter 6, and logistic classification in Chapter 7, the classification rules were constructed using criteria such as Mahalanobis distance minimization and posterior probability maximization. A basic property of these classification methods is their reliance on estimators such as sample mean vectors or sample variance-covariance matrices that summarize information contained in the observed training data from each class. Support vector machines (SVMs) proposed by Vapnik (1996, 1998), in contrast, are characterized by an internal process of constructing classification rules quite different in basic concept from those of the statistical methods. SVMs are often effective in cases for which ordinary classification methods are not effective, such as problems based on high-dimensional data and data with nonlinear structure, and they have been adapted and applied in many fields.

In this chapter, we focus on the basic concept in SVMs for the solution of recognition and classification problems and on the process of constructing SVM classifiers, and then proceed through the steps of their extension from linear to nonlinear systems.

8.1 Separating Hyperplane

Here we first establish the basic idea of the SVM using cases in which classification is actually self-evident, and then proceed to apply the SVM methodology to actual data analysis. In the following, our first objective will therefore be to construct functions that optimally separate two classes of training data in situations in which their separability is inherently obvious.

8.1.1 Linear Separability

Suppose that we have n observed training data $\boldsymbol{x}_1, \boldsymbol{x}_2, \cdots, \boldsymbol{x}_n$ relating to p variables $\boldsymbol{x} = (x_1, x_2, \cdots, x_p)^T$ that characterize (assign "features" to)

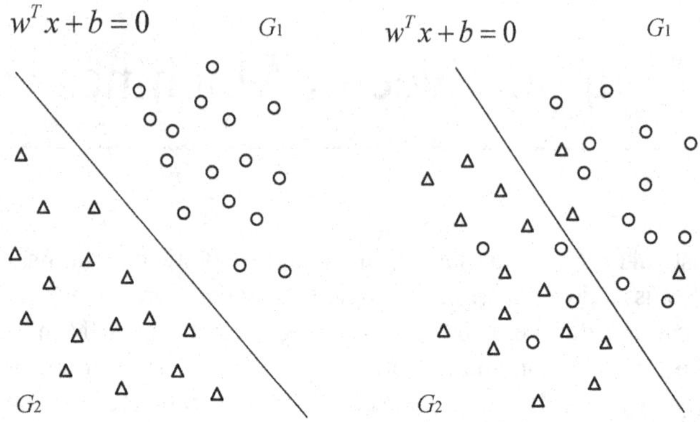

Figure 8.1 *The training data are completely separable into two classes by a hyperplane (left panel), and in contrast, separation into two classes cannot be obtained by any such linear hyperplane (right panel).*

certain individuals or objects. Here, let us assume that we already know the membership of the training data in the two classes G_1 and G_2. Our objective is to find a *hyperplane*

$$w_1 x_1 + w_2 x_2 + \cdots + w_p x_p + b = \boldsymbol{w}^T \boldsymbol{x} + b = 0, \tag{8.1}$$

which separates two classes, where $\boldsymbol{w} = (w_1, w_2, \cdots, w_p)^T$.

The meaning of inherently obvious separability in the training data is illustrated by the configuration on the left panel in Figure 8.1. As shown in the figure, the training data are completely separable into two classes by a hyperplane, which in this case is a line because the data space is two dimensional. This is known as *linear separability*. For the configuration on the right panel, in contrast, separation into two classes cannot be obtained by any such linear hyperplane.

For training data to be linearly separable by a hyperplane, all of the data must satisfy one or the other of the two inequalities $\boldsymbol{w}^T \boldsymbol{x}_i + b > 0$ and $\boldsymbol{w}^T \boldsymbol{x}_i + b < 0$. In the configuration on the left panel in Figure 8.1, the data above the separating hyperplane thus satisfy $\boldsymbol{w}^T \boldsymbol{x}_i + b > 0$, and the data below the hyperplane conversely satisfy $\boldsymbol{w}^T \boldsymbol{x}_i + b < 0$. If for example we have a straight line $x_1 + x_2 = 0$ through the origin of a two-dimensional plane, it may be readily seen that the two data at (1, 1) and (−1, −1) satisfy this relationship. It is assumed here that the training data separated into the two classes G_1 and G_2 by the hyperplane are determined as

$$\begin{aligned} G_1 &: \text{ the data set such that } \boldsymbol{w}^T\boldsymbol{x}_i + b > 0, \\ G_2 &: \text{ the data set such that } \boldsymbol{w}^T\boldsymbol{x}_i + b < 0, \end{aligned} \tag{8.2}$$

or equivalently

$$\boldsymbol{x}_i \in G_1 \Longleftrightarrow \boldsymbol{w}^T\boldsymbol{x}_i + b > 0, \qquad \boldsymbol{x}_i \in G_2 \Longleftrightarrow \boldsymbol{w}^T\boldsymbol{x}_i + b < 0. \tag{8.3}$$

To represent class membership, we introduce the class label y. If the i-th data point $\boldsymbol{x}_i$ is observed to belong to class G_1, we set $y_i = 1$, and if it is observed to belong to class G_2, we set $y_i = -1$. Accordingly, we can represent the n training data having p dimensions as follows:

$$(\boldsymbol{x}_1, y_1),\ (\boldsymbol{x}_2, y_2),\ \cdots,\ (\boldsymbol{x}_n, y_n);\quad y_i = \begin{cases} 1 & if\ \boldsymbol{x}_i \in G_1 \\ -1 & if\ \boldsymbol{x}_i \in G_2. \end{cases} \tag{8.4}$$

Since the data of class G_1 have been labeled $y_i = 1$, as shown in (8.4), the direction of inequality in the expression on the left in (8.3) does not change if we multiply both sides by y_i, and we can therefore express the relationship as $y_i(\boldsymbol{w}^T\boldsymbol{x}_i + b) > 0$. In the case of data belonging to class G_2, as these have been labeled $y_i = -1$, the direction of the inequality on the right in (8.3) is reversed if we multiply both sides by −1, and this relationship can thus also be expressed as $y_i(\boldsymbol{w}^T\boldsymbol{x}_i + b) > 0$. In short, if the training data can be completely separated into two classes by the hyperplane, then the following inequality holds for all of those data

$$y_i(\boldsymbol{w}^T\boldsymbol{x}_i + b) > 0, \qquad i = 1, 2, \cdots, n. \tag{8.5}$$

The question we must now consider is how to find the optimum separation hyperplane for linearly separable training data. In essence, this is a question of what criterion to use as a basis for estimation of the coefficient vector $\boldsymbol{w}$ (weight vector) and the intercept b (bias). The criterion we shall use for this purpose is the distance between the data and the hyperplane (Figure 8.2). In general, the distance from p-dimensional data $\boldsymbol{x} = (x_1, x_2, \cdots, x_p)^T$ to the hyperplane $\boldsymbol{w}^T\boldsymbol{x} + b = 0$ is given by

$$d \equiv \frac{|w_1x_1 + w_2x_2 + \cdots + w_px_p + b|}{\sqrt{w_1^2 + w_2^2 + \cdots + w_p^2}} = \frac{|\boldsymbol{w}^T\boldsymbol{x} + b|}{||\boldsymbol{w}||}, \tag{8.6}$$

where $||\boldsymbol{w}|| = \sqrt{w_1^2 + w_2^2 + \cdots + w_p^2}$.

$$d = \frac{|w_1 x_{01} + w_2 x_{02} + b|}{\sqrt{w_1^2 + w_2^2}} = \frac{|w^T x_0 + b|}{\|w\|}$$

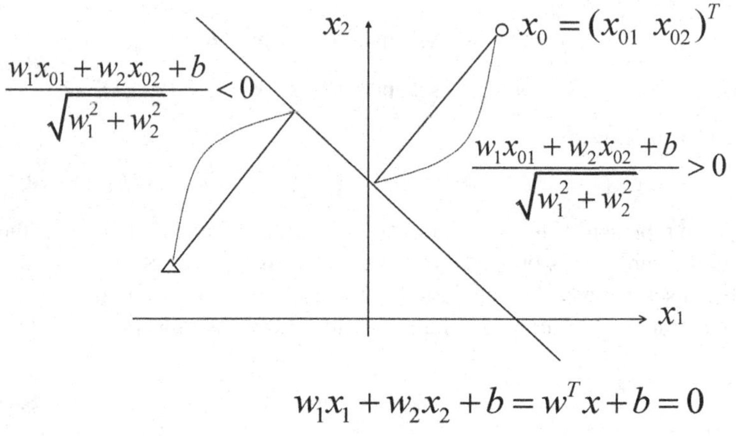

Figure 8.2 *Distance from* $\boldsymbol{x}_0 = (x_{01}\ x_{02})^T$ *to the hyperplane* $w_1 x_1 + w_2 x_2 + b = \boldsymbol{w}^T \boldsymbol{x} + b = 0$.

8.1.2 Margin Maximization

Let us now consider the method of determining the optimum hyperplane for separation of two classes, based on the distance from the data to the hyperplane. If the data are linearly separable, then a hyperplane (H) exists that separates the two classes, together with two equidistant parallel hyperplanes (H_+ and H_-) on opposite sides, as shown in Figure 8.3. Each of these two *margin-edge* hyperplanes includes at least one data point, and no data point lies between them. We may think of this as analogous to constructing a highway that separates the two data classes, with one or more data points lying on each shoulder of the highway but none lying on the highway itself; in this analogy, the highway centerline represents the separating hyperplane. For linearly separable data, as illustrated in Figure 8.4, any number of such highways can be constructed. In the SVM, the task is to build the highway that is greatest in width, with its centerline thus representing the *optimum separation hyperplane*.

The distance from data points $\boldsymbol{x}_+$ on hyperplane H_+ and $\boldsymbol{x}_-$ on hyperplane H_- to hyperplane H defines the *margin* on both sides of the

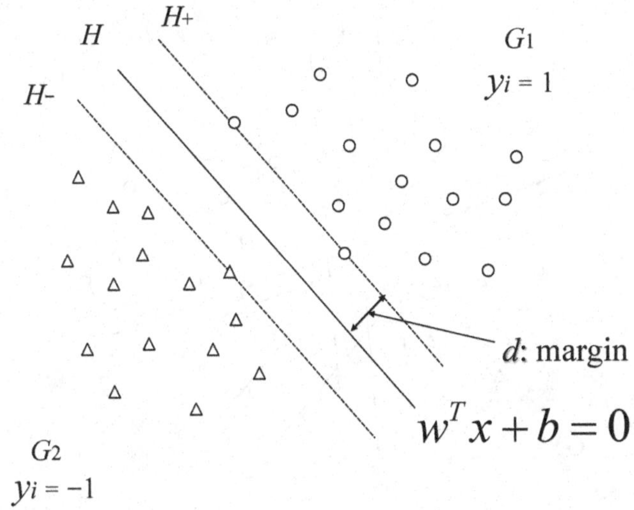

Figure 8.3 *Hyperplane (H) that separates the two classes, together with two equidistant parallel hyperplanes (H_+ and H_-) on opposite sides.*

separating hyperplane by

$$d = \frac{\boldsymbol{w}^T \boldsymbol{x}_+ + b}{||\boldsymbol{w}||} = \frac{-(\boldsymbol{w}^T \boldsymbol{x}_- + b)}{||\boldsymbol{w}||}. \tag{8.7}$$

The task of optimally separating the two classes becomes one of finding the separating hyperplane that maximizes this margin. Since we assume that no data point lies between hyperplanes H_+ and H_-, thus satisfying the condition that the distance from the separation hyperplane to any data point is at least d, the task can be formulated in terms of finding the weight vector $\boldsymbol{w}$ and bias b that maximize the margin d, and hence are expressed as

$$\max_{\boldsymbol{w},b} d \quad \text{subject to} \quad \frac{y_i(\boldsymbol{w}^T \boldsymbol{x}_i + b)}{||\boldsymbol{w}||} \geq d, \quad i = 1, 2, \cdots, n. \tag{8.8}$$

In this form, however, the problem of margin maximization is computationally difficult. As described in the following section, we therefore transform this to an optimization problem known as the quadratic programming.

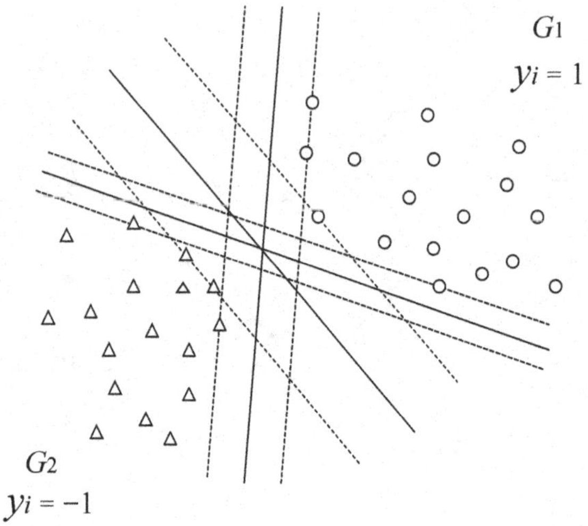

Figure 8.4 *Separating hyperplanes with different margins.*

8.1.3 Quadratic Programming and Dual Problem

In general, multiplying the coefficients on both sides by some real number $r > 0$ and rescaling the equation of the hyperplane $\boldsymbol{w}^T\boldsymbol{x} + b = 0$ does not change the hyperplane. More specifically, if we set $r\boldsymbol{w} = \boldsymbol{w}^*, rb = b^*$ in $r\boldsymbol{w}^T\boldsymbol{x} + rb = 0$, then the resulting hyperplane $\boldsymbol{w}^{*T}\boldsymbol{x} + b^* = 0$ is no different from the original. Let us now focus on the margin maximization condition

$$\frac{y_i(\boldsymbol{w}^T\boldsymbol{x}_i + b)}{||\boldsymbol{w}||} \geq d, \qquad i = 1, 2, \cdots, n, \tag{8.9}$$

with equality holding for data on the hyperplanes H_+ and H_- located on opposite sides of the separating hyperplane. If we perform the rescaling by dividing both sides of this by the margin d, we then have

$$y_i\left(\frac{1}{d||\boldsymbol{w}||}\boldsymbol{w}^T\boldsymbol{x}_i + \frac{1}{d||\boldsymbol{w}||}b\right) \geq 1, \quad i = 1, 2, \cdots, n, \tag{8.10}$$

and taking $r = 1/d||\boldsymbol{w}||$ yields

$$y_i(\boldsymbol{w}^{*T}\boldsymbol{x}_i + b^*) \geq 1, \qquad i = 1, 2, \cdots, n, \tag{8.11}$$

where

$$w^* = \frac{1}{d\|w\|}w, \qquad b^* = \frac{1}{d\|w\|}b. \tag{8.12}$$

Accordingly, all of the linearly separable data clearly satisfy the above inequality constraints, and in particular the equality in this expression applies for the data on hyperplanes H_+ and H_-. Therefore, the distance (e.g., the margin) from the data points on hyperplane H_+ or H_- to the separating hyperplane H can be expressed as

$$d^* = \frac{y_i(w^{*T}x_i + b^*)}{\|w^*\|} = \frac{1}{\|w^*\|}. \tag{8.13}$$

For the rescaled separating hyperplane $w^{*T}x_i + b^* = 0$, all of the linearly separable training data therefore satisfy the inequality constraint $y_i(w^{*T}x_i + b^*) \geq 1$. Based on this constraint, the problem now becomes one of finding w^* and b^* that maximize the margin $d^* = 1/\|w^*\|$. Furthermore, as maximizing margin d^* is equivalent to minimizing the square of the weight vector norm $\|w\|^{*2}$ in the denominator, the problem of margin maximization can be formulated as follows. In all subsequent discussion, let us therefore simply denote w^*, b^*, and d^* as w, b, and d.

Primal problem: Margin maximization

$$\min_{w} \frac{1}{2}\|w\|^2 \quad \text{subject to } y_i(w^T x_i + b) \geq 1, \quad i = 1, 2, \cdots, n. \tag{8.14}$$

In mathematical programming, the problem of maximizing the margin under the inequality constraint is known as the *quadratic programming*, and several mathematical programming procedures have been proposed for this purpose. The SVM enables us to take equation (8.14) as the *primal problem*, and use the Lagrangian function to transform it to an optimization problem known as the *dual problem*. Further information on replacement of the primal problem by the dual problem may be found in Appendices B.2 and B.3.

We first consider the following Lagrangian function, using Lagrange multipliers $\alpha_1, \alpha_2, \cdots, \alpha_n$ ($\alpha_i \geq 0$; $i = 1, 2, \cdots, n$) corresponding to the constraint number n,

$$L(w, b, \alpha_1, \alpha_2, \cdots, \alpha_n) = \frac{1}{2}\|w\|^2 - \sum_{i=1}^{n} \alpha_i \left\{ y_i(w^T x_i + b) - 1 \right\}. \tag{8.15}$$

By differentiating the Lagrangian function with respect to the weight vector $\boldsymbol{w}$ and the bias b and setting the results to zero, we have

$$\frac{\partial L(\boldsymbol{w}, b, \alpha_1, \alpha_2, \cdots, \alpha_n)}{\partial \boldsymbol{w}} = \boldsymbol{w} - \sum_{i=1}^{n} \alpha_i y_i \boldsymbol{x}_i = \boldsymbol{0},$$
$$\frac{\partial L(\boldsymbol{w}, b, \alpha_1, \alpha_2, \cdots, \alpha_n)}{\partial b} = -\sum_{i=1}^{n} \alpha_i y_i = 0. \tag{8.16}$$

Here we used $\partial \|\boldsymbol{w}\|^2 / \partial \boldsymbol{w} = \partial \boldsymbol{w}^T \boldsymbol{w} / \partial \boldsymbol{w} = 2\boldsymbol{w}$ and $\partial \boldsymbol{w}^T \boldsymbol{x}_i / \partial \boldsymbol{w} = \boldsymbol{x}_i$. We thus obtain

$$\boldsymbol{w} = \sum_{i=1}^{n} \alpha_i y_i \boldsymbol{x}_i, \qquad \sum_{i=1}^{n} \alpha_i y_i = 0. \tag{8.17}$$

Substituting these equations back into the Lagrangian function in (8.15) gives the Lagrangian dual objective function

$$L_D(\alpha_1, \alpha_2, \cdots, \alpha_n) = \sum_{i=1}^{n} \alpha_i - \frac{1}{2} \sum_{i=1}^{n} \sum_{j=1}^{n} \alpha_i \alpha_j y_i y_j \boldsymbol{x}_i^T \boldsymbol{x}_j. \tag{8.18}$$

In this way, we transform the quadratic programming problem of (8.14), which is a primal problem, to the following dual problem.

Dual problem

$$\max_{\alpha_1,\cdots,\alpha_n} L_D(\alpha_1, \alpha_2, \cdots, \alpha_n) = \max_{\alpha_1,\cdots,\alpha_n} \left\{ \sum_{i=1}^{n} \alpha_i - \frac{1}{2} \sum_{i=1}^{n} \sum_{j=1}^{n} \alpha_i \alpha_j y_i y_j \boldsymbol{x}_i^T \boldsymbol{x}_j \right\}, \tag{8.19}$$
$$\text{subject to} \quad \alpha_i \geq 0 \quad (i = 1, 2, \cdots, n) \text{ and } \sum_{i=1}^{n} \alpha_i y_i = 0.$$

Designating the solution to this dual problem $\hat{\alpha}_1, \hat{\alpha}_2, \cdots, \hat{\alpha}_n$ and substituting the Lagrange multipliers in the weight vector $\boldsymbol{w}$ of (8.17) for solution of the dual problem, we obtain the optimum solution

$$\hat{\boldsymbol{w}} = \sum_{i=1}^{n} \hat{\alpha}_i y_i \boldsymbol{x}_i. \tag{8.20}$$

To obtain the optimum solution for the bias b, we recall that the following equations hold for data in the hyperplanes H_+ and H_- lying on opposite sides of separating hyperplane $\hat{\boldsymbol{w}}^T \boldsymbol{x} + b = 0$

$$H_+ : \hat{\boldsymbol{w}}^T \boldsymbol{x}_+ + b = 1, \qquad H_- : \hat{\boldsymbol{w}}^T \boldsymbol{x}_- + b = -1. \tag{8.21}$$

The optimum solution for b directly follows from these two equations and is given by

$$\hat{b} = -\frac{1}{2}(\hat{\boldsymbol{w}}^T \boldsymbol{x}_+ + \hat{\boldsymbol{w}}^T \boldsymbol{x}_-). \tag{8.22}$$

For linearly separable training data, in summary, we can in this way obtain the hyperplane that yields the largest margin and thus best separates the two classes as follows:

$$\hat{\boldsymbol{w}}^T \boldsymbol{x} + \hat{b} = \sum_{i=1}^{n} y_i \hat{\alpha}_i \boldsymbol{x}_i^T \boldsymbol{x} + \hat{b} = 0. \tag{8.23}$$

Accordingly, the classification rule for the two classes G_1 and G_2 with the optimum separating hyperplane as the discriminant function is

$$\hat{\boldsymbol{w}}^T \boldsymbol{x} + \hat{b} = \sum_{i=1}^{n} y_i \hat{\alpha}_i \boldsymbol{x}_i^T \boldsymbol{x} + \hat{b} = \begin{cases} \geq 0 & \Longrightarrow \quad G_1 \\ < 0 & \Longrightarrow \quad G_2. \end{cases} \tag{8.24}$$

This is a key property of the discriminant functions constructed by SVMs, as distinguished from those of statistical discriminant functions in which the coefficient vectors of linear discriminant functions are estimated on the basis of sample mean vectors and sample variance-covariance matrices obtained from all of the training data. In SVMs, in contrast, the discriminant function is substantially determined by the data in the hyperplanes H_+ and H_- lying on opposite sides of the optimum separating hyperplane, and thus solely by the data that satisfy the equation

$$y_i(\hat{\boldsymbol{w}}^T \boldsymbol{x}_i + \hat{b}) = 1. \tag{8.25}$$

These data are referred to as *support vectors*, and it is for this reason that the classification machines they compose are called *support vector machines* (see Figure 8.5). If we denote the index set of the support vectors by S, the classification rule in (8.24) may then be expressed as

$$\hat{\boldsymbol{w}}^T \boldsymbol{x} + \hat{b} = \sum_{i \in S} y_i \hat{\alpha}_i \boldsymbol{x}_i^T \boldsymbol{x} + \hat{b} = \begin{cases} \geq 0 & \Longrightarrow \quad G_1 \\ < 0 & \Longrightarrow \quad G_2. \end{cases} \tag{8.26}$$

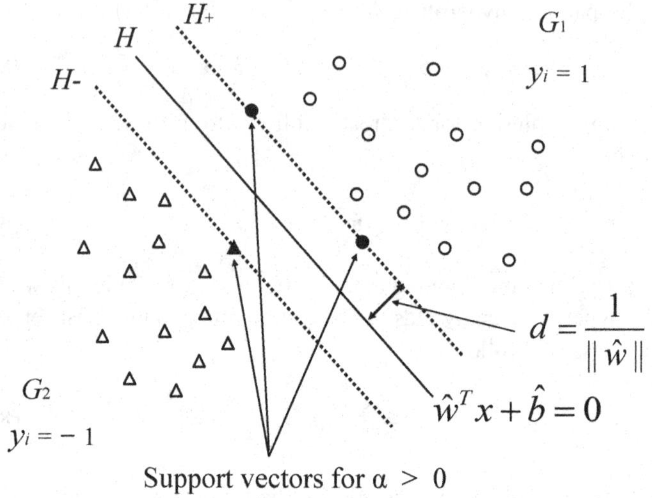

Figure 8.5 *Optimum separating hyperplane and support vectors represented by the black solid dots and triangle on the hyperplanes H_+ and H_-.*

It is known that this key property of SVMs can be derived from the *Karush-Kuhn-Tucker conditions*.

Karush-Kuhn-Tucker conditions Consider the following optimization problem:

$$\min_{\boldsymbol{w}} f(\boldsymbol{w}), \quad \text{subject to } g_i(\boldsymbol{w}) \leq 0, \quad i = 1, 2, \cdots, n, \tag{8.27}$$

where $f(w_1, w_2, \cdots, w_p) = f(\boldsymbol{w})$ is the p-dimensional objective function and $g_i(\boldsymbol{w}) \leq 0$, $i = 1, 2, \cdots, n$ are the inequality constraint functions. We define the Lagrangian function

$$L(\boldsymbol{w}, \boldsymbol{\alpha}) = f(\boldsymbol{w}) + \sum_{i=1}^{n} \alpha_i g_i(\boldsymbol{w}), \tag{8.28}$$

using Lagrange multipliers $\boldsymbol{\alpha} = (\alpha_1, \alpha_2, \cdots, \alpha_n)^T$ corresponding to the constraint number n. The solution to the inequality-constrained optimization problem is then obtained by finding the points $\boldsymbol{w}$ and $\boldsymbol{\alpha}$ satisfying the following four conditions:

$$
\begin{aligned}
&(1)\ \frac{\partial L(\boldsymbol{w}, \boldsymbol{\alpha})}{\partial w_i} = 0, \qquad i = 1, 2, \cdots, p, \\
&(2)\ \frac{\partial L(\boldsymbol{w}, \boldsymbol{\alpha})}{\partial \alpha_i} = g_i(\boldsymbol{w}) \le 0, \qquad i = 1, 2, \cdots, n, \\
&(3)\ \alpha_i \ge 0, \qquad i = 1, 2, \cdots, n, \\
&(4)\ \alpha_i g_i(\boldsymbol{w}) = 0, \qquad i = 1, 2, \cdots, n.
\end{aligned}
\tag{8.29}
$$

Support vectors In the Lagrangian function (8.15) used to obtain the optimum separating hyperplane, i.e., the one that maximizes the margin, we set the objective function $f(\boldsymbol{w})$ and the constraint equation $g_i(\boldsymbol{w})$ as follows:

$$
f(\boldsymbol{w}) = \frac{1}{2}||\boldsymbol{w}||^2, \ \ g_i(\boldsymbol{w}) = -\{y_i(\boldsymbol{w}^T \boldsymbol{x}_i + b) - 1\}, \ \ i = 1, 2, \cdots, n. \tag{8.30}
$$

Then the Karush-Kuhn-Tucker condition (4) implies that

$$
\alpha_i\{y_i(\boldsymbol{w}^T \boldsymbol{x}_i + b) - 1\} = 0, \qquad i = 1, 2, \cdots, n. \tag{8.31}
$$

In accordance with the above equation, any data that are not support vectors result in $y_i(\boldsymbol{w}^T \boldsymbol{x}_i + b) > 1$ and therefore in $\alpha_i = 0$. For any data that are support vectors, in contrast, $y_i(\boldsymbol{w}^T \boldsymbol{x}_i + b) = 1$ holds and therefore $\alpha_i > 0$. Accordingly, as illustrated in Figure 8.5, the data that determine the discriminant function are only those data that are actual support vectors for which $\alpha_i > 0$.

In this section, in summary, we have discussed the method of solution in which the quadratic programming problem, through incorporation of the Lagrangian function, is transformed to an optimization problem known as the dual problem. For more detailed theoretical discussion of this duality, refer to Cristianini and Shawe-Taylor (2000). For more information on Lagrange's method of undetermined multipliers as a method of finding stationary points in real-valued multivariable functions under equality and inequality constraints on variables, refer to Appendix B and also to Cristianini and Shawe-Taylor (2000).

8.2 Linearly Nonseparable Case

In the previous section, we focused on cases of complete separability, in which $\boldsymbol{w}^T \boldsymbol{x}_i + b > 0$ for data $\boldsymbol{x}_i \in G_1$ $(y_i = 1)$ and $\boldsymbol{w}^T \boldsymbol{x} + b < 0$ for data $x_i \in G_2$ ($y_i = -1$), with the discussion centering on the method of finding the optimum separating plane by maximizing the margin for data that were linearly separable by the hyperplane $w^T x + b = 0$. Ultimately, however, the real need is an SVM embodying a classification method that is effective for cases in which the two classes of training data cannot be separated by a hyperplane. We can now address this need, by using the basic concept of the technique developed for linearly separable data, to formulate a method of classification for training data that is not linearly separable.

8.2.1 Soft Margins

In Section 8.1.3 we showed that in cases of linear separability the following inequality holds for all of the data:

$$y_i(w^T x_i + b) \geq 1, \qquad i = 1, 2, \cdots, n. \tag{8.32}$$

In cases of linear nonseparability, however, some of the data do not satisfy this inequality constraint. No matter where we draw the hyperplane for separation of the two classes and the accompanying hyperplanes for the margin, some of the data inevitably lie beyond them, as illustrated in Figure 8.6. Some data from classes G_2 and G_1 lie beyond hyperplanes H_+ and H_-, respectively, and some data from both classes lie within the region bounded by H_+ and H_-.

In basic concept, soft margins resolve this difficulty by slightly relaxing or "softening" the inequality constraint for data that are otherwise unable to satisfy the inequality constraint equation. In the example shown in Figure 8.7, the class G_1 data at (0, 0) and (0, 1) relative to hyperplane $x_1 + x_2 - 1 = 0$ provide an illustration. Neither of these data points satisfies the original constraint $x_1 + x_2 - 1 \geq 1$. If, however, we soften this constraint to $x_1 + x_2 - 1 \geq 1 - 2$ for data (0, 0) and $x_1 + x_2 - 1 \geq 1 - 1$ for (0, 1) by subtracting 2 and 1, respectively, it is then possible for these data to satisfy these new inequality constraint equations. In Figure 8.8 the class G_2 data (1, 1) and (0, 1) are similarly unable to satisfy the original constraint, but if the restraint is softened to $-(x_1 + x_2 - 1) \geq 1 - 2$ and $-(x_1 + x_2 - 1) \geq 1 - 1$ by subtracting 2 and 1, respectively, each of these data can then satisfy its new inequality constraint equation.

In short, for training data that are not linearly separable, we introduce a non-negative real number ξ_i into the constraint equation, thereby enabling those data to satisfy the softened inequality constraint

$$y_i(w^T x_i + b) \geq 1 - \xi_i, \qquad i = 1, 2, \cdots, n. \tag{8.33}$$

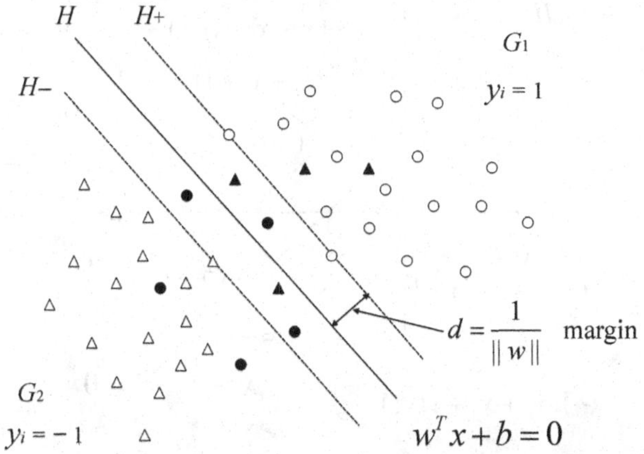

Figure 8.6 *No matter where we draw the hyperplane for separation of the two classes and the accompanying hyperplanes for the margin, some of the data (the black solid dots and triangles) do not satisfy the inequality constraint.*

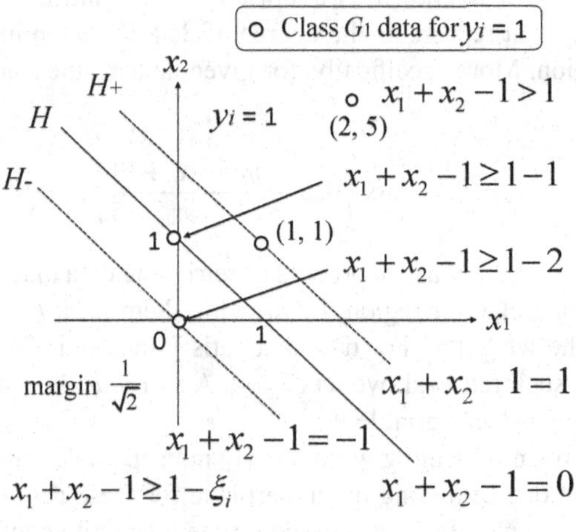

Figure 8.7 *The class G_1 data at (0, 0) and (0, 1) do not satisfy the original constraint $x_1 + x_2 - 1 \geq 1$. We soften this constraint to $x_1 + x_2 - 1 \geq 1 - 2$ for data (0, 0) and $x_1 + x_2 - 1 \geq 1 - 1$ for (0, 1) by subtracting 2 and 1, respectively; each of these data can then satisfy its new inequality constraint equation.*

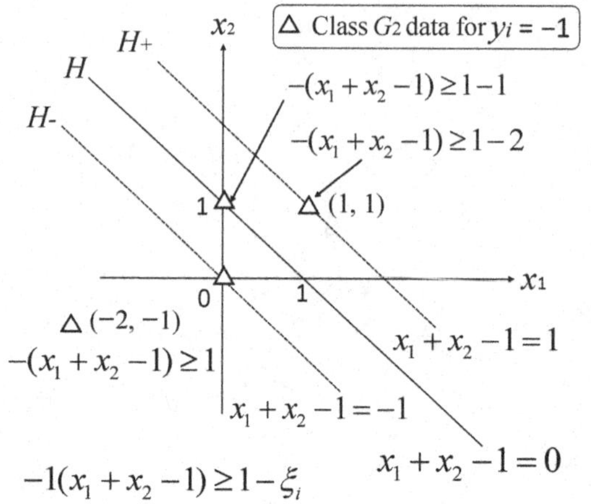

Figure 8.8 *The class G_2 data (1, 1) and (0, 1) are unable to satisfy the constraint, but if the restraint is softened to $-(x_1 + x_2 - 1) \geq 1 - 2$ and $-(x_1 + x_2 - 1) \geq 1 - 1$ by subtracting 2 and 1, respectively, each of these data can then satisfy its new inequality constraint equation.*

The term $\xi_i \geq 0$ is known as a *slack variable*, and may be regarded as a quantity that represents the degree of data intrusion into the wrong class or region. More specifically, for given data $\boldsymbol{x}_i$, the slack variable is defined by

$$\xi_i = \max\left\{0, d - \frac{y_i(\boldsymbol{w}^T\boldsymbol{x}_i + b)}{||\boldsymbol{w}||}\right\}. \tag{8.34}$$

This can be thought of as the process of retrieving data that have intruded into the wrong class or region, by drawing them back $\xi_i/||\boldsymbol{w}||$ in the direction of the weight $\boldsymbol{w}$. For data that satisfy the inequality (8.32) with no need for such retrieval, we let $\xi_i = 0$. A *soft margin* is thus a margin incorporating a slack variable.

The problem of dealing with linearly nonseparable training data has now become one of finding the hyperplane for these data that gives the largest possible margin $1/||\boldsymbol{w}||$, based on the inequality constraint (8.33). Here, however, a new difficulty arises. As illustrated in Figure 8.9, a large margin tends to increase the number of data that intrude into the other class region or into the region between hyperplanes H_+ and H_-, and the sum of the slack variables thus tends to become large. Conversely, as shown in Figure 8.10, the number of intruding data and the sum of the

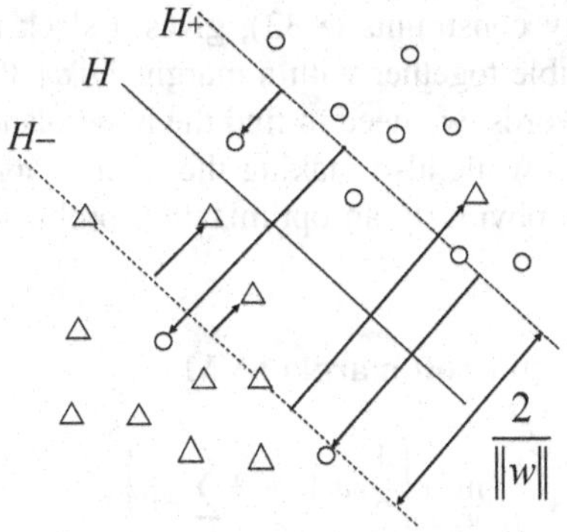

$$y_i(w^T x_i + b) \geq 1 - \xi_i, \quad i = 1, 2, \cdots, n$$

Figure 8.9 *A large margin tends to increase the number of data that intrude into the other class region or into the region between hyperplanes* H_+ *and* H_-.

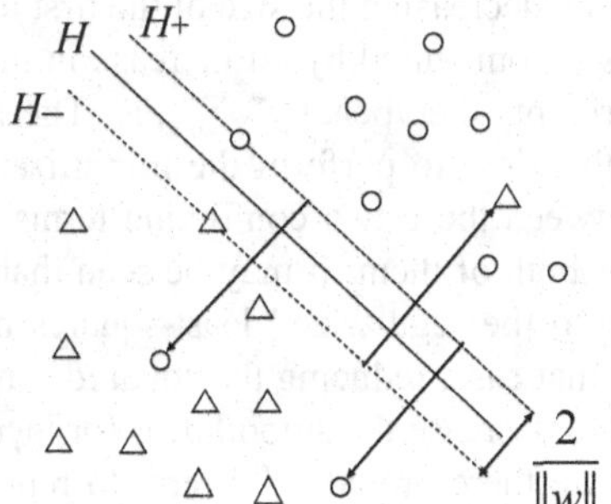

$$y_i(w^T x_i + b) \geq 1 - \xi_i, \quad i = 1, 2, \cdots, n$$

Figure 8.10 *A small margin tends to decrease the number of data that intrude into the other class region or into the region between hyperplanes* H_+ *and* H_-.

slack variables can be reduced by reducing the size of the margin. In short, the sum of the slack variables may be thought of as a penalty for intrusion into another class region, and thus preferably reduced in size, but reducing their sum also carries the penalty of reduced margin size.

It naturally follows that we would like to find the hyperplane that, under the inequality constraints (8.33), gives a slack margin sum that is as small as possible together with a margin $1/||\boldsymbol{w}||$ that is as large as possible. In other words, we need to find the hyperplane that takes $||\boldsymbol{w}||^2$ as small as possible while also making the slack margin sum as small as possible. This is obviously an optimization problem, which may be expressed as follows:

Primal problem for the soft margin SVM

$$\min_{\boldsymbol{w},\, \boldsymbol{\xi}} \left\{ \frac{1}{2}||\boldsymbol{w}||^2 + \lambda \sum_{i=1}^{n} \xi_i \right\}, \tag{8.35}$$

$$\text{subject to} \quad y_i(\boldsymbol{w}^T \boldsymbol{x}_i + b) \geq 1 - \xi_i, \quad \xi_i \geq 0, \quad i = 1, 2, \cdots, n,$$

where λ (> 0) controls the trade-off between margin maximization and constraints.

In this equation, decreasing the size of the first term $||\boldsymbol{w}||^2/2$ to obtain a large margin is accompanied by an increase in the size of the second term, which represents the "penalty" $\sum_{i=1}^{n} \xi_i$. The method of resolving this problem is therefore to perform the minimization while maintaining a balance between these two conflicting terms using the parameter λ, which controls both of them. It may be seen that this method is conceptually similar to the regularized least-squares method described in Section 3.4.1. In that case, reducing the squared sum of errors in the fitted curve tended to decrease the smoothness (or increase the roughness) excessively, and we therefore added a term to represent the amount of roughness of the curve as a penalty term and on this basis obtain a balance between the two terms. The term that we introduced to control the smoothness of the curve and the degree of fit to data was the regularization parameter.

8.2.2 From Primal Problem to Dual Problem

Like the optimization problem in the linearly separable case, the optimization problem (8.35) for linearly nonseparable training data of two classes consists of a quadratic programming as the primal problem that can be transformed into a dual problem by incorporating Lagrange multipliers, thus enabling its solution (see Appendix B.3). The constraints in the primal problem (8.35) consist of the constraint equation applied to the training data and the constraint condition applied to the slack variables. We therefore introduce the Lagrange multipliers $\boldsymbol{\alpha} = (\alpha_1, \alpha_2, \cdots, \alpha_n)^T$ ($\alpha_i \geq 0$) and $\boldsymbol{\beta} = (\beta_1, \beta_2, \cdots, \beta_n)^T$ ($\beta_i \geq 0$) to obtain the Lagrangian function, which may be expressed as

$$L(\boldsymbol{w}, b, \boldsymbol{\xi}, \boldsymbol{\alpha}, \boldsymbol{\beta}) \\ = \frac{1}{2}||\boldsymbol{w}||^2 + \lambda \sum_{i=1}^{n} \xi_i - \sum_{i=1}^{n} \alpha_i \{y_i(\boldsymbol{w}^T \boldsymbol{x}_i + b) - 1 + \xi_i\} - \sum_{i=1}^{n} \beta_i \xi_i, \tag{8.36}$$

where $\boldsymbol{\xi} = (\xi_1, \xi_2, \cdots, \xi_n)^T$ ($\xi_i \geq 0$) for slack variables.

We begin by minimizing the Lagrangian function with respect to $\boldsymbol{w}$, b, and $\boldsymbol{\xi}$. Differentiating the Lagrangian function with respect to these variables yields

$$\begin{aligned} \frac{\partial L(\boldsymbol{w}, b, \boldsymbol{\xi}, \boldsymbol{\alpha}, \boldsymbol{\beta})}{\partial \boldsymbol{w}} &= \boldsymbol{w} - \sum_{i=1}^{n} y_i \alpha_i \boldsymbol{x}_i = \boldsymbol{0}, \\ \frac{\partial L(\boldsymbol{w}, b, \boldsymbol{\xi}, \boldsymbol{\alpha}, \boldsymbol{\beta})}{\partial b} &= -\sum_{i=1}^{n} \alpha_i y_i = 0, \\ \frac{\partial L(\boldsymbol{w}, b, \boldsymbol{\xi}, \boldsymbol{\alpha}, \boldsymbol{\beta})}{\partial \xi_i} &= \lambda - \alpha_i - \beta_i = 0, \quad i = 1, 2, \cdots, n. \end{aligned} \tag{8.37}$$

We then obtain the conditional equations

$$\boldsymbol{w} = \sum_{i=1}^{n} y_i \alpha_i \boldsymbol{x}_i, \quad \sum_{i=1}^{n} \alpha_i y_i = 0, \quad \lambda = \alpha_i + \beta_i, \quad i = 1, 2, \cdots, n. \tag{8.38}$$

By substituting these three conditional equations into the Lagrangian function (8.36) and rearranging the equation, we obtain the function of Lagrange multipliers

$$L_D(\boldsymbol{\alpha}) = \sum_{i=1}^{n} \alpha_i - \frac{1}{2} \sum_{i=1}^{n} \sum_{j=1}^{n} \alpha_i \alpha_j y_i y_j \boldsymbol{x}_i^T \boldsymbol{x}_j. \tag{8.39}$$

It follows from the third condition in (8.38) that $\beta_i = \lambda - \alpha_i \geq 0$, and so $\lambda \geq \alpha_i$. Since the Lagrange multipliers are non-negative real numbers, we also obtain $0 \leq \alpha_i \leq \lambda$. For the primal problem based on the soft margin, the dual problem in (8.35) can therefore be given by the following equation.

Dual problem for the soft margin SVM

$$\max_{\alpha} L_D(\alpha) = \max_{\alpha} \left\{ \sum_{i=1}^{n} \alpha_i - \frac{1}{2} \sum_{i=1}^{n} \sum_{j=1}^{n} \alpha_i \alpha_j y_i y_j \boldsymbol{x}_i^T \boldsymbol{x}_j \right\} \tag{8.40}$$

$$\text{subject to} \quad \sum_{i=1}^{n} \alpha_i y_i = 0, \qquad 0 \leq \alpha_i \leq \lambda, \qquad i = 1, 2, \cdots, n.$$

For linearly separable training data, the constraint on the Lagrange multipliers was $0 \leq \alpha_i$ in the dual problem (8.19) giving the optimum separating hyperplane. The dual problem for linearly nonseparable data differs from that problem only in its constraint $0 \leq \alpha_i \leq \lambda$ for the soft margin, and thus in its incorporation of λ as the sum of the slack variables.

Let the solutions for the dual problem in (8.40) be $\hat{\alpha}_1, \hat{\alpha}_2, \cdots, \hat{\alpha}_n$. By substituting these solutions into the Lagrange multipliers for the weight vector $\boldsymbol{w}$ in (8.38), we then have the optimum weight vector solution

$$\hat{\boldsymbol{w}} = \sum_{i=1}^{n} \hat{\alpha}_i y_i \boldsymbol{x}_i. \tag{8.41}$$

For the data lying on hyperplanes H_+ and H_- on opposite sides of the separating hyperplane, moreover, the value for the slack variables is 0, that is,

$$H_+ : \hat{\boldsymbol{w}}^T \boldsymbol{x}_+ + b = 1, \qquad H_- : \hat{\boldsymbol{w}}^T \boldsymbol{x}_- + b = -1, \tag{8.42}$$

and the optimum solution for the bias b can therefore be given by

$$\hat{b} = -\frac{1}{2}(\hat{\boldsymbol{w}}^T \boldsymbol{x}_+ + \hat{\boldsymbol{w}}^T \boldsymbol{x}_-). \tag{8.43}$$

We thus find that, for linearly nonseparable training data, the hyperplane separating the two classes with the largest margin is

$$\hat{\boldsymbol{w}}^T \boldsymbol{x} + \hat{b} = \sum_{i=1}^{n} y_i \hat{\alpha}_i \boldsymbol{x}_i^T \boldsymbol{x} + \hat{b} = 0. \tag{8.44}$$

Accordingly, in the linearly nonseparable case the classification rule for the two classes G_1 and G_2 with the optimum separating hyperplane as the discriminant function is given by

$$\hat{\boldsymbol{w}}^T \boldsymbol{x} + \hat{b} = \sum_{i=1}^{n} y_i \hat{\alpha}_i \boldsymbol{x}_i^T \boldsymbol{x} + \hat{b} = \begin{cases} \geq 0 & \Longrightarrow \quad G_1 \\ < 0 & \Longrightarrow \quad G_2. \end{cases} \tag{8.45}$$

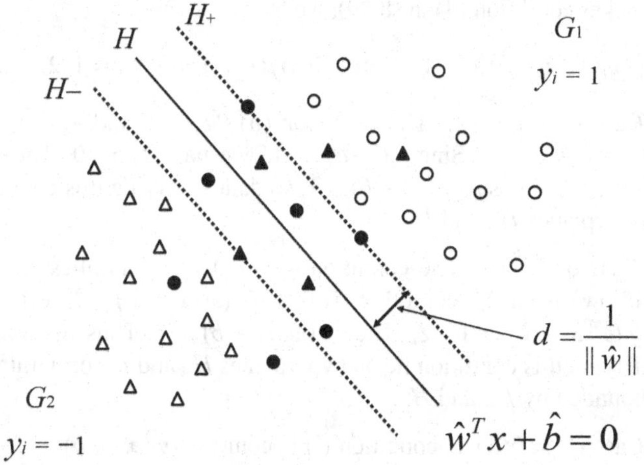

Figure 8.11 *Support vectors in a linearly nonseparable case: Data corresponding to the Lagrange multipliers such that* $0 < \hat{\alpha}_i \leq \lambda$ *(the black solid dots and triangles).*

The discriminant function for linearly separable training data was composed solely from the support vectors satisfying

$$y_i(\hat{\boldsymbol{w}}^T \boldsymbol{x}_i + \hat{b}) = 1 \tag{8.46}$$

which lie on hyperplanes H_+ and H_- on opposite sides of the optimum separating hyperplane. In the linearly nonseparable case, however, in addition to these data, all of the data indicated by the black solid symbols in Figure 8.11 are also support vectors.

If we denote the index set of the support vectors by S, then the classification rule (8.45) based on the soft margin may be expressed as

$$\hat{\boldsymbol{w}}^T \boldsymbol{x} + \hat{b} = \sum_{i \in S} y_i \hat{\alpha}_i \boldsymbol{x}_i^T \boldsymbol{x} + \hat{b} = \begin{cases} \geq 0 & \Longrightarrow \quad G_1 \\ < 0 & \Longrightarrow \quad G_2. \end{cases} \tag{8.47}$$

For the Lagrange multipliers representing the dual problem solution, the support vectors are the data set corresponding to the Lagrange multipliers such that $0 < \hat{\alpha}_i \leq \lambda$. This property of SVMs can be derived from the Karush-Kuhn-Tucker conditions as follows.

Support vectors for linearly nonseparable case From the Karush-Kuhn-Tucker condition (4) in (8.29), we have

$$\text{(i) } \alpha_i\{-y_i(\boldsymbol{w}^T\boldsymbol{x}_i + b) - \xi_i + 1\} = 0, \quad \text{(ii) } \beta_i(-\xi_i) = 0, \quad i = 1, 2, \cdots, n.$$

(a) Case $0 < \alpha_i < \lambda$: The condition (ii) $\beta_i\xi_i = 0$ and $\lambda = \alpha_i + \beta_i$ imply $(\lambda - \alpha_i)\xi_i = 0$. Since $\lambda - \alpha_i > 0$, we have $\xi_i = 0$. Then the condition (i) becomes $y_i(\boldsymbol{w}^T\boldsymbol{x}_i + b) = 1$, so data satisfying this condition lie on hyperplanes H_+ and H_-.

(b) Case $\alpha_i = \lambda$: The condition $(\lambda - \alpha_i)\xi_i = 0$ implies $\xi_i \geq 0$, and it follows from the condition (i) that $-y_i(\boldsymbol{w}^T\boldsymbol{x}_i + b) - \xi_i + 1 = 0$, that is, $y_i(\boldsymbol{w}^T\boldsymbol{x}_i + b) = 1 - \xi_i$. Then $y_i(\boldsymbol{w}^T\boldsymbol{x}_i + b) \leq 1$. Thus we see that data satisfying this condition lie on hyperplanes H_+ and H_- or within the region bounded by H_+ and H_-.

(c) Case $\alpha_i = 0$: The condition (i) becomes $y_i(\boldsymbol{w}^T\boldsymbol{x}_i + b) > 1 - \xi_i$, and $(\lambda - \alpha_i)\xi_i = 0$ implies $\xi_i = 0$. Then we have that $y_i(\boldsymbol{w}^T\boldsymbol{x}_i + b) > 1$. Consequently, we see that data satisfying this condition give linearly separable data.

8.3 From Linear to Nonlinear

In Section 8.1, we focused on the method of finding the optimum separating hyperplane and thus maximizing the margin for linearly separable training data. In Section 8.2, we then turned to the method of finding the optimum separating plane for linearly nonseparable training data by adopting soft margins that can tolerate the presence of a few erroneously discriminated data. In both of these methods, the quadratic programming was identified as the primal problem and replaced with the dual problem by incorporating a Lagrange function, and we were thereby able to obtain the optimum separating hyperplane and thus the linear discriminant function.

The question we now address is how to build on and extend these basic concepts from linear to nonlinear systems. It is known that the training data of two classes generally approach linear separability in a higher-dimensional space as the number of their dimensions increases, and in particular that the data become linearly separable if the data dimensionality exceeds the number of data. We can utilize this property to resolve the problem of nonlinear systems. In basic concept, this involves mapping the observed data to a higher-dimensional feature space using a nonlinear function, and then applying the linear SVM algorithm developed in the preceding sections to find the separating hyperplane in the high-dimensional feature space. This separating hyperplane obtained in a feature space is then applied to construct the nonlinear discriminant function in the original input space.

8.3.1 Mapping to Higher-Dimensional Feature Space

Let us first consider the process of mapping the data of an input space into a higher-dimensional feature space with a nonlinear function. Assume, for this discussion, that the two-dimensional data $\boldsymbol{x} = (x_1, x_2)^T$ has been observed in an input space and that the mapping will be performed with a function containing terms of up to two degrees as follows:

$$z = (x_1^2, x_2^2, x_1x_2, x_1, x_2)^T. \tag{8.48}$$

We are thus transforming two-dimensional data to five-dimensional data. The three two-dimensional data points (1, 2), (0, 1), and (2, 5), for example, will be transformed to the five-dimensional data points (1, 4, 2, 1, 2), (0, 1, 0, 0, 1), and (4, 25, 10, 2, 5), respectively. The nonlinear function used to transform p-dimensional data $\boldsymbol{x} = (x_1, x_2, \cdots, x_p)^T$ in an input space to a higher r-dimensional feature space is generally expressed as

$$z = \boldsymbol{\Phi}(\boldsymbol{x}) \equiv (\phi_1(\boldsymbol{x}), \phi_2(\boldsymbol{x}), \cdots, \phi_r(\boldsymbol{x}))^T, \tag{8.49}$$

where each element $\phi_j(\boldsymbol{x}) = \phi_j(x_1, x_2, \cdots, x_p)$ denotes a p-dimensional real-valued function. With this r-dimensional vector as the nonlinear function, all of the n originally observed p-dimensional training data are mapped into an r-dimensional feature space. We represent them as

$$(z_1, y_1), \quad (z_2, y_2), \quad \cdots, \quad (z_n, y_n), \tag{8.50}$$

where $z_i = \boldsymbol{\Phi}(\boldsymbol{x}_i) = (\phi_1(\boldsymbol{x}_i), \phi_2(\boldsymbol{x}_i), \cdots, \phi_r(\boldsymbol{x}_i))^T$ and y_i are the class labels. With the n r-dimensional data mapped to this feature space, the methods of the preceding sections may then be applied to construct its separating hyperplane (Figure 8.12).

For linearly separable training data, as described in the previous sections, we were able to find the separating hyperplane that maximizes the margin by transforming the primal problem to a dual problem. We can now apply the same concept to find the separating hyperplane for the data (8.50) in the feature space formed by mapping the training data. The problem thus becomes one of solving the following dual problem for the data z_i in the feature space that have replaced $\boldsymbol{x}_i$ of equation (8.19):

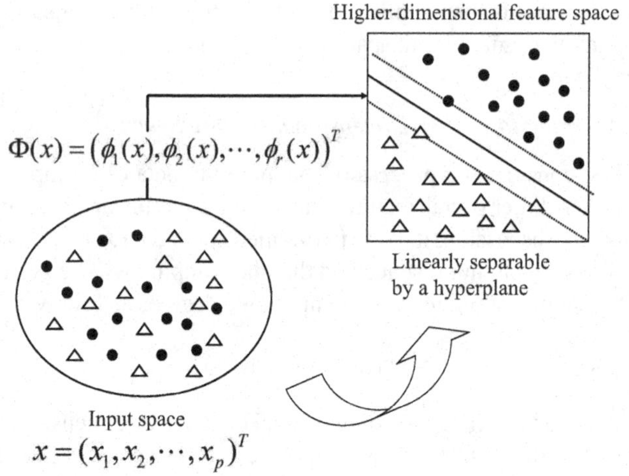

Figure 8.12 *Mapping the data of an input space into a higher-dimensional feature space with a nonlinear function.*

Dual problem in the higher-dimensional feature space

$$\max_{\alpha} L_D(\alpha) = \max_{\alpha} \left\{ \sum_{i=1}^{n} \alpha_i - \frac{1}{2} \sum_{i=1}^{n} \sum_{j=1}^{n} \alpha_i \alpha_j y_i y_j z_i^T z_j \right\} \tag{8.51}$$

$$\text{subject to} \quad \alpha_i \geq 0, \ \ i = 1, 2, \cdots, n, \qquad \sum_{i=1}^{n} \alpha_i y_i = 0.$$

For the solution $\hat{\alpha} = (\hat{\alpha}_1, \hat{\alpha}_2, \cdots, \hat{\alpha}_n)^T$ of this dual problem, as described in Section 8.1.3, the separating hyperplane in the higher-dimension feature space can be given by

$$h_f(z) = \hat{w}^T z + \hat{b} = \sum_{i \in S} \hat{\alpha}_i y_i z_i^T z + \hat{b} = 0, \tag{8.52}$$

where S denotes the set of support vector indices in the feature space, and the bias $\hat{b}$ is given by $\hat{b} = -(\hat{w}^T z_+ + w^T z_-)/2$ for any two support vectors z_+ and z_-.

It is characteristic of SVMs that a separating hyperplane obtained in this way in a feature space yields a nonlinear discriminant function in the input space. That is, using the relation $z = \Phi(x) = (\phi_1(x), \phi_2(x), \cdots, \phi_r(x))^T$, the equation in (8.52) can be expressed as

$$\begin{aligned} h_f(z) = h_f(\Phi(x)) &= \hat{w}^T \Phi(x) + \hat{b} \\ &= \sum_{i \in S} \hat{\alpha}_i y_i \Phi(x_i)^T \Phi(x) + \hat{b} \qquad (8.53) \\ &= \sum_{i \in S} \hat{\alpha}_i y_i \left\{ \sum_{j=1}^{r} \phi_j(x_i) \phi_j(x) \right\} + \hat{b}, \end{aligned}$$

which gives a nonlinear function in the input space.

In mapping two-dimensional data of an input space, for example, suppose that we map the training data $x = (x_1, x_2)^T$ into a five-dimensional feature space as follows:

$$\begin{aligned} \Phi(x) &= (\phi_1(x_1, x_2), \phi_2(x_1, x_2), \phi_3(x_1, x_2), \phi_4(x_1, x_2), \phi_5(x_1, x_2))^T \\ &= (x_1^2, x_2^2, x_1 x_2, x_1, x_2)^T \equiv (z_1, z_2, \cdots, z_5)^T. \end{aligned}$$

Then the separating hyperplane based on the variables $z_1, z_2, \cdots, z_5$ obtained in the feature space will yield a two-dimensional discriminant function for variables x_1 and x_2 in the input space, as shown by the relational expression

$$\begin{aligned} h_f(z_1, z_2, z_3, z_4, z_5) &= \hat{w}_1 z_1 + \hat{w}_2 z_2 + \hat{w}_3 z_3 + \hat{w}_4 z_4 + \hat{w}_5 z_5 + \hat{b} \\ &= \hat{w}_1 x_1^2 + \hat{w}_2 x_2^2 + \hat{w}_3 x_1 x_2 + \hat{w}_4 x_1 + \hat{w}_5 x_2 + \hat{b} \\ &= h(x_1, x_2). \end{aligned}$$

As illustrated in Figure 8.13, the separating hyperplane obtained by mapping the two-dimensional data of the input space to the higher-dimensional feature space then yields a nonlinear discriminant function in the input space.

In summary, the p-dimensional observed data x in the input space is mapped by the r-dimensional nonlinear function $\Phi(x) = z$ to a higher-dimensional feature space, in which the linear SVM is applied to the mapped data. It is thus possible to obtain a separating hyperplane by margin maximization with the mapped data in the feature space. This separation hyperplane then yields a nonlinear function in the input space, and hence a nonlinear decision boundary.

A difficulty arises, however, if the dimensionality r of the feature

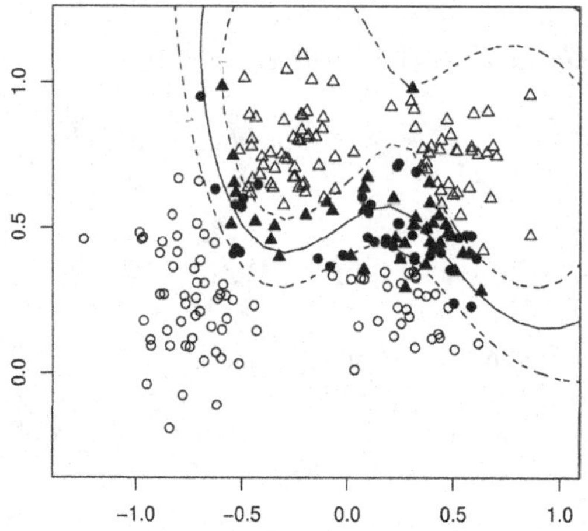

Figure 8.13 *The separating hyperplane obtained by mapping the two-dimensional data of the input space to the higher-dimensional feature space yields a nonlinear discriminant function in the input space. The black solid data indicate support vectors.*

space is very large. This results in an extremely large increase in the dimensionality of the nonlinear function $z = \boldsymbol{\Phi}(\boldsymbol{x})$, which makes it difficult to compute the inner products $z^T z$ of the dual problem in the feature space. As described in the following section, a technique known as the *kernel methods* is used to resolve this difficulty.

8.3.2 Kernel Methods

If the dimensionality of the data observed in an input space is high, the dimensionality of the feature space in which the SVM will be applied may then be expected to become extremely high. If, for example, a 256-dimensional observation in the input space is mapped into a feature space by a polynomial equation in which the polynomial ranges up to the fifth degree, the dimensionality in the feature space rises to approximately 10^{10}. In short, the feature space dimensionality r is generally far higher than the dimensionality p of the input space, and in some cases may become infinite. With very high dimensionality, it may become impossible, as a practical matter, to compute the inner products $z_i^T z_j = \boldsymbol{\Phi}(\boldsymbol{x}_i)^T \boldsymbol{\Phi}(\boldsymbol{x}_j)$ of the dual problem (8.51) for the training data as mapped to the feature space. The *kernel methods*, which are commonly known as the *kernel trick* in the field of SVMs, was conceived as a tool of resolving this problem.

The kernel methods provide a tool of resolving the extremely large computing burden for inner products by substituting the following kernel function:

$$K(\boldsymbol{x}_i, \boldsymbol{x}_j) = \boldsymbol{\Phi}(\boldsymbol{x}_i)^T \boldsymbol{\Phi}(\boldsymbol{x}_j) = \sum_{k=1}^{r} \phi_k(\boldsymbol{x}_i)\phi_k(\boldsymbol{x}_j). \tag{8.54}$$

In effect, the inner products of the right side of this equation are found by obtaining the values of the kernel function on the left side. For example, the Gaussian kernel is given by

$$K(\boldsymbol{x}_i, \boldsymbol{x}_j) = \exp\left\{-\frac{\|\boldsymbol{x}_i - \boldsymbol{x}_j\|^2}{2\sigma^2}\right\} = \sum_{k=1}^{r} \phi_k(\boldsymbol{x}_i)\phi_k(\boldsymbol{x}_j), \tag{8.55}$$

where $\|\boldsymbol{x}_i - \boldsymbol{x}_j\|^2 = (\boldsymbol{x}_i - \boldsymbol{x}_j)^T(\boldsymbol{x}_i - \boldsymbol{x}_j)$. Even in cases where extremely high dimensionality makes it difficult to compute the inner products on the right side, computation of the left side is relatively easy, as it involves only the inner products of the p-dimensional data in the input space. As an example, let us consider a case in which mapping 256-dimensional data of the input space to the feature space results in a dimensionality of $r = 10^{10}$ in the feature space. Computing the inner products on the right side in (8.55) would indeed be difficult, but it is possible to compute the kernel function values on the left side for the inner products of just the original 256-dimensional data.

The following functions are also typically used as kernel functions.

$$\begin{aligned} &\text{Polynomial kernel}: \quad && K(\boldsymbol{x}_i, \boldsymbol{x}_j) = (\boldsymbol{x}_i^T \boldsymbol{x}_j + c)^d, \\ &\text{Sigmoid kernel}: \quad && K(\boldsymbol{x}_i, \boldsymbol{x}_j) = \tanh(b\boldsymbol{x}_i^T \boldsymbol{x}_j + c). \end{aligned} \tag{8.56}$$

As a simple example of replacing inner product computation with a polynomial kernel function, let us consider the mapping of a two-dimensional original input space into a three-dimensional feature space.

$$\begin{aligned} (x_1, x_2) &\Rightarrow \boldsymbol{\Phi}(\boldsymbol{x}) = (x_1^2, \sqrt{2}x_1x_2, x_2^2)^T, \\ (y_1, y_2) &\Rightarrow \boldsymbol{\Phi}(\boldsymbol{y}) = (y_1^2, \sqrt{2}y_1y_2, y_2^2)^T. \end{aligned}$$

The three-dimensional inner products are then expressed as

$$\boldsymbol{\Phi}(\boldsymbol{x})^T\boldsymbol{\Phi}(\boldsymbol{y}) = (x_1^2,\ \sqrt{2}x_1x_2, x_2^2)^T(y_1^2,\ \sqrt{2}y_1y_2, y_2^2) = (\boldsymbol{x}^T\boldsymbol{y})^2 = K(\boldsymbol{x}, \boldsymbol{y}).$$

The inner products $\boldsymbol{x}^T\boldsymbol{y}$ in the input space are thus represented by the square of a polynomial kernel with $c = 0$ and $d = 2$ in (8.56).

A guarantee that the kernel function $K(\boldsymbol{x}_i, \boldsymbol{x}_j)$ for two p-dimensional data $\boldsymbol{x}_i$ and $\boldsymbol{x}_j$ in an input space can represent the inner products $\boldsymbol{\Phi}(\boldsymbol{x}_i)^T\boldsymbol{\Phi}(\boldsymbol{x}_j)$ for the r-dimensional data $\boldsymbol{\Phi}(\boldsymbol{x}_i)$ and $\boldsymbol{\Phi}(\boldsymbol{x}_j)$ of the feature space formed by their mapping, which establishes the relationship expressed by (8.54), is provided by *Mercer's theorem.*

Mercer's theorem Suppose that $K(\boldsymbol{x}_i, \boldsymbol{x}_j)$ is a continuous and symmetric $(K(\boldsymbol{x}_i, \boldsymbol{x}_j) = K(\boldsymbol{x}_j, \boldsymbol{x}_i))$ real-valued function on $\mathcal{R}^p \times \mathcal{R}^p$. Then for real numbers $a_k (> 0)$, the necessary and sufficient condition that guarantees the expression

$$K(\boldsymbol{x}_i, \boldsymbol{x}_j) = \sum_{k=1}^{r} a_k\phi_k(\boldsymbol{x}_i)\phi_k(\boldsymbol{x}_j) \tag{8.57}$$

is that $K(\boldsymbol{x}_i, \boldsymbol{x}_j)$ is a nonnegative-definite kernel, that is, for any n points $\boldsymbol{x}_1, \boldsymbol{x}_2, \cdots, \boldsymbol{x}_n$ and any real numbers $c_1, c_2, \cdots, c_n$, the following condition is satisfied

$$\sum_{i=1}^{n}\sum_{j=1}^{n} c_ic_jK(\boldsymbol{x}_i, \boldsymbol{x}_j) \geq 0. \tag{8.58}$$

Research is in progress on many fronts relating to the kernel methods and their use in generalizing various multivariate analysis techniques to nonlinear applications. In Section 9.4, we discuss the use of the kernel methods for non-linear principal component analysis.

8.3.3 Nonlinear Classification

As we have seen, the kernel method can effectively resolve the problem of extreme computational volume for inner products $\boldsymbol{z}_i^T\boldsymbol{z}_j = \boldsymbol{\Phi}(\boldsymbol{x}_i)^T\boldsymbol{\Phi}(\boldsymbol{x}_j)$ in (8.51) formed by mapping the p-dimensional data observed in the input space. With the application of the kernel method, the dual problem based on training data as mapped into a feature space may be given as follows.

Kernel dual problem

$$\max_{\alpha} L_D(\alpha) = \max_{\alpha} \left\{ \sum_{i=1}^{n} \alpha_i - \frac{1}{2} \sum_{i=1}^{n} \sum_{j=1}^{n} \alpha_i \alpha_j y_i y_j K(\boldsymbol{x}_i, \boldsymbol{x}_j) \right\}, \quad (8.59)$$

$$\text{subject to} \quad \alpha_i \geq 0 \;\; i = 1, 2, \cdots, n, \qquad \sum_{i=1}^{n} \alpha_i y_i = 0.$$

For the solution $\hat{\alpha} = (\hat{\alpha}_1, \hat{\alpha}_2, \cdots, \hat{\alpha}_n)^T$ of this dual problem the separating hyperplane in the higher-dimension feature space can be given by

$$h_f(\boldsymbol{z}) = \hat{\boldsymbol{w}}^T \boldsymbol{z} + \hat{b} = \sum_{i \in S} \hat{\alpha}_i y_i \boldsymbol{z}_i^T \boldsymbol{z} + \hat{b} = 0, \quad (8.60)$$

where S denotes the set of support vector indices in the feature space. By replacing the inner product $\boldsymbol{z}_i^T \boldsymbol{z} = \boldsymbol{\Phi}(\boldsymbol{x}_i)^T \boldsymbol{\Phi}(\boldsymbol{x})$ with the kernel function $K(\boldsymbol{x}_i, \boldsymbol{x})$, we obtain the nonlinear discriminant function in the input space as follows:

$$h(\boldsymbol{x}) = \sum_{i \in S} \hat{\alpha}_i y_i K(\boldsymbol{x}_i, \boldsymbol{x}) + \hat{b}. \quad (8.61)$$

By mapping the observed data by nonlinear functions into a higher-dimensional space known as a feature space, applying the kernel method to resolve the inner product computation problem, and using the linear SVM algorithm to the feature space, we are able to obtain the separating hyperplane in that space. In the input space, this separating hyperplane in the feature space then yields the following nonlinear classification rule;

$$h(\boldsymbol{x}) = \sum_{i \in S} \hat{\alpha}_i y_i K(\boldsymbol{x}_i, \boldsymbol{x}) + \hat{b} \begin{cases} \geq 0 & \Longrightarrow \;\; +1;\; G_1 \\ < 0 & \Longrightarrow \;\; -1;\; G_2. \end{cases} \quad (8.62)$$

In Section 8.2, we described the soft margin method incorporating slack variables and its application to find the separating hyperplane in an input space containing linearly nonseparable variables. We can now apply the same method to data in the feature space obtained by the mapping of training data observed in an input space, to find the separating hyperplane in the feature space by solution of the following dual problem

$$\max_{\alpha} L_D(\alpha) = \max_{\alpha} \left\{ \sum_{i=1}^{n} \alpha_i - \frac{1}{2} \sum_{i=1}^{n} \sum_{j=1}^{n} \alpha_i \alpha_j y_i y_j \boldsymbol{z}_i^T \boldsymbol{z}_j \right\}, \quad (8.63)$$

$$\text{subject to} \quad \sum_{i=1}^{n} \alpha_i y_i = 0, \qquad 0 \le \alpha_i \le \lambda, \qquad i = 1, 2, \cdots, n.$$

The problem of computing the inner products for this dual problem can be resolved by substitution of the kernel function, and the soft-margin separating hyperplane for the training data as mapped into the feature space can then be obtained by solving the dual problem given as follows.

Kernel dual problem for the soft margin SVM

$$\max_{\alpha} L_D(\alpha) = \max_{\alpha} \left\{ \sum_{i=1}^{n} \alpha_i - \frac{1}{2} \sum_{i=1}^{n} \sum_{j=1}^{n} \alpha_i \alpha_j y_i y_j K(\boldsymbol{x}_i, \boldsymbol{x}_j) \right\}, \quad (8.64)$$

$$\text{subject to} \quad \sum_{i=1}^{n} \alpha_i y_i = 0, \qquad 0 \le \alpha_i \le \lambda, \qquad i = 1, 2, \cdots, n.$$

For the solution $\hat{\alpha} = (\hat{\alpha}_1, \hat{\alpha}_2, \cdots, \hat{\alpha}_n)^T$ of this kernel dual problem, by replacing the inner product $\boldsymbol{z}_i^T \boldsymbol{z} = \boldsymbol{\Phi}(\boldsymbol{x}_i)^T \boldsymbol{\Phi}(\boldsymbol{x})$ with the kernel function $K(\boldsymbol{x}_i, \boldsymbol{x})$, we obtain the nonlinear discriminant function in the input space

$$\begin{aligned} h_f(\boldsymbol{z}) = \hat{\boldsymbol{w}}^T \boldsymbol{z} + \hat{b} &= \sum_{i \in S} \hat{\alpha}_i y_i \boldsymbol{z}_i^T \boldsymbol{z} + \hat{b} \\ &= \sum_{i \in S} \hat{\alpha}_i y_i K(\boldsymbol{x}_i, \boldsymbol{x}) + \hat{b} \equiv h(\boldsymbol{x}). \end{aligned} \quad (8.65)$$

Hence, in the input space, this separating hyperplane in the feature space yields the following nonlinear classification rule;

$$h(\boldsymbol{x}) = \sum_{i \in S} \hat{\alpha}_i y_i K(\boldsymbol{x}_i, \boldsymbol{x}) + \hat{b} \begin{cases} \ge 0 & \Longrightarrow \quad +1;\ G_1 \\ < 0 & \Longrightarrow \quad -1;\ G_2. \end{cases} \quad (8.66)$$

We have now seen that it is possible to solve the dual problem for data mapped to a feature space using the kernel method, and apply the result to construct a nonlinear discriminant function in the input space. We are left, however, with the key question of how to decide on the value of the parameters to be used in the kernel function, such as the unknown parameter σ in the Gaussian kernel (8.55) and the two unknown parameters c and d in the polynomial kernel (8.56). These parameters are involved in the nonlinearity of the decision boundary. The degree of nonlinearity in the discriminant function varies with the parameter value, and also affects the prediction error.

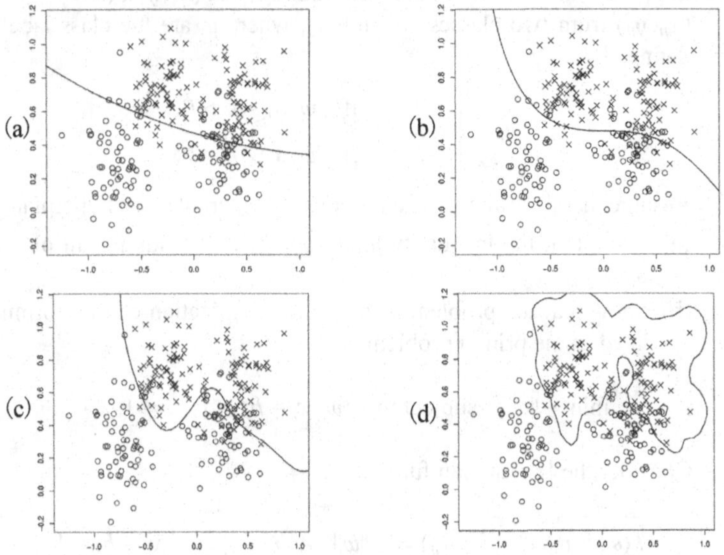

Figure 8.14 *Nonlinear decision boundaries in the input space vary with different values* σ *in the Gaussian kernel; (a)* $\sigma = 10$, *(b)* $\sigma = 1$, *(c)* $\sigma = 0.1$, *and (d)* $\sigma = 0.01$.

Figure 8.14 provides an example of the marked change that occurs in the boundary of the discriminant function in the input space with different values σ in the Gaussian kernel; (a) $\sigma = 10$, (b) $\sigma = 1$, (c) $\sigma = 0.1$, and (d) $\sigma = 0.01$. As shown in the figure, decreasing the value of σ in the Gaussian kernel substantially increases the nonlinearity. It is therefore essential to determine the optimum value of the parameters based on an appropriate criterion. One method that may be considered for this purpose is cross-validation, but it involves computational difficulty. Research is currently in progress on many fronts in regard to this issue.

Exercises

8.1 Show that the distance from p-dimensional data $\boldsymbol{x}$ to the hyperplane $\boldsymbol{w}^T\boldsymbol{x} + b = 0$ is given by

$$d \equiv \frac{|\boldsymbol{w}^T\boldsymbol{x} + b|}{||\boldsymbol{w}||}.$$

8.2 Suppose that we have n observed training data $(\boldsymbol{x}_1, y_1)$, $(\boldsymbol{x}_2, y_2)$, $\cdots$, $(\boldsymbol{x}_n, y_n)$ from two classes G_1 and G_2, where y_i are the class labels defined by

$$y_i = \begin{cases} 1 & \text{if} \quad \boldsymbol{w}^T\boldsymbol{x}_i + b > 0, \\ -1 & \text{if} \quad \boldsymbol{w}^T\boldsymbol{x}_i + b < 0. \end{cases}$$

Assume that the training data are linearly separable by a hyperplane.

(a) Show that the inequality $y_i(\boldsymbol{w}^T\boldsymbol{x}_i + b) > 0$ holds for all of the data.

(b) Show that the problem of margin maximization can be formulated as the primal problem

$$\min_{\boldsymbol{w}} \frac{1}{2}||\boldsymbol{w}||^2 \text{ subject to } y_i(\boldsymbol{w}^T\boldsymbol{x}_i + b) \geq 1, \quad i = 1, 2, \cdots, n.$$

8.3 Consider the Lagrangian function

$$L(\boldsymbol{w}, b, \alpha_1, \alpha_2, \cdots, \alpha_n) = \frac{1}{2}||\boldsymbol{w}||^2 - \sum_{i=1}^{n} \alpha_i \left\{ y_i(\boldsymbol{w}^T\boldsymbol{x}_i + b) - 1 \right\},$$

where $\alpha_1, \alpha_2, \cdots, \alpha_n$ ($\alpha_i \geq 0$; $i = 1, 2, \cdots, n$) are Lagrange multipliers corresponding to the constraint number n (see (8.15)).

(a) Derive the conditions

$$\boldsymbol{w} = \sum_{i=1}^{n} \alpha_i y_i \boldsymbol{x}_i, \qquad \sum_{i=1}^{n} \alpha_i y_i = 0.$$

(b) Show that substituting the conditions in (a) into the Lagrangian function yields the dual problem given by equation (8.19).

8.4 Assume that the training data are linearly separable by a hyperplane. Show that by the Karush-Kuhn-Tucker condition (4) in (8.29), any data that are not support vectors result in $\alpha_i = 0$, and any data that are support vectors in $\alpha_i > 0$.

8.5 For linearly nonseparable training data of two classes, consider the Lagrangian function

$$L(\boldsymbol{w}, b, \boldsymbol{\xi}, \boldsymbol{\alpha}, \boldsymbol{\beta})$$

$$= \frac{1}{2}||\boldsymbol{w}||^2 + \lambda \sum_{i=1}^{n} \xi_i - \sum_{i=1}^{n} \alpha_i \{y_i(\boldsymbol{w}^T\boldsymbol{x}_i + b) - 1 + \xi_i\} - \sum_{i=1}^{n} \beta_i \xi_i,$$

where $\boldsymbol{\xi} = (\xi_1, \xi_2, \cdots, \xi_n)^T$ ($\xi_i \geq 0$) are slack variables (see (8.36)).

(a) Derive the conditions

$$\boldsymbol{w} = \sum_{i=1}^{n} y_i \alpha_i \boldsymbol{x}_i, \quad \sum_{i=1}^{n} \alpha_i y_i = 0, \quad \lambda = \alpha_i + \beta_i, \quad i = 1, 2, \cdots, n.$$

(b) Show that substituting the conditions in (a) into the Lagrangian function yields the dual problem for the soft margin SVM given by (8.40).

8.6 Describe how to construct the nonlinear discriminant function in the original input space based on the kernel function defined by (8.54).

Chapter 9

Principal Component Analysis

Principal component analysis (PCA) is a method for consolidating the mutually correlated variables of multidimensional observed data into new variables by linear combinations of the original variables with minimal loss of the information in the observed data. Information of interest can be extracted from the data through this merging of the multiple variables that characterize individuals. PCA can also be used as a technique for visual apprehension of data structures by consolidating high-dimensional data into a smaller number of variables, performing dimension reduction (reduction of dimensionality, also known as dimensional compression), and projecting the results onto a one-dimensional line, two-dimensional plane, or three-dimensional space.

In this chapter, we first discuss the basic concept and purpose of PCA in the context of linearity. We then consider an application of PCA to the process of image decompression (and thus reconstruction) from dimensionally compressed transmitted data. We also discuss singular value decomposition of data matrices, which shares a number of aspects with PCA, and the relation between PCA and image compression. We then proceed to a discussion of nonlinear PCA using the kernel method for dimensionally reduced structure searches and information extraction from multidimensional data involving complex nonlinear structures.

9.1 Principal Components

PCA is a useful technique for extracting information contained in multidimensional observed data from the combined new variables. For simplicity, we begin by explaining the process of obtaining new variables through the linear combination of two variables.

9.1.1 Basic Concept

Figure 9.1 shows a plot of the mathematics (x_1) and English (x_2) test scores of 25 students. Projecting these two-dimensional data onto a sin-

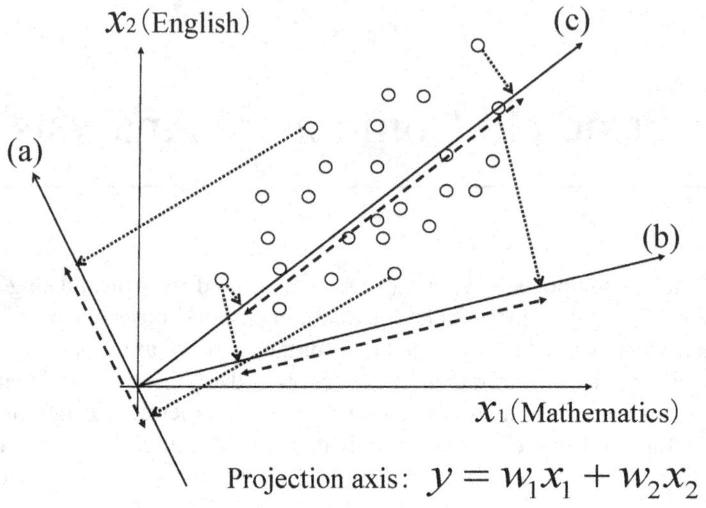

Figure 9.1 *Projection onto three different axes, (a), (b), and (c) and the spread of the data.*

gle axis by substituting the scores for both subjects in the following equation transforms them into one-dimensional data.

$$y = w_1x_1 + w_2x_2. \tag{9.1}$$

Figure 9.1 shows the projection onto three different axes, (a), (b), and (c). The spread of the data (as a measure of variance) for each axis is indicated by a double-headed arrow $\leftarrow - - \rightarrow$.

This figure shows that the variance is greater when projected onto axis (b) than onto axis (a), and greater onto axis (c) than onto axis (b). The axis with the greatest variance most clearly shows separation between the data. The question immediately arises of how best to find the projection axis that yields the maximum variance. In PCA, we use a sequence of projection axes for this purpose. We first find the projection axis, known as the first principal component, which maximizes the overall variance. We next find the projection axis that maximizes variance under the constraint of orthogonality to the first principal component. This axis is known as the second principal component. Let us now apply this to two-dimensional data.

We denote the n observed data for the two variables $\boldsymbol{x} = (x_1, x_2)^T$ as

$$\boldsymbol{x}_1 = \begin{pmatrix} x_{11} \\ x_{12} \end{pmatrix}, \quad \boldsymbol{x}_2 = \begin{pmatrix} x_{21} \\ x_{22} \end{pmatrix}, \quad \cdots, \quad \boldsymbol{x}_n = \begin{pmatrix} x_{n1} \\ x_{n2} \end{pmatrix}. \tag{9.2}$$

These n two-dimensional data are projected onto $y = w_1 x_1 + w_2 x_2$, and then expressed as

$$y_i = w_1 x_{i1} + w_2 x_{i2} = \boldsymbol{w}^T \boldsymbol{x}_i, \qquad i = 1, 2, \cdots, n, \tag{9.3}$$

where $\boldsymbol{w} = (w_1, w_2)^T$ represents a coefficient vector. The mean of the data $y_1, y_2, \cdots, y_n$ that are projected onto the projection axis is

$$\overline{y} = \frac{1}{n}\sum_{i=1}^{n} y_i = \frac{1}{n}\sum_{i=1}^{n}(w_1 x_{i1} + w_2 x_{i2}) = w_1\overline{x}_1 + w_2\overline{x}_2 = \boldsymbol{w}^T\overline{\boldsymbol{x}}, \tag{9.4}$$

where $\overline{\boldsymbol{x}} = (\overline{x}_1, \overline{x}_2)^T$ is the sample mean vector having as its components the sample mean $\overline{x}_j = n^{-1}\sum_{i=1}^{n} x_{ij}$ $(j = 1, 2)$ of each variable. The variance may be given as

$$\begin{aligned}
s_y^2 &= \frac{1}{n}\sum_{i=1}^{n}(y_i - \overline{y})^2 \\
&= \frac{1}{n}\sum_{i=1}^{n}\{w_1(x_{i1} - \overline{x}_1) + w_2(x_{i2} - \overline{x}_2)\}^2 \\
&= w_1^2 \frac{1}{n}\sum_{i=1}^{n}(x_{i1} - \overline{x}_1)^2 \\
&\qquad + 2w_1 w_2 \frac{1}{n}\sum_{i=1}^{n}(x_{i1} - \overline{x}_1)(x_{i2} - \overline{x}_2) + w_2^2 \frac{1}{n}\sum_{i=1}^{n}(x_{i2} - \overline{x}_2)^2 \\
&= w_1^2 s_{11} + 2w_1 w_2 s_{12} + w_2^2 s_{22} \\
&= \boldsymbol{w}^T S \boldsymbol{w},
\end{aligned} \tag{9.5}$$

where S is the sample variance-covariance matrix defined by

$$S = \begin{pmatrix} s_{11} & s_{12} \\ s_{21} & s_{22} \end{pmatrix}, \quad s_{jk} = \frac{1}{n}\sum_{i=1}^{n}(x_{ij} - \overline{x}_j)(x_{ik} - \overline{x}_k), \quad j, k = 1, 2. \tag{9.6}$$

The problem of finding the coefficient vector $\boldsymbol{w} = (w_1, w_2)^T$, which corresponds to the maximum variance for the n two-dimensional data as projected onto $y = w_1 x_1 + w_2 x_2$ becomes the maximization problem of the variance $s_y^2 = \boldsymbol{w}^T S \boldsymbol{w}$ in (9.5) under the constraint $\boldsymbol{w}^T \boldsymbol{w} = 1$. This constraint is applied because $||\boldsymbol{w}||$ would be infinitely large without it and the variance would thus diverge. The problem of variance maximization under this constraint can be solved by Lagrange's method of undetermined multipliers, by finding the stationary point (at which the derivative becomes 0) of the Lagrangian function

$$L(\boldsymbol{w}, \lambda) = \boldsymbol{w}^T S \boldsymbol{w} + \lambda(1 - \boldsymbol{w}^T \boldsymbol{w}) \tag{9.7}$$

with λ as the Lagrange multiplier (Appendix B.1).

When setting the equation obtained by partial differentiation of the Lagrangian function with respect to $\boldsymbol{w}$ to $\boldsymbol{0}$, we obtain the equation

$$S\boldsymbol{w} = \lambda \boldsymbol{w}. \tag{9.8}$$

Here, we used the vector differential formula $\partial(\boldsymbol{w}^T S \boldsymbol{w})/\partial \boldsymbol{w} = 2S\boldsymbol{w}$, $\partial(\boldsymbol{w}^T \boldsymbol{w})/\partial \boldsymbol{w} = 2\boldsymbol{w}$. This solution is the eigenvector $\boldsymbol{w}_1 = (w_{11}, w_{12})^T$ corresponding to the maximum eigenvalue λ_1 obtained by solving the characteristic equation for the sample variance-covariance matrix S. Accordingly, the first principal component is given by

$$y_1 = w_{11}x_1 + w_{12}x_2 = \boldsymbol{w}_1^T \boldsymbol{x}. \tag{9.9}$$

The variance of the first principal component y_1 is attained with $\boldsymbol{w} = \boldsymbol{w}_1$, which maximizes the variance of (9.5), and is therefore

$$s_{y_1}^2 = \boldsymbol{w}_1^T S \boldsymbol{w}_1 = \lambda_1. \tag{9.10}$$

Here we have utilized the fact that, in the relation $S\boldsymbol{w}_1 = \lambda_1 \boldsymbol{w}_1$ between the eigenvalue and the eigenvector of the symmetric matrix S, left multiplying both sides by the vector $\boldsymbol{w}_1$ ($\boldsymbol{w}_1^T \boldsymbol{w}_1 = 1$) normalized to length 1 results in $\boldsymbol{w}_1^T S \boldsymbol{w}_1 = \lambda_1 \boldsymbol{w}_1^T \boldsymbol{w}_1 = \lambda_1$.

The second principal component, having the two-dimensional data projected onto $y = w_1 x_1 + w_2 x_2$ under the normalization constraint $\boldsymbol{w}^T \boldsymbol{w} = 1$ together with orthogonality to the first principal component, is given as the coefficient vector that maximizes the variance $s_{y_2}^2 = \boldsymbol{w}^T S \boldsymbol{w}$ with $\boldsymbol{w}_1^T \boldsymbol{w} = 0$. The solution, in the same way as for the first principal component, is given as the stationary point with respect to $\boldsymbol{w}$ for the Lagrangian function

$$L(\boldsymbol{w}, \lambda, \gamma) = \boldsymbol{w}^T S \boldsymbol{w} + \lambda(1 - \boldsymbol{w}^T \boldsymbol{w}) + \gamma \boldsymbol{w}_1^T \boldsymbol{w}, \tag{9.11}$$

where λ, γ are Lagrange multipliers.

When we partially differentiate the above equation with respect to $\boldsymbol{w}$ and set the result equal to $\mathbf{0}$, we obtain the equation

$$2S\boldsymbol{w} - 2\lambda\boldsymbol{w} + \gamma\boldsymbol{w}_1 = \mathbf{0}. \tag{9.12}$$

If we then left multiply by the eigenvector $\boldsymbol{w}_1$ corresponding to the maximum eigenvalue λ_1, we have

$$2\boldsymbol{w}_1^T S\boldsymbol{w} - 2\lambda\boldsymbol{w}_1^T\boldsymbol{w} + \gamma\boldsymbol{w}_1^T\boldsymbol{w}_1 = 0. \tag{9.13}$$

From the requirement of orthogonality $\boldsymbol{w}_1^T\boldsymbol{w} = 0$ to the first principal component and $\boldsymbol{w}_1^T\boldsymbol{w}_1 = 1$, the left-hand side of the above equation becomes $2\boldsymbol{w}_1^T S\boldsymbol{w} + \gamma = 0$. Moreover, when we apply the relation $S\boldsymbol{w}_1 = \lambda_1\boldsymbol{w}_1$ ($\boldsymbol{w}_1^T S = \lambda_1\boldsymbol{w}_1^T$) between the eigenvalue and the eigenvector, we have $\boldsymbol{w}_1^T S\boldsymbol{w} = \lambda_1\boldsymbol{w}_1^T\boldsymbol{w} = 0$ and thus $\gamma = 0$. Equation (9.12) therefore becomes $S\boldsymbol{w} = \lambda\boldsymbol{w}$, and finding the solution $\boldsymbol{w}$ again becomes an eigenvalue problem $S\boldsymbol{w} = \lambda\boldsymbol{w}$ under the length 1 constraint. This solution is the eigenvector $\boldsymbol{w}_2 = (w_{21}, w_{22})^T$ corresponding to the second eigenvalue λ_2 of the sample variance-covariance matrix S. The second principal component can therefore be given by the following equation:

$$y_2 = w_{21}x_1 + w_{22}x_2 = \boldsymbol{w}_2^T\boldsymbol{x}. \tag{9.14}$$

From the relation between the eigenvalue and the eigenvector of the symmetric matrix S, the variance of the second principal component y_2 becomes $s_{y_2}^2 = \boldsymbol{w}_2^T S\boldsymbol{w}_2 = \lambda_2$ in the same manner as in (9.10).

In summary, PCA based on the two-dimensional data in (9.2) is essentially a problem of finding the eigenvalues and eigenvectors of the sample variance-covariance matrix S.

Eigenvalues, eigenvectors, and principal components We denote the eigenvalues of the sample variance-covariance matrix S in descending order according to magnitude as $\lambda_1 \geq \lambda_2 \geq 0$, and the corresponding mutually orthogonal eigenvectors normalized to length 1 as $\boldsymbol{w}_1 = (w_{11}, w_{12})^T$ and $\boldsymbol{w}_2 = (w_{21}, w_{22})^T$. The first and second principal components and their respective variances are then given as follows (Figure 9.2):

$$\begin{aligned} y_1 &= w_{11}x_1 + w_{12}x_2 = \boldsymbol{w}_1^T\boldsymbol{x}, \qquad \mathrm{var}(y_1) = \lambda_1, \\ y_2 &= w_{21}x_1 + w_{22}x_2 = \boldsymbol{w}_2^T\boldsymbol{x}, \qquad \mathrm{var}(y_2) = \lambda_2. \end{aligned} \tag{9.15}$$

Lagrange's method of undetermined multipliers, as used in this subsection, provides a tool of finding the stationary points of multivariate real-valued functions under equality and inequality constraints on the variables. For details on this method, refer to Appendix B.

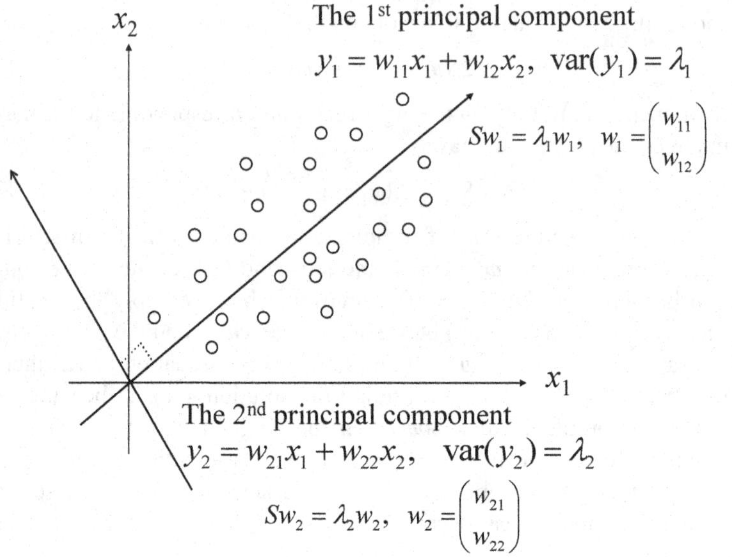

Figure 9.2 *Eigenvalue problem and the first and second principal components.*

9.1.2 Process of Deriving Principal Components and Properties

In general, we denote as $\boldsymbol{x} = (x_1, x_2, \cdots, x_p)^T$ the p variables representing the characteristics of the individuals. Based on the p-dimensional data $\boldsymbol{x}_1, \boldsymbol{x}_2, \cdots, \boldsymbol{x}_n$ observed for these p variables, we obtain the sample variance-covariance matrix

$$S = (s_{jk}) = \frac{1}{n} \sum_{i=1}^{n} (\boldsymbol{x}_i - \overline{\boldsymbol{x}})(\boldsymbol{x}_i - \overline{\boldsymbol{x}})^T, \tag{9.16}$$

where $\overline{\boldsymbol{x}}$ is the p-dimensional sample mean vector and $s_{jk} = \sum_{i=1}^{n}(x_{ij} - \overline{x}_j)(x_{ik} - \overline{x}_k)/n$.

Following the basic concept of the derivation of principal components as described for two-dimensional data in the preceding subsection, we first project the n observed p-dimensional data onto the projection axis

$$y = w_1 x_1 + w_2 x_2 + \cdots + w_p x_p = \boldsymbol{w}^T \boldsymbol{x}, \tag{9.17}$$

and then obtain the one-dimensional data $y_i = \boldsymbol{w}^T \boldsymbol{x}_i$ $(i = 1, 2, \cdots, n)$.

Since the mean of the projected data is $\overline{y} = n^{-1} \sum_{i=1}^{n} \boldsymbol{w}^T \boldsymbol{x}_i = \boldsymbol{w}^T \overline{\boldsymbol{x}}$, the variance can be expressed as

$$\begin{aligned} s_y^2 &= \frac{1}{n} \sum_{i=1}^{n} (y_i - \overline{y})^2 \\ &= \frac{1}{n} \sum_{i=1}^{n} (\boldsymbol{w}^T \boldsymbol{x}_i - \boldsymbol{w}^T \overline{\boldsymbol{x}})^2 = \boldsymbol{w}^T \frac{1}{n} \sum_{i=1}^{n} (\boldsymbol{x}_i - \overline{\boldsymbol{x}})(\boldsymbol{x}_i - \overline{\boldsymbol{x}})^T \boldsymbol{w} = \boldsymbol{w}^T S \boldsymbol{w}. \end{aligned} \tag{9.18}$$

Accordingly, in the same manner as applied for two dimensions, the coefficient vector that maximizes the variance of the projected data can be given as the eigenvector $\boldsymbol{w}_1$ corresponding to the maximum eigenvalue λ_1 of the sample variance-covariance matrix S. The projection axis $y_1 = \boldsymbol{w}_1^T \boldsymbol{x}$ having $\boldsymbol{w}_1$ as its coefficient vector is then the first principal component. In the same way, again, the variance on this projection axis is the maximum eigenvalue λ_1.

The second principal component is the axis that, under the requirement of orthogonality to the first principal component, maximizes the variance of the projected p-dimensional data, and is thus the projection axis generated by the eigenvector $\boldsymbol{w}_2$, which corresponds to the second-largest eigenvalue λ_2 of the matrix S. Continuing in the same manner, the third principal component is defined as the axis which, under the requirement of orthogonality to the first and second principal components, maximizes the variance of the projected p-dimensional data. By successive repetitions of this process, we can in principal derive p principal components for the linear combinations of the original variables. As a result, PCA thus becomes the eigenvalue problem of the sample variance-covariance matrix S, as next described.

Eigenvalue problem of the sample variance-covariance matrix and principal components Let S be a sample variance-covariance matrix based on n observed p-dimensional data. As seen from the definition in (9.16), it is a symmetric matrix of order p. We then denote the p eigenvalues

$$\lambda_1 \geq \lambda_2 \geq \cdots \geq \lambda_i \geq \cdots \geq \lambda_p \geq 0, \tag{9.19}$$

given as the solution to the characteristic equation of S, $|S - \lambda I_p| = 0$. Further, we denote the p-dimensional eigenvectors normalized to length

1 corresponding to these eigenvalues as

$$\boldsymbol{w}_1 = \begin{pmatrix} w_{11} \\ w_{12} \\ \vdots \\ w_{1p} \end{pmatrix}, \quad \boldsymbol{w}_2 = \begin{pmatrix} w_{21} \\ w_{22} \\ \vdots \\ w_{2p} \end{pmatrix}, \quad \cdots, \quad \boldsymbol{w}_p = \begin{pmatrix} w_{p1} \\ w_{p2} \\ \vdots \\ w_{pp} \end{pmatrix}. \tag{9.20}$$

For these eigenvectors, normalization $\boldsymbol{w}_i^T \boldsymbol{w}_i = 1$ to vector length 1 and orthogonality $\boldsymbol{w}_i^T \boldsymbol{w}_j = 0$ $(i \neq j)$ are thus established. The p principal components and their variance expressed in terms of the linear combination of the original variables may now be given in order as follows:

$$\begin{aligned} y_1 &= w_{11}x_1 + w_{12}x_2 + \cdots + w_{1p}x_p = \boldsymbol{w}_1^T \boldsymbol{x}, \quad && \mathrm{var}(y_1) = \lambda_1, \\ y_2 &= w_{21}x_1 + w_{22}x_2 + \cdots + w_{2p}x_p = \boldsymbol{w}_2^T \boldsymbol{x}, \quad && \mathrm{var}(y_2) = \lambda_2, \\ &\vdots \\ y_p &= w_{p1}x_1 + w_{p2}x_2 + \cdots + w_{pp}x_p = \boldsymbol{w}_p^T \boldsymbol{x}, \quad && \mathrm{var}(y_p) = \lambda_p. \end{aligned} \tag{9.21}$$

In applying PCA, it is possible to reduce the dimensionality from the n observed p-dimensional data $\{\boldsymbol{x}_i = (x_{i1}, x_{i2}, \cdots, x_{ip})^T;\ i = 1, 2, \cdots, n\}$ for the p original variables to a smaller number, by using only the first several principal components; e.g., to the two-dimensional data $\{(y_{i1}, y_{i2});\ i = 1, 2, \cdots, n\}$ by using only the first and second principal components, where $y_{i1} = \boldsymbol{w}_1^T \boldsymbol{x}_i$, $y_{i2} = \boldsymbol{w}_2^T \boldsymbol{x}_i$. By thus projecting the data set of the higher-dimensional space onto a two-dimensional plane, we can visually apprehend the data structure. By finding the meaning of the new variables combined as linear combinations of the original variables, moreover, we can extract useful information.

The meaning of the principal components can be understood in terms of the magnitude and sign of the coefficients w_{ij} of each variable. Also the correlation between the principal components and the variables as a quantitative indicator is highly useful for identifying the variables that influence the principal components. The correlation between the i-th principal component y_i and the j-th variable x_j is given by

$$r_{y_i, x_j} = \frac{\mathrm{cov}(y_i, x_j)}{\sqrt{\mathrm{var}(y_i)}\sqrt{\mathrm{var}(x_j)}} = \frac{\lambda_i w_{ij}}{\sqrt{\lambda_i}\sqrt{s_{jj}}} = \frac{\sqrt{\lambda_i} w_{ij}}{\sqrt{s_{jj}}}, \tag{9.22}$$

where λ_i is the variance of the i-th principal component, w_{ij} is the coefficient of the variable x_j for the i-th principal component, and s_{jj} is the variance of variable x_j.

Eigenvalues and eigenvectors of a symmetric matrix The sample variance-covariance matrix S given by (9.16) is a symmetric matrix of order p, with the following relation between its eigenvalues and eigenvectors:

$$S\boldsymbol{w}_i = \lambda_i \boldsymbol{w}_i, \quad \boldsymbol{w}_i^T \boldsymbol{w}_i = 1, \quad \boldsymbol{w}_i^T \boldsymbol{w}_j = 0 \quad (i \neq j) \tag{9.23}$$

for $i, j = 1, 2, \cdots, p$. We denote as W the matrix of order p having p eigenvectors as columns, and as Λ the matrix of order p having the eigenvalues as its diagonal elements, that is,

$$W = (\boldsymbol{w}_1, \boldsymbol{w}_2, \cdots, \boldsymbol{w}_p), \qquad \Lambda = \begin{pmatrix} \lambda_1 & 0 & \cdots & 0 \\ 0 & \lambda_2 & \cdots & 0 \\ \vdots & \vdots & \ddots & \vdots \\ 0 & 0 & \cdots & \lambda_p \end{pmatrix}. \tag{9.24}$$

The relation between the eigenvalues and eigenvectors of the sample variance-covariance matrix S given by (9.23) may then be expressed as follows:

$$\begin{aligned} &(1)\ SW = W\Lambda, \qquad W^T W = I_p. \\ &(2)\ W^T S W = \Lambda. \\ &(3)\ S = W\Lambda W^T = \lambda_1 \boldsymbol{w}_1 \boldsymbol{w}_1^T + \lambda_2 \boldsymbol{w}_2 \boldsymbol{w}_2^T + \cdots + \lambda_p \boldsymbol{w}_p \boldsymbol{w}_p^T. \\ &(4)\ \mathrm{tr}\, S = \mathrm{tr}(W\Lambda W^T) = \mathrm{tr}\, \Lambda = \lambda_1 + \lambda_2 + \cdots + \lambda_p. \end{aligned} \tag{9.25}$$

Equation (2) shows that the symmetric matrix S can be diagonalized by the orthogonal matrix W, and (3) is known as the *spectral decomposition* of symmetric matrix S. Equation (4) shows that the sum, tr$S$ $= s_{11} + s_{22} + \cdots + s_{pp}$, of the variances of the original variables $x_1, x_2, \cdots, x_p$ equals the sum, tr$\Lambda = \lambda_1 + \lambda_2 + \cdots + \lambda_p$, of the variances of the constructed p principal components.

Standardization and sample correlation matrix Let us assume that, in addition to scores in mathematics, English, science, and Japanese, we add as a fifth variable the proportion of correct answers given for 100 calculation problems measuring calculating ability and perform PCA based on the resulting five-dimensional data. In this case, the units of measurement differ substantially between the proportion of correct answers and the subject test scores, and it is therefore necessary to standardize the observed data.

For the n observed p-dimensional data

$$\boldsymbol{x}_i = (x_{i1}, x_{i2}, \cdots, x_{ip})^T, \qquad i = 1, 2, \cdots, n, \tag{9.26}$$

we first obtain the sample mean vector $\overline{\boldsymbol{x}} = (\overline{x}_1, \overline{x}_2, \cdots, \overline{x}_p)^T$ and the sample variance-covariance matrix $S = (s_{jk})$. We standardize the p-dimensional data in (9.26) such that

$$\boldsymbol{z}_i = (z_{i1}, z_{i2}, \cdots, z_{ip})^T, \quad z_{ij} = \frac{x_{ij} - \overline{x}_j}{\sqrt{s_{jj}}}, \quad j = 1, 2, \cdots, p. \tag{9.27}$$

The sample variance (s^*_{jj}) and the sample covariance (s^*_{jk}) based on these standardized p-dimensional data are then given by

$$s^*_{jk} = \frac{1}{n}\sum_{i=1}^{n} z_{ij} z_{ik} = \frac{1}{n}\sum_{i=1}^{n} \frac{(x_{ij} - \overline{x}_j)(x_{ik} - \overline{x}_k)}{\sqrt{s_{jj}}\sqrt{s_{kk}}} = \frac{s_{jk}}{\sqrt{s_{jj}}\sqrt{s_{kk}}} \equiv r_{jk}$$

for $j, k = 1, 2, \cdots, p$. In the sample variance-covariance matrix based on the standardized data, all of the diagonal elements r_{jj} are accordingly 1, and we obtain the $p \times p$ matrix R with r_{jk} as the non-diagonal elements, the sample correlation coefficients between the j-th and k-th variables, which is known as the sample correlation matrix.

PCA starting with standardized multidimensional data thus becomes a problem of finding the eigenvalues and eigenvectors of the sample correlation matrix R, which is performed by the same process of principal component derivation as that of PCA starting with a sample variance-covariance matrix.

9.1.3 Dimension Reduction and Information Loss

As previously noted, PCA is a method of consolidating mutually correlated variables to a smaller number of variables, known as principal components, by linear combination with minimal loss of information contained in the observed data. The principal components are mutually uncorrelated, and are constructed in order of decreasing variance. From the principal components, the dimensionality of data scattered in a high-dimensional space can be reduced to a one-dimensional line, two-dimensional plane, or three-dimensional space, thus enabling intuitive, visual apprehension of structures embodied by the higher-dimensional data. However, information loss may occur in such a dimension reduction, and so a quantitative measure of this loss is necessary.

As a simple example of such information loss, the projection of two

two-dimensional data points consisting of the mathematics and English score pairs (80, 60) and (60, 80) onto the projection axis $y = x_1 + x_2$ results in the data point 140 for both. Data that are distinguishable on a two-dimensional plane may become indistinguishable when projected into one dimension, resulting in information loss even in a dimension reduction from two to one. To minimize the loss of information in dimension reduction, it is therefore essential to find the projection axis that provides the greatest dispersion of the projected data.

In PCA, the variance provides a measure of information, and information loss can therefore be quantitatively estimated from the relative sizes of variances of principal components. As shown in (9.21), the variance of principal component is given by the eigenvalue of the sample variance-covariance matrix. In short, we can utilize $\lambda_1/(\lambda_1+\lambda_2+\cdots+\lambda_p)$ to assess what proportion of the information contained in the p original variables is present in the first principal component y_1. In general the following equation is used as the measure of the information present in the i-th principal component y_i:

$$\frac{\lambda_i}{\lambda_1 + \lambda_2 + \cdots + \lambda_p}. \tag{9.28}$$

Similarly, the percentage of variance accounted for by the first k principal components is given by

$$\frac{\lambda_1 + \lambda_2 + \cdots + \lambda_k}{\lambda_1 + \lambda_2 + \cdots + \lambda_k + \cdots + \lambda_p}. \tag{9.29}$$

As may be seen from the relation $\mathrm{tr}S = \mathrm{tr}\Lambda$ between the eigenvalues and eigenvectors of the symmetric matrix given in (9.25), the p principal components $y_1, y_2, \cdots, y_p$ taken together contain all of the information in the original variables $x_1, x_2, \cdots, x_p$ (i.e., the sum of their variances) but information loss occurs in dimension reduction, which reduces the number of principal components that are actually used. In conclusion, the quantitative measure of the information contained in the principal components that are actually used is given by the ratio of the sum of the variances of those principal components to the sum of the variances of all the principal components, which thus serves as the measure of information loss.

9.1.4 Examples

We illustrate the dimension reduction, information loss, and the use of PCA through some examples.

Example 9.1 (Relation between correlation structure and information loss) Suppose that we have the sample correlation matrix based on observed two-dimensional data for variables x_1 and x_2 as follows:

$$R = \begin{pmatrix} 1 & r \\ r & 1 \end{pmatrix}, \tag{9.30}$$

where r (> 0) is the sample correlation coefficient between two variables x_1 and x_2. The characteristic equation of the sample correlation matrix R is

$$|R - \lambda I_2| = (1 - \lambda)^2 - r^2 = \lambda^2 - 2\lambda + 1 - r^2 = 0. \tag{9.31}$$

Its solutions are $\lambda_1 = 1 + r$ and $\lambda_2 = 1 - r$, and the corresponding eigenvectors are then $\boldsymbol{w}_1 = (1/\sqrt{2}, 1/\sqrt{2})^T$ and $\boldsymbol{w}_2 = (-1/\sqrt{2}, 1/\sqrt{2})^T$, respectively. The first and second principal components obtained from the sample correlation matrix are therefore given by

$$y_1 = \frac{1}{\sqrt{2}}x_1 + \frac{1}{\sqrt{2}}x_2, \qquad y_2 = -\frac{1}{\sqrt{2}}x_1 + \frac{1}{\sqrt{2}}x_2. \tag{9.32}$$

The contributions of the first and second principal components are, respectively,

$$\frac{\lambda_1}{\lambda_1 + \lambda_2} = \frac{1 + r}{2}, \qquad \frac{\lambda_2}{\lambda_1 + \lambda_2} = \frac{1 - r}{2}. \tag{9.33}$$

As shown in Figure 9.3, the contribution of the first principal component, and thus the proportion that can be explained by the first principal component alone, increases with increasing correlation between the two variables.

Example 9.2 (Analysis of test-score data) Let us consider the test scores in a class of 50 students, in which the sample variance-covariance matrix computed from these four-dimensional test-score data for Japanese (x_1), science (x_2), English (x_3), and mathematics (x_4) is given by

$$S = \begin{pmatrix} 622 & 112 & 329 & 57 \\ 112 & 137 & 290 & 142 \\ 329 & 290 & 763 & 244 \\ 57 & 142 & 244 & 514 \end{pmatrix}. \tag{9.34}$$

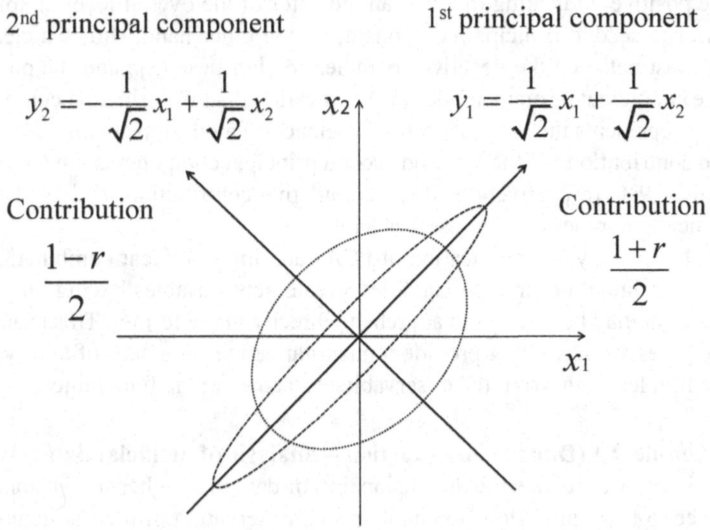

Figure 9.3 *Principal components based on the sample correlation matrix and their contributions: The contribution of the first principal component increases with increasing correlation between the two variables.*

The eigenvectors corresponding to the eigenvalues of this sample variance-covariance matrix are summarized as follows:

Eigenvalues	1218	509	291	19
Eigenvectors				
Japanese(x_1)	0.495	-0.690	0.527	0.021
Science(x_2)	0.296	0.120	-0.158	0.934
English(x_3)	0.736	0.074	-0.581	-0.341
Mathematics(x_4)	0.355	0.710	0.600	-0.102

The first and second principal components, having eigenvectors corresponding to the first two eigenvalues 1218 and 509 as their coefficients, are then given by

$$\begin{aligned} y_1 &= 0.495x_1 + 0.296x_2 + 0.736x_3 + 0.355x_4, \\ y_2 &= -0.690x_1 + 0.120x_2 + 0.074x_3 + 0.710x_4. \end{aligned} \tag{9.35}$$

In the first principal component, all of the coefficients of the variables are positive, indicating that it is an indicator of the overall level of ability. The second principal component, on the other hand, shows a clear contrast between the negative coefficient of Japanese (x_1) and the positive coefficient of mathematics (x_4), indicating that this principal component represents the level of ability in science-related subjects. Moreover, the contributions of the first and second principal components are 59.8% and 24.9%, respectively, and the cumulative contribution, if these two principal components are used, is 84.7%.

In this way, the application of PCA can, through linear combination of the original observable variables, create new variables having meanings that may be difficult to apprehend directly in the former. This example gives variables that provide a quantitative measurement of achievement in learning, from the observable test scores in the four subjects.

Example 9.3 (Dimension reduction – analysis of artificial data) We apply PCA to reduce the dimensionality of data in a higher-dimensional space to two dimensions to enable visual observation of their structure. For this purpose, we generated three classes, each composed of 21-dimensional data of size 100, using the algorithm:

$$\begin{aligned} \text{Class 1}: \ x_{ij} &= U(0,1)h_1(j) + \{1 - U(0,1)\}h_2(j) + \varepsilon_j, \\ \text{Class 2}: \ x_{ij} &= U(0,1)h_1(j) + \{1 - U(0,1)\}h_3(j) + \varepsilon_j, \\ \text{Class 3}: \ x_{ij} &= U(0,1)h_2(j) + \{1 - U(0,1)\}h_3(j) + \varepsilon_j, \end{aligned} \tag{9.36}$$

where $j = 1, 2, \cdots, 21$, $U(0,1)$ is a uniform distribution on (0,1), ε_j are standard normal variates, and $h_i(j)$ are the shifted triangular waveforms given by

$$h_1(j) = \max(6 - |j - 11|, 0), \quad h_2(j) = h_1(j-4), \quad h_3(j) = h_1(j+4).$$

This algorithm for generating artificial data sets, known as waveform data, was developed by Breiman et al. (1984) and is widely used to investigate the effectiveness of classification methods (e.g., Hastie et al., 2009, p. 451).

For the 21-dimensional data set of size 300 generated by the algorithm, we obtained the sample variance-covariance matrix and applied PCA. Figure 9.4 shows a plot of the values of the first and second principal components. The application of PCA for projection of a higher-dimensional space data set onto a two-dimensional plane enables visual observation of the dispersion patterns of the three different data sets and their positional relation.

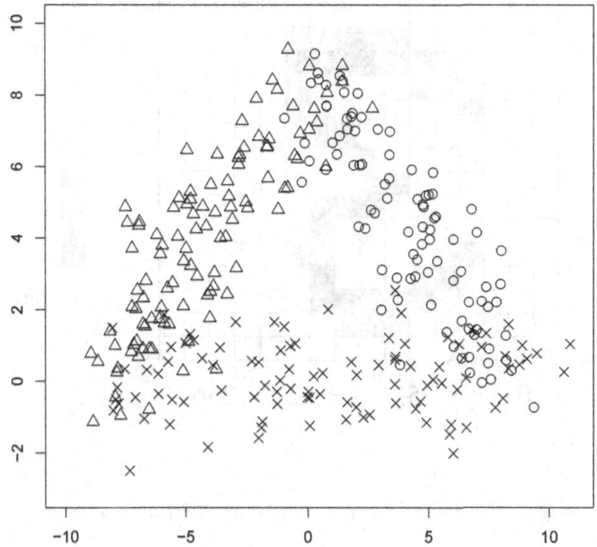

Figure 9.4 *Two-dimensional view of the 21-dimensional data set, projected onto the first (x) and second (y) principal components.*

9.2 Image Compression and Decompression

Dimension reduction by PCA can be used in compression and transmission of image data expressed in the form of multidimensional data. If the coefficient vectors of the principal components are known, then it can also be used to decompress the transmitted compressed data. In the following, we discuss the basic concept of this image compression and decompression.

Figure 9.5 shows an example of image digitization of characters, image digitization in which the handwritten character region has been divided into 16 × 16 pixels. The data are thus 256-dimensional data that have been digitized by pixel correspondence to a grey scale of 0 to 255. We denote the digitized p-dimensional data as $\boldsymbol{x}_1, \boldsymbol{x}_2, \cdots, \boldsymbol{x}_n$ and obtain sample variance-covariance matrix S to derive the principal components. The principal components obtained by solving the eigenvalue problem of the symmetric matrix S are given as shown in (9.21).

In general, the i-th image data $\boldsymbol{x}_i = (x_{i1}, x_{i2}, \cdots, x_{ip})^T$ $(i = 1, 2, \cdots, n)$

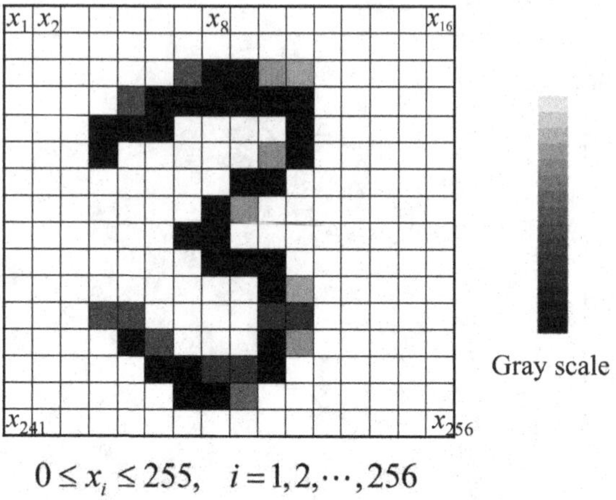

Figure 9.5 *Image digitization of a handwritten character.*

is transformed to principal components as follows:

$$
\begin{aligned}
y_{i1} &= w_{11}x_{i1} + w_{12}x_{i2} + \cdots + w_{1p}x_{ip} = \boldsymbol{w}_1^T \boldsymbol{x}_i, \\
y_{i2} &= w_{21}x_{i1} + w_{22}x_{i2} + \cdots + w_{2p}x_{ip} = \boldsymbol{w}_2^T \boldsymbol{x}_i, \\
&\vdots \\
y_{ip} &= w_{p1}x_{i1} + w_{p2}x_{i2} + \cdots + w_{pp}x_{ip} = \boldsymbol{w}_p^T \boldsymbol{x}_i.
\end{aligned}
\tag{9.37}
$$

We express these equations in matrix and vector form

$$
\boldsymbol{y}_i = W^T \boldsymbol{x}_i, \qquad W = (\boldsymbol{w}_1, \boldsymbol{w}_2, \cdots, \boldsymbol{w}_p), \tag{9.38}
$$

where $\boldsymbol{y}_i = (y_{i1}, y_{i2}, \cdots, y_{ip})^T$ and W is a $p \times p$ orthogonal matrix with the principal component coefficient vectors as column vectors.

If we have access to the orthogonal matrix W and the transmitted $\boldsymbol{y}_i$ represents the i-th p-dimensional image data $\boldsymbol{x}_i$ transformed to all principal components, we can then obtain

$$
\boldsymbol{y}_i = W^T \boldsymbol{x}_i \quad \Longrightarrow \quad W\boldsymbol{y}_i = WW^T \boldsymbol{x}_i = \boldsymbol{x}_i, \tag{9.39}
$$

and therefore completely decompress and thereby reconstruct the entire image. If, for example, only the first and second principal component values (y_{i1}, y_{i2}) have been transmitted, then a question arises concerning the degree to which the image can be reconstructed by decompression. This is key to PCA-based image data compression/transmission and decompression. The process can in fact be regarded as follows.

From (9.39) and (9.37), we have

$$\begin{aligned} \boldsymbol{x}_i &= W\boldsymbol{y}_i \\ &= y_{i1}\boldsymbol{w}_1 + y_{i2}\boldsymbol{w}_2 + \cdots + y_{ip}\boldsymbol{w}_p \\ &= (\boldsymbol{x}_i^T\boldsymbol{w}_1)\boldsymbol{w}_1 + (\boldsymbol{x}_i^T\boldsymbol{w}_2)\boldsymbol{w}_2 + \cdots + (\boldsymbol{x}_i^T\boldsymbol{w}_p)\boldsymbol{w}_p. \end{aligned} \tag{9.40}$$

We now need to subtract the sample mean vector from the image data and perform data centering. It follows from (9.40) that

$$\begin{aligned} \overline{\boldsymbol{x}} &= \frac{1}{n}\sum_{i=1}^{n}\boldsymbol{x}_i \\ &= \frac{1}{n}\sum_{i=1}^{n}\left\{(\boldsymbol{x}_i^T\boldsymbol{w}_1)\boldsymbol{w}_1 + (\boldsymbol{x}_i^T\boldsymbol{w}_2)\boldsymbol{w}_2 + \cdots + (\boldsymbol{x}_i^T\boldsymbol{w}_p)\boldsymbol{w}_p\right\} \\ &= (\overline{\boldsymbol{x}}^T\boldsymbol{w}_1)\boldsymbol{w}_1 + (\overline{\boldsymbol{x}}^T\boldsymbol{w}_2)\boldsymbol{w}_2 + \cdots + (\overline{\boldsymbol{x}}^T\boldsymbol{w}_p)\boldsymbol{w}_p. \end{aligned} \tag{9.41}$$

The centering of the i-th data $\boldsymbol{x}_i$ is then given by

$$\begin{aligned} &\boldsymbol{x}_i - \overline{\boldsymbol{x}} \\ &= \left\{(\boldsymbol{x}_i - \overline{\boldsymbol{x}})^T\boldsymbol{w}_1\right\}\boldsymbol{w}_1 + \left\{(\boldsymbol{x}_i - \overline{\boldsymbol{x}})^T\boldsymbol{w}_2\right\}\boldsymbol{w}_2 + \cdots + \left\{(\boldsymbol{x}_i - \overline{\boldsymbol{x}})^T\boldsymbol{w}_p\right\}\boldsymbol{w}_p. \end{aligned} \tag{9.42}$$

If the image data $\boldsymbol{x}_i$ for all principal components are transmitted, complete decompression and reproduction can thereby be performed as follows:

$$\boldsymbol{x}_i = \overline{\boldsymbol{x}} + \sum_{j=1}^{p}\left\{(\boldsymbol{x}_i - \overline{\boldsymbol{x}})^T\boldsymbol{w}_j\right\}\boldsymbol{w}_j. \tag{9.43}$$

If we reduce the dimensionality of the p-dimensional image data by using just m $(< p)$ principal components in the image transmission, the transmitted i-th image is then decompressed as

$$\boldsymbol{x}_i^{(m)} = \overline{\boldsymbol{x}} + \sum_{j=1}^{m}\left\{(\boldsymbol{x}_i - \overline{\boldsymbol{x}})^T\boldsymbol{w}_j\right\}\boldsymbol{w}_j. \tag{9.44}$$

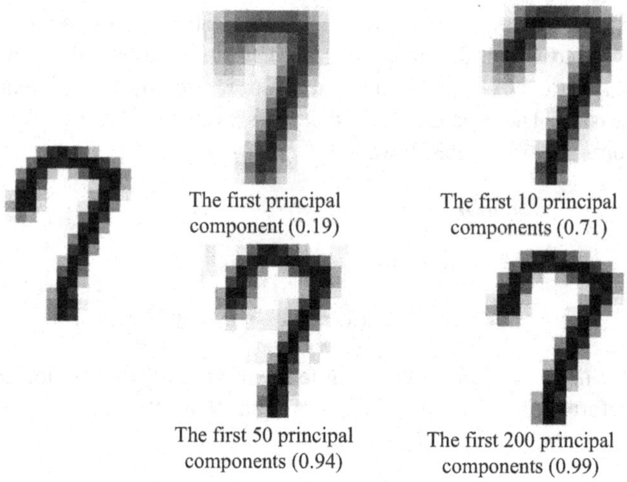

Figure 9.6 *The images obtained by first digitizing and compressing the leftmost image 7 and then decompressing transmitted data using a successively increasing number of principal components. The number in parentheses shows the cumulative contribution rate in each case.*

This constitutes image compression and decompression based on the information contained in the first m principal components.

Example 9.4 (Compression and decompression of character image) A character image region is divided into 16×16 pixels in correspondence with a grey scale of 0 to 255. We have 646 256-dimensional data (LeCun et al., 1998; Bishop, 2006). In this example, we obtained the eigenvectors corresponding to the eigenvalues of the sample variance-covariance matrix based on these data.

Figure 9.6 shows the images obtained by first digitizing and compressing the leftmost image 7 and then decompressing transmitted data using a successively increasing number of principal components. The number in parentheses shows the cumulative contribution rate in each case. The image obtained by decompression of the transmitted data using just one principal component is quite blurry, but when the first 50 principal components are used, the cumulative contribution rate rises to 94% and the original image is reproduced very closely. As evident from this figure, increasing the number of principal components tends to increase the cumulative contribution, with a corresponding increase in the amount of information transmitted.

PCA is used as a method of dimension reduction in fields such as pattern recognition, image analysis, and signal processing, where it is referred to as the Karhunen-Loève expansion. In the next subsection, we discuss singular value decomposition of data matrices, which shares several aspects with PCA, and its relation to PCA and image compression discussed thus far.

9.3 Singular Value Decomposition

We describe the singular value decomposition of a matrix. Let us denote as $\overline{\boldsymbol{x}} = n^{-1} \sum_{i=1}^{n} \boldsymbol{x}_i$ the sample mean vector obtained from n observed p-dimensional data, and as

$$X = (\boldsymbol{x}_1 - \overline{\boldsymbol{x}},\ \boldsymbol{x}_2 - \overline{\boldsymbol{x}},\ \cdots,\ \boldsymbol{x}_n - \overline{\boldsymbol{x}})^T \tag{9.45}$$

the $n \times p$ matrix obtained by subtracting the sample mean vector from the data. Matrix X is an $n \times p$ matrix containing the centered data as rows, that is, the origin becomes the sample mean vector corresponding to the centroid of the data set. Singular value decomposition of data matrix X can be stated as follows.

Singular value decomposition Let data matrix X be an $n \times p$ matrix of rank q $(q \leq p)$. Matrix X can then be decomposed using the $n \times q$ matrix V, $q \times q$ diagonal matrix L, and $p \times q$ matrix W, as

$$X = VLW^T, \tag{9.46}$$

where V, W, and L are matrices having the following properties:

(1) $V = (\boldsymbol{v}_1, \boldsymbol{v}_2, \cdots, \boldsymbol{v}_q)$ is a $n \times q$ matrix composed of mutually orthogonal column vectors of length 1, thus satisfying $\boldsymbol{v}_i^T \boldsymbol{v}_i = 1$ and $\boldsymbol{v}_i^T \boldsymbol{v}_j = 0$, $V^T V = I_q$.

(2) $W = (\boldsymbol{w}_1, \boldsymbol{w}_2, \cdots, \boldsymbol{w}_q)$ is a $p \times q$ matrix composed of mutually orthogonal column vectors of length 1, thus satisfying $\boldsymbol{w}_i^T \boldsymbol{w}_i = 1$, $\boldsymbol{w}_i^T \boldsymbol{w}_j = 0$ and $W^T W = I_q$.

(3) L is a $q \times q$ diagonal matrix.

The matrices that express the singular value decomposition of the data matrix X are obtained as follows. We first perform spectral decomposition of X^TX, noting that the $p \times p$ symmetric matrix X^TX is of rank q, as

$$X^TX = \ell_1 \boldsymbol{w}_1 \boldsymbol{w}_1^T + \ell_2 \boldsymbol{w}_2 \boldsymbol{w}_2^T + \cdots + \ell_q \boldsymbol{w}_q \boldsymbol{w}_q^T, \tag{9.47}$$

where $\ell_1 \geq \ell_2 \geq \cdots \geq \ell_q > 0$ are eigenvalues of X^TX in descending order, and $\boldsymbol{w}_1, \boldsymbol{w}_2, \cdots, \boldsymbol{w}_q$ are the corresponding normalized eigenvectors of length 1. As the rank of the $p \times p$ matrix X^TX is q, $(p-q)$ of the p eigenvalues are therefore 0. The matrices W $(p \times q)$, V $(n \times q)$, and L $(q \times q)$ based on this spectral decomposition are defined as

$$\begin{aligned} W &= (\boldsymbol{w}_1, \boldsymbol{w}_2, \cdots, \boldsymbol{w}_q), \\ V &= (\boldsymbol{v}_1, \boldsymbol{v}_2, \cdots, \boldsymbol{v}_q) = \left(\frac{1}{\sqrt{\ell_1}} X\boldsymbol{w}_1, \frac{1}{\sqrt{\ell_2}} X\boldsymbol{w}_2, \cdots, \frac{1}{\sqrt{\ell_q}} X\boldsymbol{w}_q \right), \\ L &= \mathrm{diag}\left(\sqrt{\ell_1}, \sqrt{\ell_2}, \cdots, \sqrt{\ell_q} \right). \end{aligned} \tag{9.48}$$

We first observe that, from the spectral decomposition of X^TX, the columns of matrix W are of length 1 and are mutually orthogonal, and thus satisfy property (2). We next observe in regard to matrix V that, since

$$\boldsymbol{v}_i^T \boldsymbol{v}_j = \frac{1}{\sqrt{\ell_i}\sqrt{\ell_j}} \boldsymbol{w}_i X^T X \boldsymbol{w}_j = \frac{\ell_j}{\sqrt{\ell_i}\sqrt{\ell_j}} \boldsymbol{w}_i^T \boldsymbol{w}_j = \begin{cases} 1 & i = j \\ 0 & i \neq j, \end{cases} \tag{9.49}$$

its column vectors are mutually orthogonal and of length 1 and thus satisfy property (1). From the three matrices defined by (9.48), we then have

$$VLW^T = \sum_{i=1}^{q} \boldsymbol{v}_i \sqrt{\ell_i} \boldsymbol{w}_i^T = \sum_{i=1}^{q} \frac{1}{\sqrt{\ell_i}} X\boldsymbol{w}_i \sqrt{\ell_i} \boldsymbol{w}_i^T = \sum_{i=1}^{q} X\boldsymbol{w}_i \boldsymbol{w}_i^T. \tag{9.50}$$

Since the rank of the $p \times p$ matrix X^TX is q and $(p-q)$ of the p eigenvalues are 0, we then have $X^TX\boldsymbol{w}_i = \boldsymbol{0}$ for the corresponding eigenvectors $\boldsymbol{w}_i$ $(i = q+1, q+2, \cdots, p)$. We thus obtain $X\boldsymbol{w}_i = \boldsymbol{0}$, and (9.50) can be written as

$$VLW^T = \sum_{i=1}^{q} X\boldsymbol{w}_i \boldsymbol{w}_i^T = \sum_{i=1}^{p} X\boldsymbol{w}_i \boldsymbol{w}_i^T = X \sum_{i=1}^{p} \boldsymbol{w}_i \boldsymbol{w}_i^T. \tag{9.51}$$

Finally, since the matrix having as its columns the eigenvectors $\boldsymbol{w}_i$ $(i =$

$1, 2, \cdots, p$) corresponding to all of the eigenvalues of the $p \times p$ symmetric matrix $X^T X$ is an orthogonal matrix, it follows from $\sum_{i=1}^{p} \boldsymbol{w}_i \boldsymbol{w}_i^T = I_p$ that $VLW^T = X$.

Relation between PCA and image compression When data matrix X is decomposed by singular value decomposition into the three matrices V, L, and W given in (9.48), each of these is closely related to the eigenvalues and eigenvectors of PCA. They are also related to the image decompression discussed in the previous subsection. To observe this relationship, let us assume that the rank, q, of data matrix X is equal to the dimension, p, of each data.

Using the data matrix X, the sample variance-covariance matrix can be expressed as

$$S = \frac{1}{n} \sum_{i=1}^{n} (\boldsymbol{x}_i - \overline{\boldsymbol{x}})(\boldsymbol{x}_i - \overline{\boldsymbol{x}})^T = \frac{1}{n} X^T X. \tag{9.52}$$

Also, by singular value decomposition letting $q = p$, we obtain

$$X^T X = W L^2 W^T. \tag{9.53}$$

We therefore have the following results:

(1) The j-th eigenvalue λ_j of the sample variance-covariance matrix S is given by $\lambda_j = n^{-1} \ell_j$.

(2) The p eigenvectors of the sample variance-covariance matrix S are given as the columns of matrix W.

(3) Since the j-th column of the $n \times p$ matrix V can be expressed as

$$\boldsymbol{v}_j = \frac{1}{\sqrt{\ell_j}} X \boldsymbol{w}_j = \frac{1}{\sqrt{n \lambda_j}} X \boldsymbol{w}_j, \tag{9.54}$$

it gives the score of the principal component standardized to $1/n$ when the n centered data are projected onto the j-th principal component.

(4) Using the first m principal components, any given data $\boldsymbol{x}_i$ can be approximated as

$$\boldsymbol{x}_i \approx \boldsymbol{x}_i^{(m)} = \overline{\boldsymbol{x}} + \sum_{k=1}^{m} \{(\boldsymbol{x}_i - \overline{\boldsymbol{x}})^T \boldsymbol{w}_k\} \boldsymbol{w}_k. \tag{9.55}$$

The above results (1) and (2) are immediately obvious from (9.52) and (9.53). The result (3) clearly follows from the fact that the variance of the j-th principal component is λ_j and that the score of the j-th principal component for the data is given by $(\boldsymbol{x}_i - \overline{\boldsymbol{x}})^T \boldsymbol{w}_j$. The result (4) can be readily obtained by substituting matrix V defined by (9.48) into $X^T = WLV^T$ as transformed from the singular value decomposition (9.46) and comparing the i-th column on both sides. It may be seen that these equations correspond to the completely decompressed data of (9.43) discussed in Section 9.2 on image data compression and decompression and to the decompressed data using the first m principal components of (9.44).

It may also be noted that by singular value decomposition we obtain $XX^T = VL^2V^T$ and that the columns of matrix V are eigenvectors corresponding to the non-zero eigenvalues of XX^T. This constitutes a PCA in which the roles of the variables and the data are seen in reverse. It is used as a method of searching for relations between variables, in which case it is referred to as the biplot method. For further information on *biplots*, see, for example, Johnson and Wichern (2007, p. 726) and Seber (1984, p. 204).

9.4 Kernel Principal Component Analysis

PCA is a useful method for finding structures in high-dimensional data by projecting them onto several principal components expressed as linear combinations of the variables. We have seen in particular that it is useful for extraction of information from data in linear structures. The question that naturally next arises is how to search for structures that are nonlinear in data dispersed in a high-dimensional space. This leads us to nonlinear PCA, which is a method of apprehending nonlinear structures in high-dimensional data. In this section, we discuss the basic concept of kernel-based PCA, which is one approach with this as its purpose.

9.4.1 Data Centering and Eigenvalue Problem

Let us assume that we have observed n p-dimensional data $\boldsymbol{x}_1, \boldsymbol{x}_2, \cdots, \boldsymbol{x}_n$. By calculating the sample mean vector $\overline{\boldsymbol{x}} = n^{-1}\sum_{i=1}^n \boldsymbol{x}_i$, which corresponds to the data centroid, we first perform data centering $\boldsymbol{x}_i^* = \boldsymbol{x}_i - \overline{\boldsymbol{x}}$ $(i = 1, 2, \cdots, n)$. The sample mean vector for the centered data is then $n^{-1}\sum_{i=1}^n \boldsymbol{x}_i^* = n^{-1}\sum_{i=1}^n (\boldsymbol{x}_i - \overline{\boldsymbol{x}}) = \boldsymbol{0}$. With the centroid of the p-dimensional data moved to the origin, we can now omit the asterisk (*) and again write

$$\boldsymbol{x}_1,\ \boldsymbol{x}_2,\ \cdots,\ \boldsymbol{x}_n; \qquad \sum_{i=1}^n \boldsymbol{x}_i = \boldsymbol{0}. \tag{9.56}$$

The PCA based on the centered data thus becomes a problem of finding the eigenvalues and eigenvectors of the sample variance-covariance matrix

$$S = \frac{1}{n}\sum_{i=1}^{n} \boldsymbol{x}_i \boldsymbol{x}_i^T. \tag{9.57}$$

The basic idea of kernel PCA, as described below, is to replace the eigenvalue problem of matrix S with the eigenvalue problem of a new matrix defined by the inner products of the data. Let us first consider this basic idea in the context of observed data. Then, in the next subsection, we shall apply it to transformed data in a higher-dimensional space.

We first obtain the eigenvalues of the sample variance-covariance matrix S in (9.57) based on the centered data, together with the corresponding mutually orthogonal normalized eigenvectors of length 1, and arrange them as follows:

$$\begin{array}{ccccccccc} \lambda_1 & \geq & \lambda_2 & \geq & \cdots & \geq & \lambda_p & > & 0, \\ \boldsymbol{w}_1, & & \boldsymbol{w}_2, & & \cdots & , & \boldsymbol{w}_p. & & \end{array} \tag{9.58}$$

It follows from the relation between the eigenvalues and the eigenvectors,

$$S\boldsymbol{w}_\alpha = \lambda_\alpha \boldsymbol{w}_\alpha, \qquad \alpha = 1, 2, \cdots, p \tag{9.59}$$

that the left-hand side may be written as

$$S\boldsymbol{w}_\alpha = \frac{1}{n}\sum_{i=1}^{n} \boldsymbol{x}_i \boldsymbol{x}_i^T \boldsymbol{w}_\alpha = \frac{1}{n}\sum_{i=1}^{n} (\boldsymbol{x}_i^T \boldsymbol{w}_\alpha)\boldsymbol{x}_i. \tag{9.60}$$

The relation between the eigenvalues of the sample variance-covariance matrix S and the eigenvectors can therefore be rewritten as

$$\frac{1}{n}\sum_{i=1}^{n} (\boldsymbol{x}_i^T \boldsymbol{w}_\alpha)\boldsymbol{x}_i = \lambda_\alpha \boldsymbol{w}_\alpha. \tag{9.61}$$

The α-th eigenvector $\boldsymbol{w}_\alpha$, if its eigenvalue is not 0, can then be expressed as a linear combination of the p-dimensional data, as follows:

$$\boldsymbol{w}_\alpha = \sum_{i=1}^{n} c_{i\alpha} \boldsymbol{x}_i, \qquad c_{i\alpha} = \frac{1}{n\lambda_\alpha} \boldsymbol{x}_i^T \boldsymbol{w}_\alpha, \qquad i = 1, 2, \cdots, n. \tag{9.62}$$

We next organize all of the above in matrix and vector form. For this purpose, we denote as X the $n \times p$ data matrix having the centered n p-dimensional data as rows. X and its transposed matrix X^T are then given as

$$X = \begin{pmatrix} \boldsymbol{x}_1^T \\ \boldsymbol{x}_2^T \\ \vdots \\ \boldsymbol{x}_n^T \end{pmatrix}, \qquad X^T = (\boldsymbol{x}_1, \boldsymbol{x}_2, \cdots, \boldsymbol{x}_n). \tag{9.63}$$

The sample variance-covariance matrix is, using the data matrix, expressed as

$$S = \frac{1}{n}\sum_{i=1}^{n} \boldsymbol{x}_i \boldsymbol{x}_i^T = \frac{1}{n} X^T X, \tag{9.64}$$

and the eigenvector $\boldsymbol{w}_\alpha$ in (9.62) may be written as

$$\boldsymbol{w}_\alpha = X^T \boldsymbol{c}_\alpha, \tag{9.65}$$

where $\boldsymbol{c}_\alpha$ is the n-dimensional coefficient vector $\boldsymbol{c}_\alpha = (c_{1\alpha}, c_{2\alpha}, \cdots, c_{n\alpha})^T$.

When we substitute the eigenvector $\boldsymbol{w}_\alpha$ representing the linear combination of the data into (9.59), we then have

$$\frac{1}{n} X^T X X^T \boldsymbol{c}_\alpha = \lambda_\alpha X^T \boldsymbol{c}_\alpha. \tag{9.66}$$

If we left multiply both sides of this equation by data matrix X, we have

$$X X^T X X^T \boldsymbol{c}_\alpha = n\lambda_\alpha X X^T \boldsymbol{c}_\alpha. \tag{9.67}$$

It may be seen from (9.63) that XX^T is an n-dimensional symmetric matrix with elements comprising the inner products of the data as follows:

$$K \equiv XX^T = \begin{pmatrix} \boldsymbol{x}_1^T\boldsymbol{x}_1 & \boldsymbol{x}_1^T\boldsymbol{x}_2 & \cdots & \boldsymbol{x}_1^T\boldsymbol{x}_n \\ \boldsymbol{x}_2^T\boldsymbol{x}_1 & \boldsymbol{x}_2^T\boldsymbol{x}_2 & \cdots & \boldsymbol{x}_2^T\boldsymbol{x}_n \\ \vdots & \vdots & \ddots & \vdots \\ \boldsymbol{x}_n^T\boldsymbol{x}_1 & \boldsymbol{x}_n^T\boldsymbol{x}_2 & \cdots & \boldsymbol{x}_n^T\boldsymbol{x}_n \end{pmatrix}. \tag{9.68}$$

Equation (9.67) may accordingly be expressed as

$$K^2 \boldsymbol{c}_\alpha = n\lambda_\alpha K \boldsymbol{c}_\alpha. \tag{9.69}$$

The problem of finding the eigenvector $\boldsymbol{c}_\alpha$ corresponding to the α-th non-zero eigenvalue λ_α is thus reduced to the following eigenvalue problem:

$$K\boldsymbol{c}_\alpha = n\lambda_\alpha \boldsymbol{c}_\alpha. \tag{9.70}$$

When we substitute the eigenvector $\boldsymbol{c}_\alpha$ corresponding to the non-zero eigenvalue λ_α of the symmetric matrix K into (9.62) or (9.65), we obtain the coefficient vector $\boldsymbol{w}_\alpha = X^T \boldsymbol{c}_\alpha$ for the α-th principal component. Since $\boldsymbol{w}_\alpha$ is an eigenvector of length 1, it is assumed that the eigenvector $\boldsymbol{c}_\alpha$ satisfies the constraint

$$1 = \boldsymbol{w}_\alpha^T \boldsymbol{w}_\alpha = \boldsymbol{c}_\alpha^T X X^T \boldsymbol{c}_\alpha = n\lambda_\alpha \boldsymbol{c}_\alpha^T \boldsymbol{c}_\alpha. \tag{9.71}$$

In summary, the above shows that we can obtain the principal components by replacing the eigenvalue problem of the sample variance-covariance matrix S with the eigenvalue problem of matrix K composed of the inner products of the data. In the next subsection, we apply this eigenvalue problem replacement to data mapped to a higher-dimensional space.

9.4.2 *Mapping to a Higher-Dimensional Space*

Nonlinearization of the support vector machine, as discussed in Section 8.3, was achieved by mapping the observed data with nonlinear structure to a higher-dimensional space known as the feature space, applying the linear support vector theory to obtain the separating hyperplane, and constructing a nonlinear discriminant function in the original input space. In this subsection, we now apply this method of finding nonlinear structures to PCA. Essentially, this consists of mapping n p-dimensional observed data $\boldsymbol{x}_1, \boldsymbol{x}_2, \cdots, \boldsymbol{x}_n$ in an input space to a higher-dimensional space, where PCA is performed with linear combinations of variables (Figure 9.7).

In general, a nonlinear function that transforms p-dimensional data $\boldsymbol{x} = (x_1, x_2, \cdots, x_p)^T$ in an input space to a higher r-dimensional space is expressed in vector form as

$$\boldsymbol{\Phi}(\boldsymbol{x}) \equiv (\phi_1(\boldsymbol{x}), \phi_2(\boldsymbol{x}), \cdots, \phi_r(\boldsymbol{x}))^T, \tag{9.72}$$

where the components $\phi_j(\boldsymbol{x})$ are real-valued functions of p variables. The data as mapped from the n p-dimensional observed data to the r-dimensional feature space by the vector $\boldsymbol{\Phi}(\boldsymbol{x})$ with nonlinear functions $\phi_j(\boldsymbol{x})$ are then

$$\boldsymbol{\Phi}(\boldsymbol{x}_1),\ \boldsymbol{\Phi}(\boldsymbol{x}_2),\ \cdots,\ \boldsymbol{\Phi}(\boldsymbol{x}_n), \tag{9.73}$$

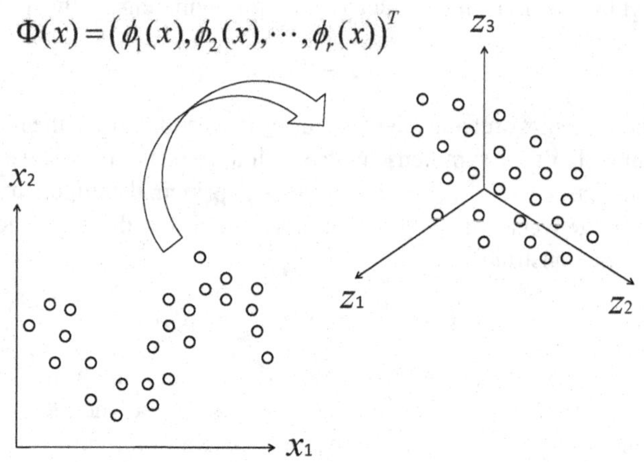

Figure 9.7 *Mapping the observed data with nonlinear structure to a higher-dimensional feature space, where PCA is performed with linear combinations of variables* z_1, z_2, z_3.

where $\boldsymbol{\Phi}(\boldsymbol{x}_i) = (\phi_1(\boldsymbol{x}_i), \phi_2(\boldsymbol{x}_i), \cdots \phi_r(\boldsymbol{x}_i))^T$. We perform the PCA for the n r-dimensional data in this feature space.

First, we move the origin to the centroid of the r-dimensional data dispersed in the feature space. For this purpose, we obtain the sample mean vector

$$\overline{\boldsymbol{\Phi}} = \frac{1}{n}\sum_{i=1}^{n} \boldsymbol{\Phi}(\boldsymbol{x}_i), \tag{9.74}$$

and perform the data centering as follows:

$$\boldsymbol{\Phi}_c(\boldsymbol{x}_i) = \boldsymbol{\Phi}(\boldsymbol{x}_i) - \overline{\boldsymbol{\Phi}}, \qquad i = 1, 2, \cdots, n. \tag{9.75}$$

The sample mean vector for the centered data is thus $\mathbf{0}$, that is,

$$\frac{1}{n}\sum_{i=1}^{n} \boldsymbol{\Phi}_c(\boldsymbol{x}_i) = \frac{1}{n}\sum_{i=1}^{n} \left\{\boldsymbol{\Phi}(\boldsymbol{x}_i) - \overline{\boldsymbol{\Phi}}\right\} = \mathbf{0}. \tag{9.76}$$

We next denote the $n \times r$ data matrix having the centered data as row vectors as

$$Z_c = (\boldsymbol{\Phi}_c(\boldsymbol{x}_1), \boldsymbol{\Phi}_c(\boldsymbol{x}_2), \cdots, \boldsymbol{\Phi}_c(\boldsymbol{x}_n))^T. \tag{9.77}$$

The sample variance-covariance matrix is then given by

$$S_c = \frac{1}{n}\sum_{i=1}^{n} \boldsymbol{\Phi}_c(\boldsymbol{x}_i)\boldsymbol{\Phi}_c(\boldsymbol{x}_i)^T = \frac{1}{n}Z_c^T Z_c. \tag{9.78}$$

By the method described in the previous subsection, we next replace the problem of finding eigenvalues of the sample variance-covariance matrix S_c with the problem of finding eigenvalues of the matrix based on the inner products of the data in the feature space.

In general, from the relationship between the α-th eigenvalue λ_α^F of matrix S_c and the corresponding eigenvector $\boldsymbol{w}_\alpha^F$,

$$S_c \boldsymbol{w}_\alpha^F = \lambda_\alpha^F \boldsymbol{w}_\alpha^F, \tag{9.79}$$

the eigenvector on the basis of the centered data in the feature space can be expressed as

$$\boldsymbol{w}_\alpha^F = Z_c^T \boldsymbol{c}_\alpha^F, \quad Z_c^T = (\boldsymbol{\Phi}_c(\boldsymbol{x}_1), \boldsymbol{\Phi}_c(\boldsymbol{x}_2), \cdots, \boldsymbol{\Phi}_c(\boldsymbol{x}_n)), \tag{9.80}$$

where $\boldsymbol{c}_\alpha^F$ denotes the coefficient vector (see (9.62) and (9.65)).

When we substitute the eigenvector expressed in this manner into (9.79), we obtain

$$\frac{1}{n} Z_c^T Z_c Z_c^T \boldsymbol{c}_\alpha^F = \lambda_\alpha^F Z_c^T \boldsymbol{c}_\alpha^F. \tag{9.81}$$

If we left multiply both sides of this equation by the data matrix Z_c, we then have

$$Z_c Z_c^T Z_c Z_c^T \boldsymbol{c}_\alpha^F = n\lambda_\alpha^F Z_c Z_c^T \boldsymbol{c}_\alpha^F. \tag{9.82}$$

It may be seen that $Z_c Z_c^T$ is an n-dimensional symmetric matrix composed of the inner products of the centered data in the feature space, expressed as

$$K_c \equiv Z_c Z_c^T \tag{9.83}$$

$$= \begin{pmatrix} \boldsymbol{\Phi}_c(\boldsymbol{x}_1)^T\boldsymbol{\Phi}_c(\boldsymbol{x}_1) & \boldsymbol{\Phi}_c(\boldsymbol{x}_1)^T\boldsymbol{\Phi}_c(\boldsymbol{x}_2) & \cdots & \boldsymbol{\Phi}_c(\boldsymbol{x}_1)^T\boldsymbol{\Phi}_c(\boldsymbol{x}_n) \\ \boldsymbol{\Phi}_c(\boldsymbol{x}_2)^T\boldsymbol{\Phi}_c(\boldsymbol{x}_1) & \boldsymbol{\Phi}_c(\boldsymbol{x}_2)^T\boldsymbol{\Phi}_c(\boldsymbol{x}_2) & \cdots & \boldsymbol{\Phi}_c(\boldsymbol{x}_2)^T\boldsymbol{\Phi}_c(\boldsymbol{x}_n) \\ \vdots & \vdots & \ddots & \vdots \\ \boldsymbol{\Phi}_c(\boldsymbol{x}_n)^T\boldsymbol{\Phi}_c(\boldsymbol{x}_1) & \boldsymbol{\Phi}_c(\boldsymbol{x}_n)^T\boldsymbol{\Phi}_c(\boldsymbol{x}_2) & \cdots & \boldsymbol{\Phi}_c(\boldsymbol{x}_n)^T\boldsymbol{\Phi}_c(\boldsymbol{x}_n) \end{pmatrix},$$

Equation (9.82) can therefore be expressed as

$$K_c^2 \boldsymbol{c}_\alpha^F = n\lambda_\alpha^F K_c \boldsymbol{c}_\alpha^F, \qquad \alpha = 1, 2, \cdots, n. \tag{9.84}$$

In this way, the problem of finding the eigenvector $\boldsymbol{c}_\alpha^F$ corresponding to the α-th-largest eigenvalue λ_α^F ($\neq 0$) thus becomes the problem of finding the eigenvalue of the n-dimensional symmetric matrix K_c as follows:

$$K_c \boldsymbol{c}_\alpha^F = n\lambda_\alpha^F \boldsymbol{c}_\alpha^F. \tag{9.85}$$

By substituting the eigenvector $\boldsymbol{c}_\alpha^F$ into (9.80), we obtain the eigenvector corresponding to the α-th non-zero eigenvalue of the sample variance-covariance matrix S_c based on the centered data as follows:

$$\boldsymbol{w}_\alpha^F = Z_c^T \boldsymbol{c}_\alpha^F = \sum_{i=1}^n c_{i\alpha}^F \boldsymbol{\Phi}_c(\boldsymbol{x}_i). \tag{9.86}$$

Here, we require that the eigenvectors of matrix K_c satisfy the constraint

$$1 = \boldsymbol{w}_\alpha^{F^T} \boldsymbol{w}_\alpha^F = (Z_c^T \boldsymbol{c}_\alpha^F)^T (Z_c^T \boldsymbol{c}_\alpha^F) = \boldsymbol{c}_\alpha^{F^T} Z_c Z_c^T \boldsymbol{c}_\alpha^F = n\lambda_\alpha^F \boldsymbol{c}_\alpha^{F^T} \boldsymbol{c}_\alpha^F. \tag{9.87}$$

From the above, it follows that the α-th principal component constructed on the basis of the centered data in the feature space is given by

$$y_\alpha^F = \boldsymbol{w}_\alpha^{F^T} \boldsymbol{\Phi}_c(\boldsymbol{x}) = \sum_{i=1}^n c_{i\alpha}^F \boldsymbol{\Phi}_c(\boldsymbol{x}_i)^T \boldsymbol{\Phi}_c(\boldsymbol{x}). \tag{9.88}$$

In PCA, as also described for support vectors in Section 8.3.2, high-dimensional data in the input space requires data mapping to an extremely high-dimensional feature space, and it is therefore difficult to compute the inner products $\boldsymbol{\Phi}_c(\boldsymbol{x}_j)^T \boldsymbol{\Phi}_c(\boldsymbol{x}_k)$ of the data in (9.83). The kernel method provides an effective technique of circumventing this difficulty.

9.4.3 Kernel Methods

With the nonlinearization of principal components, it becomes essential to replace the matrix defined by the inner products of the data with a kernel function. In short, we circumvent the problem of computing inner products of the transformed extremely high-dimensional data by using a kernel function as follows:

$$K_c(\boldsymbol{x}_j, \boldsymbol{x}_k) = \boldsymbol{\Phi}_c(\boldsymbol{x}_j)^T \boldsymbol{\Phi}_c(\boldsymbol{x}_k). \tag{9.89}$$

In effect, we obtain the value of the inner product on the right-hand side as the computable value of the kernel function on the left-hand side. If, for example, we use the Gaussian kernel, which is a highly representative kernel function, the expression is then

$$K_c(\boldsymbol{x}_j, \boldsymbol{x}_k) = \exp\left\{-\frac{\|\boldsymbol{x}_j - \boldsymbol{x}_k\|^2}{2\sigma^2}\right\} = \boldsymbol{\Phi}_c(\boldsymbol{x}_j)^T \boldsymbol{\Phi}_c(\boldsymbol{x}_k). \qquad (9.90)$$

Computation of the inner products on the right-hand side of this equation would be difficult because of the extremely high dimensionalization, but the kernel function is relatively easy to compute because, as may be seen from its terms, it is based on the inner products of the p-dimensional data in the input space.

In using the kernel function, matrix K_c of (9.83) defined by the inner products of the high-dimensional data in the feature space can be expressed by the following matrix using the computable kernel function for its elements:

$$K_c \equiv Z_c Z_c^T = \begin{pmatrix} K_c(\boldsymbol{x}_1, \boldsymbol{x}_1) & K_c(\boldsymbol{x}_1, \boldsymbol{x}_2) & \cdots & K_c(\boldsymbol{x}_1, \boldsymbol{x}_n) \\ K_c(\boldsymbol{x}_2, \boldsymbol{x}_1) & K_c(\boldsymbol{x}_2, \boldsymbol{x}_2) & \cdots & K_c(\boldsymbol{x}_2, \boldsymbol{x}_n) \\ \vdots & \vdots & \ddots & \vdots \\ K_c(\boldsymbol{x}_n, \boldsymbol{x}_1) & K_c(\boldsymbol{x}_n, \boldsymbol{x}_2) & \cdots & K_c(\boldsymbol{x}_n, \boldsymbol{x}_n) \end{pmatrix}. \qquad (9.91)$$

Moreover, the α-th principal component given by (9.88) can also be expressed, in terms of K_c, as

$$y_\alpha = \boldsymbol{w}_\alpha^{F^T} \boldsymbol{\Phi}_c(\boldsymbol{x}) = \sum_{i=1}^{n} c_{i\alpha}^F \boldsymbol{\Phi}_c(\boldsymbol{x}_i)^T \boldsymbol{\Phi}_c(\boldsymbol{x}) = \sum_{i=1}^{n} c_{i\alpha}^F K_c(\boldsymbol{x}_i, \boldsymbol{x}). \qquad (9.92)$$

In the feature space, the principal components are based on linearity. In the input space, however, it may be seen that the kernel function is nonlinear and therefore produces nonlinear principal components. If, for example, we express (9.92) by using a polynomial as the kernel function, we then have

$$y_\alpha = \sum_{i=1}^{n} c_{i\alpha}^F \left(\boldsymbol{x}_i^T \boldsymbol{x} + 1\right)^5, \qquad (9.93)$$

which clearly presents data as nonlinear principal components in the observed input space.

Thus far in our discussion, we have mapped data observed in the input space onto the feature space, substituted the kernel function for the inner products of the centered data (9.75), and derived the nonlinear principal components. This leaves the problem, however, of computing the sample mean vector of (9.74), which is essential for the data centering process. To circumvent this problem, it is necessary to express the kernel function in terms of the centered data of (9.75), using the equation which replaced the inner products of the r-dimensional data $\boldsymbol{\Phi}(\boldsymbol{x}_1), \boldsymbol{\Phi}(\boldsymbol{x}_2), \cdots, \boldsymbol{\Phi}(\boldsymbol{x}_n)$ transformed to the higher-dimensional feature space with the kernel function

$$K(\boldsymbol{x}_j, \boldsymbol{x}_k) = \boldsymbol{\Phi}(\boldsymbol{x}_j)^T \boldsymbol{\Phi}(\boldsymbol{x}_k). \tag{9.94}$$

The relation is given by the following equation:

$$\begin{aligned} K_c(\boldsymbol{x}_j, \boldsymbol{x}_k) &= \boldsymbol{\Phi}_c(\boldsymbol{x}_j)^T \boldsymbol{\Phi}_c(\boldsymbol{x}_k) \\ &= \left(\boldsymbol{\Phi}(\boldsymbol{x}_j) - \frac{1}{n} \sum_{r=1}^{n} \boldsymbol{\Phi}(\boldsymbol{x}_r) \right)^T \left(\boldsymbol{\Phi}(\boldsymbol{x}_k) - \frac{1}{n} \sum_{s=1}^{n} \boldsymbol{\Phi}(\boldsymbol{x}_s) \right) \\ &= K(\boldsymbol{x}_j, \boldsymbol{x}_k) - \frac{1}{n} \sum_{r=1}^{n} K(\boldsymbol{x}_r, \boldsymbol{x}_k) - \frac{1}{n} \sum_{s=1}^{n} K(\boldsymbol{x}_j, \boldsymbol{x}_s) \\ &\quad + \frac{1}{n^2} \sum_{r=1}^{n} \sum_{s=1}^{n} K(\boldsymbol{x}_r, \boldsymbol{x}_s). \end{aligned} \tag{9.95}$$

We also need to replace matrix K_c in (9.91) defined by the inner products based on the centered data, with the matrix K having as its (j, k)-th element the kernel function defined by (9.94). For this purpose, we define the $n \times r$ data matrix

$$Z = (\boldsymbol{\Phi}(\boldsymbol{x}_1), \boldsymbol{\Phi}(\boldsymbol{x}_2), \cdots, \boldsymbol{\Phi}(\boldsymbol{x}_n))^T, \tag{9.96}$$

having the r-dimensional data of the feature space as rows. Using this data matrix, we can then express the sample mean vector as

$$\overline{\boldsymbol{\Phi}} = \frac{1}{n} \sum_{i=1}^{n} \boldsymbol{\Phi}(\boldsymbol{x}_i) = \frac{1}{n} Z^T \boldsymbol{1}_n, \tag{9.97}$$

where $\boldsymbol{1}_n$ is an n-dimensional vector with elements 1. In addition, the $n \times r$ data matrix Z_c having the centered data defined by (9.75) as rows can be expressed as

$$Z_c = \begin{pmatrix} \boldsymbol{\Phi}_c(\boldsymbol{x}_1)^T \\ \boldsymbol{\Phi}_c(\boldsymbol{x}_2)^T \\ \vdots \\ \boldsymbol{\Phi}_c(\boldsymbol{x}_n)^T \end{pmatrix} = \begin{pmatrix} \boldsymbol{\Phi}(\boldsymbol{x}_1)^T \\ \boldsymbol{\Phi}(\boldsymbol{x}_2)^T \\ \vdots \\ \boldsymbol{\Phi}(\boldsymbol{x}_n)^T \end{pmatrix} - \begin{pmatrix} 1 \\ 1 \\ \vdots \\ 1 \end{pmatrix} \overline{\boldsymbol{\Phi}}^T = Z - \frac{1}{n}\mathbf{1}_n\mathbf{1}_n^T Z. \quad (9.98)$$

Incorporating this finding, the n-dimensional symmetric matrix composed of the inner products of the centered data is then given by

$$\begin{aligned} K_c = Z_c Z_c^T &= \left(Z - \frac{1}{n}\mathbf{1}_n\mathbf{1}_n^T Z\right)\left(Z - \frac{1}{n}\mathbf{1}_n\mathbf{1}_n^T Z\right)^T \\ &= \left(I_n - \frac{1}{n}\mathbf{1}_n\mathbf{1}_n^T\right) ZZ^T \left(I_n - \frac{1}{n}\mathbf{1}_n\mathbf{1}_n^T\right)^T, \end{aligned} \quad (9.99)$$

where I_n is an n-dimensional identity matrix.

Accordingly, the n-dimensional symmetric matrix K_c having the inner products based on the centered data being replaced with the kernel function can be expressed as

$$K_c = \left(I_n - \frac{1}{n}\mathbf{1}_n\mathbf{1}_n^T\right) K \left(I_n - \frac{1}{n}\mathbf{1}_n\mathbf{1}_n^T\right), \quad (9.100)$$

where $K = ZZ^T$ is an n-dimensional symmetric matrix having the inner products based on the data transformed to the feature space replaced by the kernel function.

Nonlinear PCA incorporating the kernel method can be summarized essentially as follows:

Process of nonlinear principal component analysis Nonlinear principal components based on n p-dimensional observed data $\boldsymbol{x}_1, \boldsymbol{x}_2, \cdots, \boldsymbol{x}_n$ are constructed by the following process.

(1) Let the n r-dimensional data transformed to a higher-dimensional feature space by nonlinear mapping $\boldsymbol{\Phi}(\boldsymbol{x}) = (\phi_1(\boldsymbol{x}), \phi_2(\boldsymbol{x}), \cdots, \phi_r(\boldsymbol{x}))^T$ from the observed data in the input space be denoted as

$$\boldsymbol{\Phi}(\boldsymbol{x}_1),\ \boldsymbol{\Phi}(\boldsymbol{x}_2),\ \cdots,\ \boldsymbol{\Phi}(\boldsymbol{x}_n), \quad (9.101)$$

where $\boldsymbol{\Phi}(\boldsymbol{x}_i) = (\phi_1(\boldsymbol{x}_i), \phi_2(\boldsymbol{x}_i), \cdots, \phi_r(\boldsymbol{x}_i))^T$.

(2) Replace the inner products of the r-dimensional data in the feature space with the kernel function as follows:

$$K(\boldsymbol{x}_j, \boldsymbol{x}_k) = \boldsymbol{\Phi}(\boldsymbol{x}_j)^T \boldsymbol{\Phi}(\boldsymbol{x}_k). \quad (9.102)$$

(3) Let K be the kernel matrix of order n having the kernel function $K(\boldsymbol{x}_j, \boldsymbol{x}_k)$ as its (j,k)-th element, and obtain the kernel matrix K_c (9.91) of order n, having as its elements the inner products of the centered data in (9.75), by

$$K_c = \left(I_n - \frac{1}{n}\mathbf{1}_n\mathbf{1}_n^T \right) K \left(I_n - \frac{1}{n}\mathbf{1}_n\mathbf{1}_n^T \right). \tag{9.103}$$

(4) Solve the following eigenvalue problem for the symmetric matrix K_c of order n defined in step (3) under the constraint $n\lambda^F \boldsymbol{c}^{F^T} \boldsymbol{c}^F = 1$.

$$K_c \boldsymbol{c}^F = n\lambda^F \boldsymbol{c}^F. \tag{9.104}$$

(5) With the eigenvalues arranged in descending order according to magnitude and the eigenvector corresponding to the α-th non-zero eigenvalue denoted by $\boldsymbol{c}_\alpha^F = (c_{1\alpha}^F, c_{2\alpha}^F, \cdots, c_{n\alpha}^F)^T$, the α-th principal component is then given by

$$y_\alpha = \sum_{i=1}^{n} c_{i\alpha}^F K_c(\boldsymbol{x}_i, \boldsymbol{x}), \tag{9.105}$$

where $K_c(\boldsymbol{x}_i, \boldsymbol{x})$ having the inner products of the centered data replaced with the kernel function is, from (9.102), expressed as

$$K_c(\boldsymbol{x}_i, \boldsymbol{x}) = K(\boldsymbol{x}_i, \boldsymbol{x}) - \frac{1}{n}\sum_{r=1}^{n} K(\boldsymbol{x}_r, \boldsymbol{x}) - \frac{1}{n}\sum_{s=1}^{n} K(\boldsymbol{x}_i, \boldsymbol{x}_s) + \frac{1}{n^2}\sum_{r=1}^{n}\sum_{s=1}^{n} K(\boldsymbol{x}_r, \boldsymbol{x}_s). \tag{9.106}$$

Exercises

9.1 Show that when projecting n observed p-dimensional data onto the axis $y = \boldsymbol{w}^T\boldsymbol{x}$, the variance of the projected data is given by $s_y^2 = \boldsymbol{w}^T S \boldsymbol{w}$, where S is the sample variance-covariance matrix based on the observed data.

9.2 Show that the sum of the variances of the original variables $x_1, x_2, \cdots, x_p$ equals the sum of the variances of the constructed p principal components $y_1, y_2, \cdots, y_p$.

9.3 Show that the correlation between the i-th principal component y_i and the j-th variable x_j is given by (9.22).

9.4 Let X be an $n \times p$ data matrix of rank p. Matrix X can then be decomposed using the $n \times p$ matrix V, $p \times p$ diagonal matrix L, and $p \times p$ matrix W, as $X = VLW^T$, where V, W and L are matrices having the properties given in (9.46).

(a) Show that the j-th column of the $n \times p$ matrix V expressed as

$$\boldsymbol{v}_j = \frac{1}{\sqrt{\ell_j}} X\boldsymbol{w}_j = \frac{1}{\sqrt{n\lambda_j}} X\boldsymbol{w}_j$$

gives the score of the principal component standardized to $1/n$ when the n centered data are projected onto the j-th principal component.

(b) Show that when using the first m principal components, any given data $\boldsymbol{x}_i$ can be approximated as

$$\boldsymbol{x}_i \approx \boldsymbol{x}_i^{(m)} = \overline{\boldsymbol{x}} + \sum_{k=1}^{m} \{(\boldsymbol{x}_i - \overline{\boldsymbol{x}})^T \boldsymbol{w}_k\} \boldsymbol{w}_k.$$

9.5 Show that the $n \times r$ data matrix Z_c in (9.98) can be expressed as

$$Z_c = Z - \frac{1}{n} \mathbf{1}_n \mathbf{1}_n^T Z,$$

where Z is the $n \times r$ data matrix defined by (9.96).

9.6 Show that the n-dimensional symmetric matrix K_c in (9.100) can be expressed as

$$K_c = \left(I_n - \frac{1}{n} \mathbf{1}_n \mathbf{1}_n^T \right) K \left(I_n - \frac{1}{n} \mathbf{1}_n \mathbf{1}_n^T \right),$$

where $K = ZZ^T$ is an n-dimensional symmetric matrix having the inner products based on the data transformed to the feature space replaced by the kernel function.

Chapter 10

Clustering

Clustering, also referred to as cluster analysis, consists of partitioning a large number of objects characterized by multiple variables into groups (clusters) based on some indicator of mutual similarity between the objects. In discriminant analysis, the classification rule is constructed from data having known group memberships. In cluster analysis, in contrast, grouping data (i.e., objects) on the basis of some criterion of similarity is used to discern relationships between the objects.

Clustering may be either hierarchical or nonhierarchical. Hierarchical clustering proceeds in successive steps from smaller to larger clusters, which can be directly observed visually by humans. Nonhierarchical clustering, in contrast, consists of progressively refining the data partitions to obtain a given number of clusters. In this chapter, we first consider the process and characteristics of hierarchical clustering and the basic requirements for its application, with typical examples. We then turn to nonhierarchical clustering and the k-means, self-organizing map, and mixture-model clustering that are commonly used for this purpose in many fields, and consider the clustering procedure for each of these techniques.

10.1 Hierarchical Clustering

Hierarchical clustering essentially consists of progressively organizing all of the candidate objects into clusters comprising mutually similar objects as determined by some measure of interobject and intercluster similarity, proceeding in succession from the formation of small clusters containing just two objects to large clusters containing many objects. It is characteristic of this procedure that the clusters formed in each step can be graphically displayed in *tree diagrams* referred to as *dendrograms*. In the following, we discuss the techniques and measures of interobject and intercluster similarity, their application, and the cluster formation process.

10.1.1 Interobject Similarity

Cluster analysis begins with a quantitative measurement of similarity between candidate objects among a large number of objects characterized by several variables. Consider the objects

$$A: \boldsymbol{a} = (a_1, a_2, \cdots, a_p)^T, \qquad B: \boldsymbol{b} = (b_1, b_2, \cdots, b_p)^T \quad (10.1)$$

that have been observed in terms of p variables $\boldsymbol{x} = (x_1, x_2, \cdots, x_p)^T$. As measures of similarity (and dissimilarity) between objects A and B, the following modes of distance may be applied:

(1) Euclidean distance

$$d(A, B) = \sqrt{(a_1 - b_1)^2 + (a_2 - b_2)^2 + \cdots + (a_p - b_p)^2}.$$

(2) Standardized Euclidean distance

$$d_s(A, B) = \sqrt{\left(\frac{a_1 - b_1}{s_1}\right)^2 + \left(\frac{a_2 - b_2}{s_2}\right)^2 + \cdots + \left(\frac{a_p - b_p}{s_p}\right)^2},$$

where s_i is a standard deviation of x_i.

(3) L_1 distance

$$d(A, B) = |a_1 - b_1| + |a_2 - b_2| + \cdots + |a_p - b_p|.$$

(4) Minkowski distance

$$d(A, B) = \left\{|a_1 - b_1|^m + |a_2 - b_2|^m + \cdots + |a_p - b_p|^m\right\}^{1/m}.$$

Euclidean distance (1) usually serves as the measure of similarity, but if the variables characterizing the objects are in different units, then it is commonly replaced by standardized Euclidean distance (2), which is based on the standard deviations of these variables over all objects in the dataset. Note that Minkowski distance (4) reduces to the L_1 distance (3) when $m = 1$ and to the Euclidean distance (1) when $m = 2$. Various other modes have been proposed, such as correlation-coefficient and Mahalanobis distances, but the Euclidean distance $d(A, B)$ in (1) or squared Euclidean distance $d^2(A, B)$ is generally used as the basic quantitative measure of interobject similarity.

10.1.2 Intercluster Distance

Based on the distances found between the data in the dataset, we begin to form a single cluster containing the objects separated by the shortest interobject distance. It is then necessary to determine the distance from that cluster to the other objects in the set. In forming further clusters, moreover, it is also necessary to determine the intercluster distances. Here we describe five techniques used for this purpose: the *single linkage* (or minimum distance or nearest neighbor), *complete linkage* (or maximum distance or furthest neighbor), *average* (or average distance), *centroid*, and *median* linkage techniques.

Consider the cluster analysis of a dataset $S = \{\boldsymbol{x}_1, \boldsymbol{x}_2, \cdots, \boldsymbol{x}_n\}$ comprising n p-dimensional observed data and let

$$d_{ij} = d(\boldsymbol{x}_i, \boldsymbol{x}_j), \qquad (i \neq j), \qquad i, j = 1, 2, \cdots, n \tag{10.2}$$

denote the distance between any two objects. Let C_α and C_β be two clusters formed on the basis of distance between objects. Then $d(C_\alpha, C_\beta)$, the intercluster distance between C_α and C_β, is determined by the above-mentioned five linkage techniques as follows (see Figure 10.1):

(1) *Single linkage* (minimum distance): Calculate distance between each pair of one object from C_α and one object from C_β, and take the smallest distance as the intercluster distance

$$d(C_\alpha, C_\beta) = \min_{i,j} \left\{ d(\boldsymbol{x}_i^{(\alpha)}, \boldsymbol{x}_j^{(\beta)});\ \boldsymbol{x}_i^{(\alpha)} \in C_\alpha,\ \boldsymbol{x}_j^{(\beta)} \in C_\beta \right\}. \tag{10.3}$$

(2) *Complete linkage* (maximum distance): Calculate distance between each pair of one object from C_α and one object from C_β, and take the largest distance as the intercluster distance

$$d(C_\alpha, C_\beta) = \max_{i,j} \left\{ d(\boldsymbol{x}_i^{(\alpha)}, \boldsymbol{x}_j^{(\beta)});\ \boldsymbol{x}_i^{(\alpha)} \in C_\alpha,\ \boldsymbol{x}_j^{(\beta)} \in C_\beta \right\}. \tag{10.4}$$

(3) *Average linkage*: Calculate distance between each pair of one object from C_α and one object from C_β, and take the average of the distances as the intercluster distance

$$d(C_\alpha, C_\beta) = \frac{1}{n_\alpha n_\beta} \sum_{i=1}^{n_\alpha} \sum_{j=1}^{n_\beta} d(\boldsymbol{x}_i^{(\alpha)}, \boldsymbol{x}_j^{(\beta)}),\ \ \boldsymbol{x}_i^{(\alpha)} \in C_\alpha,\ \boldsymbol{x}_j^{(\beta)} \in C_\beta, \tag{10.5}$$

where n_α and n_β are the number of data in clusters C_α and C_β, respectively.

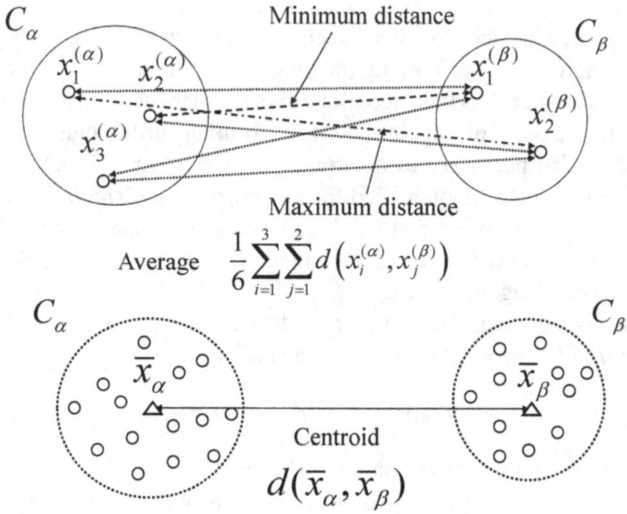

Figure 10.1 *Intercluster distances: Single linkage (minimum distance), complete linkage (maximum distance), average linkage, centroid linkage.*

(4) *Centroid linkage*: Take the distance between the cluster centroids (their sample mean vectors) as the intercluster distance

$$d(C_\alpha, C_\beta) = d(\overline{\boldsymbol{x}}_\alpha, \overline{\boldsymbol{x}}_\beta), \quad \overline{\boldsymbol{x}}_\alpha = \frac{1}{n_\alpha}\sum_{i=1}^{n_\alpha} \boldsymbol{x}_i^{(\alpha)}, \quad \overline{\boldsymbol{x}}_\beta = \frac{1}{n_\beta}\sum_{i=1}^{n_\beta} \boldsymbol{x}_i^{(\beta)}. \quad (10.6)$$

(5) *Median linkage*: As in centroid linkage, the distance between the cluster centroids is taken as the intercluster distance. However, the two techniques differ in terms of the definition of the centroid, specifically in the calculation of the centroid of a post-merger cluster. In centroid linkage, the post-merger cluster centroid can be determined from the pre-merger cluster centroids by assigning a weighting according to the number of data in the pre-merger clusters, as $(n_\alpha\overline{\boldsymbol{x}}_\alpha + n_\beta\overline{\boldsymbol{x}}_\beta)/(n_\alpha + n_\beta)$. In median linkage, the center point $(\overline{\boldsymbol{x}}_\alpha + \overline{\boldsymbol{x}}_\beta)/2$ between the pre-merger centroids is simply taken as the post-merger centroid, which in effect may be regarded as a means for preventing post-merger centroid bias in cases involving large differences in the sizes of the pre-merger clusters.

When single, complete, or average linkage is used to determine the intercluster distance, any of the distance modes described in the previous section can be used as the measure of interobject distance. If the intercluster distance is determined by centroid or median linkage, or by Ward's method described in Section 10.1.4, only the (squared) Euclidean distance can be used for the interobject distance. In the following, we describe the cluster formation procedure based on interobject and intercluster distances.

10.1.3 Cluster Formation Process

Let us denote by $D = (d_{ij})$ the $n \times n$ matrix having as its (i, j)-th element the distance $d_{ij} = d(\boldsymbol{x}_i, \boldsymbol{x}_j)$ representing the similarity measurement between given data $\boldsymbol{x}_i$ and $\boldsymbol{x}_j$ contained in the set S of n p-dimensional data. Matrix D is referred to as the *distance matrix*. It is a symmetric matrix, as each of its diagonal elements is the distance between a datum and itself, and thus $d_{ii} = 0$, and in all cases $d_{ij} = d(\boldsymbol{x}_i, \boldsymbol{x}_j) = d(\boldsymbol{x}_j, \boldsymbol{x}_i) = d_{ji}$. For simplicity, let us consider the following distance matrix constructed for five objects $\{a, b, c, d, e\}$ and showing the distance between them, such as 8 between objects a and b.

$$\begin{array}{c} \\ a \\ b \\ c \\ d \\ e \end{array} \begin{array}{c} \begin{array}{ccccc} a & b & c & d & e \end{array} \\ \begin{pmatrix} 0 & 8 & 3 & 7 & 10 \\ 8 & 0 & 6 & 5 & 12 \\ 3 & 6 & 0 & 11 & 2 \\ 7 & 5 & 11 & 0 & 9 \\ 10 & 12 & 2 & 9 & 0 \end{pmatrix} \end{array}. \qquad (10.7)$$

From this distance matrix, we implement a successive cluster formation procedure based on single linkage (minimum distance) in the following steps:

(1) Find the element in the distance matrix representing the shortest interobject distance, and form a new cluster composed of the corresponding object pair. In this example, we form the cluster (c, e) based on $d(c, e) = 2$, and enter the first step in the corresponding dendrogram with a height of 2 representing the interobject distance (Figure 10.2 (1)). The distance representing the height in the dendrogram at which a cluster is formed is referred to as the *fusion level* (or, alternatively, the *fusion distance*).

(2) Determine the distances between cluster (c, e) formed in step (1) and the remaining objects a, b, and d by single linkage, and construct a new distance matrix. Since $d((c, e), a) = 3$, $d((c, e), b) = 6$, and $d((c, e), d) = 9$, we then have the following distance matrix:

$$\begin{array}{c} \\ (c,e) \\ a \\ b \\ d \end{array} \begin{array}{c} \begin{array}{cccc} (c,e) & a & b & d \end{array} \\ \begin{pmatrix} 0 & 3 & 6 & 9 \\ 3 & 0 & 8 & 7 \\ 6 & 8 & 0 & 5 \\ 9 & 7 & 5 & 0 \end{pmatrix} \end{array}. \tag{10.8}$$

The shortest distance represented in this matrix is between cluster (c, e) and a. We therefore form the new cluster (a, c, e). In the dendrogram, as shown in Figure 10.2 (2), the new merger appears at fusion level 3.

(3) Determine the distance between cluster (a, c, e) formed in step (2) and objects b and d by single linkage and construct the corresponding distance matrix. In this example, $d((a, c, e), b) = 6$ and $d((a, c, e), d) = 7$, and we therefore have the following distance matrix:

$$\begin{array}{c} \\ (a,c,e) \\ b \\ d \end{array} \begin{array}{c} \begin{array}{ccc} (a,c,e) & b & d \end{array} \\ \begin{pmatrix} 0 & 6 & 7 \\ 6 & 0 & 5 \\ 7 & 5 & 0 \end{pmatrix} \end{array}. \tag{10.9}$$

The shortest distance represented in this matrix is 5, between objects b and d, and we therefore form the new cluster (b, d) and add the corresponding fusion level 5 in the dendrogram as shown in Figure 10.2 (3).

(4) Merge cluster (b, d) formed in step (3) and cluster (a, c, e), thus forming a final, single cluster. The fusion level of the shortest interobject distance between the two clusters is 6, as shown in the completed dendrogram in Figure 10.2 (4).

$$\begin{array}{c} \\ (a,c,e) \\ (b,d) \end{array} \begin{array}{c} \begin{array}{cc} (a,c,e) & (b,d) \end{array} \\ \begin{pmatrix} 0 & 6 \\ 6 & 0 \end{pmatrix} \end{array}. \tag{10.10}$$

This example illustrates the cluster formation procedure and the corresponding dendrogram when the intercluster distances are determined by single linkage. The procedure is essentially the same with complete,

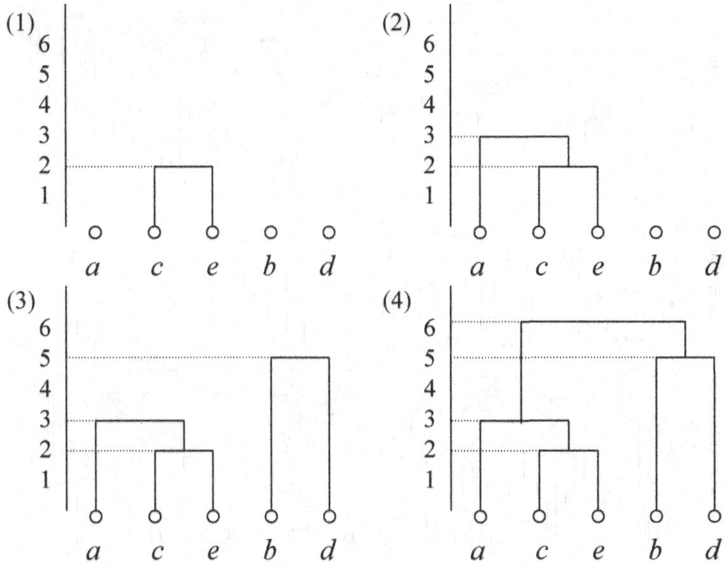

Figure 10.2 *Cluster formation process and the corresponding dendrogram based on single linkage when starting from the distance matrix in (10.7).*

average, centroid, and median linkages, but using a different linkage may result in different intercluster distances, correspondingly different clusters, and substantially different fusion levels. Figure 10.3 shows the dendrograms obtained for a single set of 72 six-dimensional data using three different linkage techniques: single, complete, and centroid linkages. The resulting differences in cluster configurations and fusion levels are easily seen. In actual analyses, it is therefore necessary to consider several different linkage techniques, one of which may be Ward's method, described in the next section, and apply them together with knowledge of the relevant field of application.

Properties of hierarchical clustering It should be kept in mind that single linkage is susceptible to a *chaining effect* in which the proximity of certain neighboring objects leads to a series of mergers that make the cluster structure in the dendrogram hard to recognize, as is shown in the circled portion of the uppermost dendrogram in Figure 10.3.

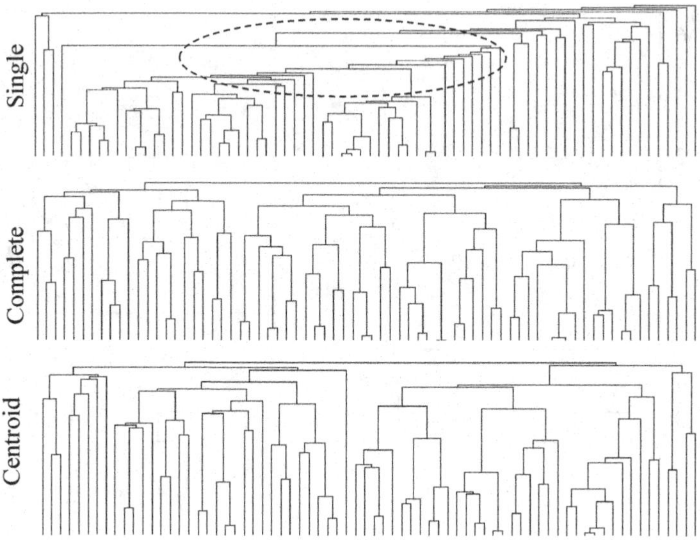

Figure 10.3 *The dendrograms obtained for a single set of 72 six-dimensional data using three different linkage techniques: single, complete, and centroid linkages. The circled portion of the dendrogram shows a chaining effect.*

A different problem may arise with centroid and median linkages, involving relative fusion distances (or levels) as illustrated in Figure 10.4. In the cluster formation procedure shown by the dendrogram on the left, the fusion distances increase monotonically in the order $\{3 < 7 < 12 < 25 < 30\}$, and thus the fusion distances increase with *fusion-distance monotonicity*. In the dendrogram on the right, however, a fusion-distance inversion has occurred in the final update of intercluster distance, from 30 in the merger of cluster $\{1, 2, 3\}$ with cluster $\{4, 5\}$, to the shorter distance of 27 in the merger of cluster $\{1, 2, 3, 4, 5\}$ with cluster $\{6\}$. Centroid and median linkages sometimes lack fusion-distance monotonicity, and are particularly susceptible to fusion-distance inversion in cases where the cluster structure is not clearly discernible.

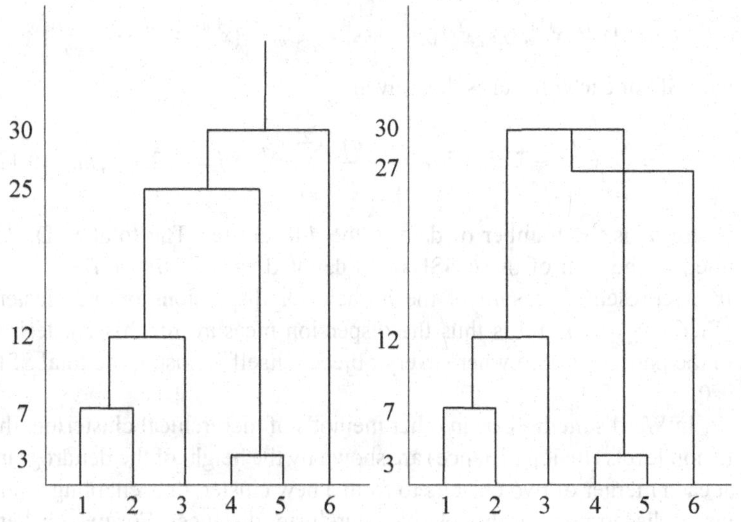

Figure 10.4 *Fusion-distance monotonicity (left) and fusion-distance inversion (right).*

10.1.4 Ward's Method

Ward's method is widely applied to categorization in many fields and regarded as a highly practical method of hierarchical clustering that is relatively effective for detection of useful patterns from datasets. In Ward's method, clustering is based on quantitative indicators of data dispersion. The underlying measure is squared Euclidean distance. In this section, we describe the essentials of Ward's method, beginning with the dispersion measures for one-dimensional and then for multidimensional datasets, and then proceed to the process of its implementation.

(1) *Dispersion measure for one-dimensional datasets* Consider the clustering of a dataset $\{x_1, x_2, \cdots, x_n\}$ comprising n one-dimensional data observed for some variable. As the indicator for their degree of dispersion, we use the following sum of squared deviations (SSD):

$$\sum_{i=1}^{n}(x_i - \overline{x})^2, \qquad \overline{x} = \frac{1}{n}\sum_{i=1}^{n} x_i. \tag{10.11}$$

If the one-dimensional dataset is partitioned into m clusters as

$$C_1 = \{x_1^{(1)}, x_2^{(1)}, \cdots, x_{n_1}^{(1)}\}, \qquad \cdots, \qquad C_m = \{x_1^{(m)}, x_2^{(m)}, \cdots, x_{n_m}^{(m)}\},$$

the SSD of each cluster is then given by

$$W_j = \sum_{i=1}^{n_j}(x_i^{(j)} - \overline{x}_j)^2 \ \text{ with } \ \overline{x}_j = \frac{1}{n_j}\sum_{i=1}^{n_j} x_i^{(j)}, \ \ j = 1, 2, \cdots, m, \quad (10.12)$$

where n_j is the number of data in the j-th cluster. The total SSD, defined as the sum of these SSDs and denoted as $W = W_1 + W_2 + \cdots + W_m$, represents the sum of the intracluster dispersions of the clusters $\{C_1, C_2, \cdots, C_m\}$ and is thus the dispersion measure for this clustering. In the particular case where every object is itself a cluster, the total SSD is 0.

In Ward's method, as in other methods of hierarchical clustering, the fusion levels (fusion distance) are shown by the height of the dendrogram at each merger of two clusters to form a new cluster, thus enabling ready visual discernment of the relative intercluster distances. For two clusters

$$C_\alpha = \{x_1^{(\alpha)}, x_2^{(\alpha)}, \cdots, x_{n_\alpha}^{(\alpha)}\}, \qquad C_\beta = \{x_1^{(\beta)}, x_2^{(\beta)}, \cdots, x_{n_\beta}^{(\beta)}\} \quad (10.13)$$

having W_α and W_β, respectively, the SSD of the cluster formed by their merger is $W_{\alpha\beta}$, and the related fusion level is defined as the increase in SSD resulting from the merger in the following:

$$d(C_\alpha, C_\beta) = W_{\alpha\beta} - W_\alpha - W_\beta, \quad (10.14)$$

the increase in SSD resulting from the merger.

(2) *Dispersion measure for multidimensional datasets* We define the measure of dispersion for a dataset $\{\boldsymbol{x}_1, \boldsymbol{x}_2, \cdots, \boldsymbol{x}_n\}$ composed of n p-dimensional data. The measure of dispersion essentially consists of the SSD matrix

$$S = (s_{jk}) = \sum_{i=1}^{n} (\boldsymbol{x}_i - \overline{\boldsymbol{x}})(\boldsymbol{x}_i - \overline{\boldsymbol{x}})^T, \qquad \overline{\boldsymbol{x}} = \frac{1}{n}\sum_{i=1}^{n} \boldsymbol{x}_i, \quad (10.15)$$

which corresponds to (10.11) in the one-dimensional SSD. The dispersion of the multidimensional data can then be determined by

$$\text{(i) tr } S = s_{11} + s_{22} + \cdots + s_{pp}, \qquad \text{(ii) } |S|. \quad (10.16)$$

The equation (i) is the total SSD of the variables and (ii) is the *generalized variance*, and either one can be used as the dispersion measure for multidimensional datasets.

For the cluster analysis of the dataset of n p-dimensional data, let us group the data into m clusters and denote the j-th cluster as $C_j = \{\boldsymbol{x}_1^{(j)}, \boldsymbol{x}_2^{(j)}, \cdots, \boldsymbol{x}_{n_j}^{(j)}\}$ $(j = 1, 2, \cdots, m)$. If we let S_j denote the SSD matrix based on the n_j p-dimensional data in this cluster, the dispersion in cluster C_j can then be measured in terms of either $W_j = \text{tr}\, S_j$ or $W_j = |S_j|$, and the measure for the dispersion of dataset S as grouped in m clusters is accordingly given by

$$\text{(i) } W = \text{tr}\, S_1 + \text{tr}\, S_2 + \cdots + \text{tr}\, S_m, \quad \text{(ii) } W = |S_1| + |S_2| + \cdots + |S_m|.$$

In the initial state, each cluster is composed of just one datum, and so we find $W = 0$.

The fusion level in the merger of two clusters C_α and C_β containing data from a multidimensional dataset is given by

$$\begin{aligned} &\text{(i) } d(C_\alpha, C_\beta) = \text{tr}\, S_{\alpha\beta} - \text{tr}\, S_\alpha - \text{tr}\, S_\beta, \\ &\text{(ii) } d(C_\alpha, C_\beta) = |S_{\alpha\beta}| - |S_\alpha| - |S_\beta|. \end{aligned} \tag{10.17}$$

This corresponds to (10.14), which is based on SSD in one-dimensional data clustering, but now with S_α and S_β as SSD matrices for the data in clusters C_α and C_β, respectively, and $S_{\alpha\beta}$ as the SSD matrix for all data in the two clusters.

(3) *Ward's hierarchical clustering method* In Ward's method, clusters are formed according to the following steps:

(i) For each datum, form one cluster containing only that datum. The dispersion measure W is then 0.

(ii) Reduce the number of clusters by one through the merger of two clusters that yields the smallest increase in dispersion measure W.

(iii) Repeat step (ii) until all clusters have been merged into a single cluster.

The following is a simple example of a cluster formation procedure by Ward's method. Figure 10.5 shows the dendrogram corresponding to the clusters formed by this procedure.

Example 10.1 (Clustering by Ward's method) For cluster analysis of a one-dimensional dataset $\{1, 2, 4, 6, 9, 12\}$, we use the SSD of (10.11) as the dispersion measure.

(1) Start with just one datum in each cluster. The number of clusters is thus 6, and the total SSD is $W = 0$.

(2) Reduce the number of clusters by one through the two-cluster merger that causes the smallest increase in W. In this example, the merger of (one-datum clusters) 1 and 2 yields an SSD of $(1-1.5)^2+(2-1.5)^2=0.5$ in the new cluster and a total SSD of $W=0.5$ for the resulting five clusters $C_1=\{1,2\}$, $C_2=4$, $C_3=6$, $C_4=9$, $C_5=12$, which is the smallest increase from $W=0$ among all possible two-cluster mergers in this step. In the dendrogram, the fusion level of this merger is $d(\{1,2\})=W_{1,2}-W_1-W_2=2-0-0=0.5$, as found from (10.14).

(3) Merge 4 and 6 as the next two-cluster merger causing the smallest increase in total SSD. The SSD of the resulting cluster {4, 6} is $(4-5)^2+(6-5)^2=2$. The merger yields $W=2.5$ as the total SSD for the four resulting clusters $C_1=\{1,2\}$, $C_2=\{4,6\}$, $C_3=9$, $C_4=12$, and a fusion level of $d(\{4,6\})=W_{4,6}-W_4-W_6=2-0-0=2$ in the dendrogram.

(4) Merge 9 and 12, as this yields a smaller increase in total SSD than would be obtained by merging 9 with cluster {4, 6} or merging cluster {1, 2} with cluster {4, 6}. The SSD of cluster {9, 12} is $(9-10.5)^2+(12-10.5)^2=4.5$. The total SSD for the resulting clusters $C_1=\{1,2\}$, $C_2=\{4,6\}$, $C_3=\{9,12\}$ is $W=7$, and the fusion level is $d(\{9,12\})=W_{9,12}-W_9-W_{12}=4.5-0-0=4.5$.

(5) Calculate the increase in total SSD for merging {1, 2} with {4, 6} and for merging {4, 6} with {9, 12}. As the increase is 14.75 for the former and 36.75 for the latter, form the new cluster {1, 2, 4, 6}. The total SSD for the two clusters $C_1=\{1,2,4,6\}$ and $C_2=\{9,12\}$ is then $W=19.25$, and by (10.14) the fusion level is $d(\{1,2,4,6\})=W_{1,2,4,6}-W_{1,2}-W_{4,6}=14.75-0.5-2=12.25$.

(6) Finally, merge all of the data into a single cluster, thus completing the procedure and obtaining a total SSD of $W=89.33$ and a fusion level of $d(\{1,2,4,6,9,12\})=W_{1,2,4,6,9,12}-W_{1,2,4,6}-W_{9,12}=89.33-14.75-4.5=70.08$.

This example and the related dendrogram in Figure 10.5 show the essentials of a stepwise cluster formation procedure by Ward's method.

10.2 Nonhierarchical Clustering

In nonhierarchical clustering, a predetermined number of clusters are formed. Various techniques may be used for this purpose. Here we describe two typical techniques, *k-means clustering* and *self-organizing map clustering*, and the basic processes of their implementation.

No. of clusters	Cluster	SSD	Fusion level
6	1, 2, 4, 6, 9, 12	W = 0	
5	{1, 2}, 4, 6, 9, 12	W = 0.5	0.5
4	{1, 2}, {4, 6}, 9, 12	W = 2.5	2
3	{1, 2}, {4, 6}, {9, 12}	W = 7	4.5
2	{1, 2, 4, 6}, {9, 12}	W = 19.25	12.25
1	{1, 2, 4, 6, 9, 12}	W = 89.3	70.08

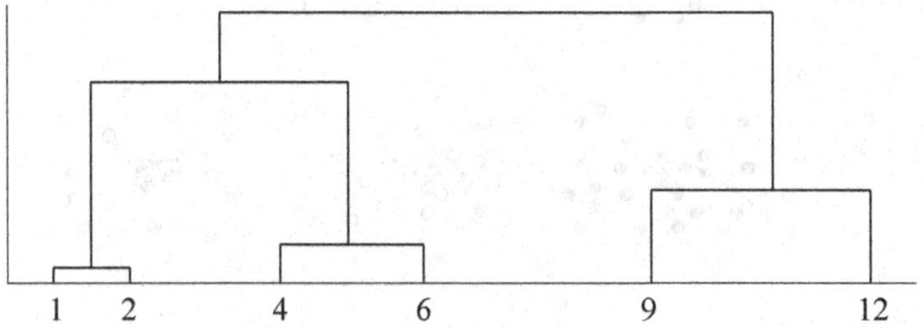

Figure 10.5 *Stepwise cluster formation procedure by Ward's method and the related dendrogram.*

10.2.1 K-Means Clustering

In k-means clustering, the objects are partitioned into k clusters, essentially as described in the following example and illustrated in Figure 10.6. In the example, the set has n objects and $k = 3$, and so the objects are partitioned into three clusters.

(1) From among the n objects, take k objects at random as *seeds* that will form the nuclei of $k = 3$ clusters, as illustrated by the three circled objects in Figure 10.6 (1).

(2) Assign each of the remaining objects to the nearest nucleus using an appropriate measure of similarity (usually squared Euclidean distance), thus partitioning the entire set into k (three) clusters. Figure 10.6 (2) shows the three resulting clusters with their object members separately indicated by open, closed, and shaded circles based on assignment determined by squared Euclidean distance.

(3) Compute the centroid (sample mean vector; $\triangle$) of each of the k clusters formed in step (2), and reassign each of the objects to the nearest resulting centroid, thus forming three new clusters with these com-

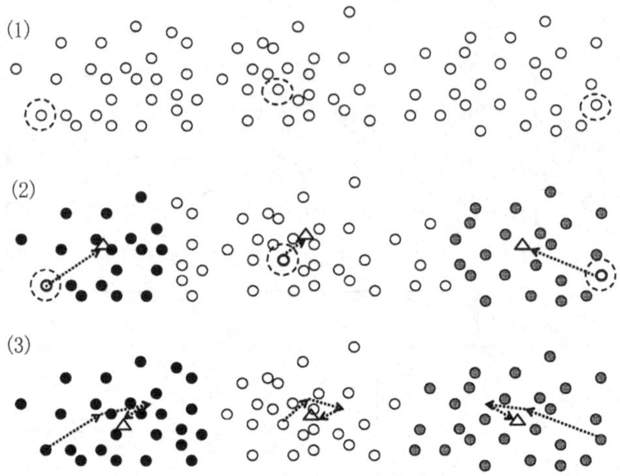

Figure 10.6 *Stepwise cluster formation process by k-means.*

puted centroids replacing the original nuclei set with the initial values.

(4) Compute a new centroid for each of the new k clusters and then reassign the objects to these new centroids, thus updating their membership in these k clusters, in the same manner as in step (3). Repeat this step until no further intercluster movement occurs for any of the objects. As illustrated in Figure 10.6 (3), note that the centroid position tends to change to a varying extent in each repetition.

It is generally recognized that the clusters ultimately formed in k-means clustering may differ with the seed selection in step (1); that is, the clusters are dependent on the initial values. One technique that has been proposed to resolve this problem consists of randomly changing the manner of seed selection, assessing the dispersion in the clusters formed in each case, and selecting the clustering that minimizes the evaluation function. As a source of explanation and commentary on both hierarchical and nonhierarchical clustering, Seber (1984) and Johnson and Wichern (2007) may be useful as a reference.

10.2.2 Self-Organizing Map Clustering

A self-organizing map is a type of neural network originally proposed by Kohonen (1990, 2001) as a learning algorithm. It involves two layers (also referred to as spaces). One is the *input layer*, in which p-dimensional data are entered as the objects of the clustering. The other is the *competitive layer*, which comprises a node array that is generally in the form of a grid. The self-organizing map results are output on the competitive layer and provide an intuitive visual display.

To illustrate the basic concept, let us assume that the competitive layer comprises an array of m nodes as shown in Figure 10.7. Each node is assigned a different weight vector $\boldsymbol{w}_j = (w_{j1}, w_{j2}, \cdots, w_{jp})^T$ $(j = 1, 2, \cdots, m)$ (also referred to as a *connection weight*) having the same dimensionality as the data. If we train the weight vectors by repeating the steps described below, a set of nodes having similar weight vectors emerges in the competitive layer. The weight vectors thus play the role of cluster seeds, with the input data each distributed around the nearest weight vector to form clusters. The training of the weight vectors in the self-organizing map is implemented in the following steps.

(1) Randomly assign different weight vectors $\boldsymbol{w}_1, \boldsymbol{w}_2, \cdots, \boldsymbol{w}_m$ to m nodes in the competitive layer.

(2) Enter the p-dimensional data $\boldsymbol{x} = (x_1, x_2, \cdots, x_p)^T$ into the input layer, and compute the Euclidean distance of each datum to the weight vector of each node as follows:

$$d_j = \sqrt{\sum_{k=1}^{p}(x_k - w_{jk})^2}, \qquad j = 1, 2, \cdots, m. \qquad (10.18)$$

Denote as c the node among the array of m nodes for which the distance is shortest, and denote as $\boldsymbol{w}_c$ the corresponding weight vector.

(3) For each of the nodes j $(j = 1, 2, \cdots, m)$ in the neighborhood of node c, update the weight vector from $w_j(t)$ to $w_j(t+1)$ by the following equation:

$$\boldsymbol{w}_j(t+1) = \boldsymbol{w}_j(t) + h_{cj}\left\{\boldsymbol{x} - \boldsymbol{w}_j(t)\right\}, \quad j = 1, 2, \cdots, m, \quad (10.19)$$

where h_{cj}, which is referred to as the *neighborhood function*, is usually in the form of a Gaussian function

$$h_{cj} = \eta(t)\exp\left\{-\frac{\|\boldsymbol{r}_c - \boldsymbol{r}_j\|^2}{2\sigma^2(t)}\right\}. \qquad (10.20)$$

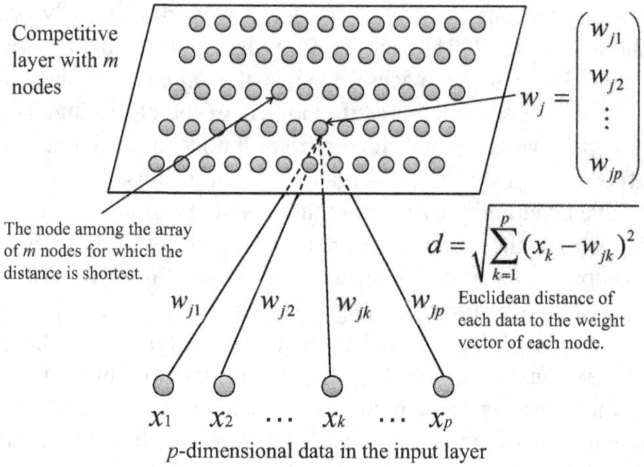

Figure 10.7 *The competitive layer comprises an array of m nodes. Each node is assigned a different weight vector* $\boldsymbol{w}_j = (w_{j1}, w_{j2}, \cdots, w_{jp})^T$ $(j = 1, 2, \cdots, m)$, *and the Euclidean distance of each p-dimensional data to the weight vector is computed.*

Here $\boldsymbol{r}_c$ and $\boldsymbol{r}_j$ are two-dimensional position vectors of nodes c and j, respectively, in the competitive layer. The $\eta(t)$ in (10.20) is the *learning coefficient*, and $\sigma^2(t)$ is a monotonically decreasing function of t that may be regarded as equivalent to the radius of the neighborhood. One example of a function that may be used for $\eta(t)$ and $\sigma^2(t)$ is $1-t/T$, in which the variable is the number of the training iteration and T is a predetermined, sufficiently large total number of training iterations.

(4) Repeat steps (2) and (3) for $t = 0, 1, 2, \cdots$ to train the weight vector.
(5) Repeat steps (2) to (4) to train the weight vectors for all of the data $\{\boldsymbol{x}_i;\ i = 1, 2, \cdots, n\}$ and thus implement the clustering.

Next, denote as $\{\boldsymbol{w}_j;\ j = 1, 2, \cdots, m\}$ the final weight vectors found by implementation of the training algorithm for self-organization, and for each of the n p-dimensional data compute the Euclidean distance to the m weight vectors and assign the distance to the cluster containing the weight vector that yields the shortest distance. In other words, construct the j-th cluster C_j as the set of data for which the Euclidean distance to the weight vector $\boldsymbol{w}_j$ is shortest, and in this manner form m clusters $C_1, C_2, \cdots, C_m$ for the n p-dimensional data.

As may be seen from this training algorithm, the self-organizing map technique resembles k-means clustering in its determination of weight vectors of nodes that yield the shortest distance from certain data, the influence of those data on the weight vectors of other nodes in the same neighborhood, and the repeated updating. Each technique, however, poses its own particular challenge. In self-organizing map clustering, the question arises of how to best set the values for the several parameters in the equation of step (3) for updating the weight vector. In k-means clustering, on the other hand, it is always necessary to bear in mind that what clusters are obtained is dependent on the initial values; that is, the clusters vary with the selection of seed data in the first step.

10.3 Mixture Models for Clustering

A mixture model is essentially a probability distribution model formed by superimposing several probability distributions by linear combination (see, e.g., McLachlan and Peel, 2000). In the following, we first consider the basic procedure of cluster analysis with mixture models and then turn to the expectation-maximization (EM) algorithm used in the model estimation.

10.3.1 Mixture Models

The histogram in Figure 10.8 is based on observed data on the speed of recession from Earth of 82 galaxies scattered in space (Roeder, 1990). It suggests that these recession speeds can be grouped into several clusters. Let us represent galactic recession speed by the variable (x); then we will assume that its values can be generated by a probability distribution model and attempt to find a model consistent with the data.

A normal distribution is often used as a probability distribution model, but would clearly not be sufficient in the case of a complex data distribution such as that shown in Figure 10.8. We will therefore attempt to construct a more appropriate model by linear combination of multiple normal distributions

$$f(x|\boldsymbol{\theta}) = \sum_{j=1}^{g} \pi_j \frac{1}{\sqrt{2\pi\sigma_j^2}} \exp\left\{-\frac{(x-\mu_j)^2}{2\sigma_j^2}\right\}, \tag{10.21}$$

where $0 \leq \pi_j \leq 1$ $(j = 1, 2, \cdots, g)$ and $\sum_{j=1}^{g} \pi_j = 1$. The parameters

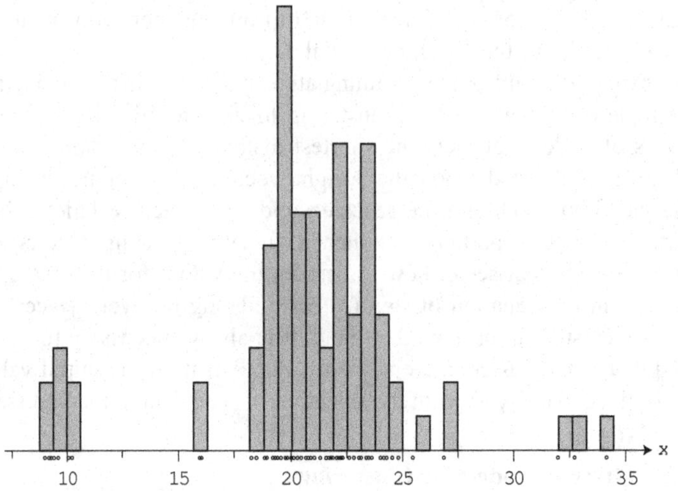

Figure 10.8 *Histogram based on observed data on the speed of recession from Earth of 82 galaxies scattered in space.*

$\{\mu_j, \sigma_j^2;\ j = 1, 2, \cdots, g\}$ and $\{\pi_j;\ j = 1, 2, \cdots, g\}$ in this probability distribution model may together be represented by a parameter vector $\boldsymbol{\theta}$. If we can estimate the model parameters from the data and appropriately select the number g of superimposed normal distributions, we can then know the number of clusters to use in categorizing the galactic recession speeds. The mixture model formed by the linear combination of probability distributions is generally referred to as *mixture model-based clustering* and may be formalized as follows.

Suppose that p-dimensional data $\boldsymbol{x}$ have been observed and conform to the following probability distribution as represented by the linear combination of g probability density functions:

$$f(\boldsymbol{x}|\boldsymbol{\theta}) = \sum_{j=1}^{g} \pi_j f_j(\boldsymbol{x}|\boldsymbol{\theta}_j), \tag{10.22}$$

where $\pi_1, \pi_2, \cdots, \pi_g$ are mixing proportions such that $0 \leq \pi_j \leq 1$ and they sum to 1, and $\boldsymbol{\theta} = \{\boldsymbol{\theta}_1, \boldsymbol{\theta}_2, \cdots, \boldsymbol{\theta}_g, \pi_1, \pi_2, \cdots, \pi_g\}$ is the distribution-regulating parameter. A Gaussian mixture model, in particular, is a distribution model representing a linear combination of $N_p(\boldsymbol{\mu}_j, \Sigma_j)$ with mean vector $\boldsymbol{\mu}_j$ and variance-covariance matrix Σ_j.

Data categorization by a mixed distribution model may also be understood in the framework of the Bayesian approach described in Chapter 7. In this case, the probability distribution $f_j(\boldsymbol{x}|\boldsymbol{\theta}_j)$ is taken as a distribution characterizing the j-th group G_j, representing the conditional likelihood when observed data $\boldsymbol{x}$ is assumed to belong to G_j, and the corresponding mixing proportion π_j is taken to represent the probability of selection of G_j, and thus the prior probability. By Bayes' theorem, the posterior probability of the observed data $\boldsymbol{x}$ being a member of group G_j is then given by

$$p(j|\boldsymbol{x}) = \frac{\pi_j f_j(\boldsymbol{x}|\boldsymbol{\theta}_j)}{\displaystyle\sum_{k=1}^{g} \pi_k f_k(\boldsymbol{x}|\boldsymbol{\theta}_k)} = \frac{\pi_j f_j(\boldsymbol{x}|\boldsymbol{\theta}_j)}{f(\boldsymbol{x}|\boldsymbol{\theta})}, \quad j = 1, 2, \cdots, g. \quad (10.23)$$

If the mixture model can be estimated, the observed data $\boldsymbol{x}$ is taken to belong to the group yielding the highest of the g posterior probabilities. Cluster analysis can thus be implemented in the Bayesian approach by estimating the model from the candidate data and categorizing the data by posterior probability maximization.

10.3.2 Model Estimation by EM Algorithm

It is necessary to use numerical optimization for estimating mixture model parameters by the maximum likelihood methods, since they cannot be expressed analytically in closed form. The estimation method by the EM algorithm is widely used for this purpose (see McLachlan and Krishnan, 2008). In the form shown in Appendix C, the EM algorithm can be applied to the estimation of the parameters of the Gaussian mixture model as follows.

Estimation of Gaussian mixture model by EM algorithm We first set the initial values of the mixing proportions, mean vectors, and variance-covariance matrices $\{(\pi_j, \boldsymbol{\mu}_j, \Sigma_j);\ j = 1, 2, \cdots, g\}$. Thereafter, the $\{(\pi_j^{(t)}, \boldsymbol{\mu}_j^{(t)}, \Sigma_j^{(t)});\ j = 1, 2, \cdots, g\}$ values in the t-th step are updated to new values in the $(t+1)$-th step by alternating repetitions of the following E and M steps until convergence:

E step Compute the posterior probability that the i-th data $\boldsymbol{x}_i$ is from group G_j as

$$p^{(t+1)}(j|\boldsymbol{x}_i) = \frac{\pi_j^{(t)} f_j(\boldsymbol{x}_i|\boldsymbol{\theta}_j^{(t)})}{\sum_{k=1}^{g} \pi_k^{(t)} f_k(\boldsymbol{x}_i|\boldsymbol{\theta}_k^{(t)})}, \quad i = 1, \cdots, n, \;\; j = 1, \cdots, g. \tag{10.24}$$

M step Update the mixing proportions, mean vectors, and variance-covariance matrices in the $(t+1)$-th step, using the equation

$$\pi_j^{(t+1)} = \frac{1}{n}\sum_{i=1}^{n} p^{(t+1)}(j|\boldsymbol{x}_i), \quad \boldsymbol{\mu}_j^{(t+1)} = \frac{1}{n\pi_j^{(t+1)}}\sum_{i=1}^{n} p^{(t+1)}(j|\boldsymbol{x}_i)\boldsymbol{x}_i,$$
$$\Sigma_j^{(t+1)} = \frac{1}{n\pi_j^{(t+1)}}\sum_{i=1}^{n} p^{(t+1)}(j|\boldsymbol{x}_i)(\boldsymbol{x}_i - \boldsymbol{\mu}_j^{(t+1)})(\boldsymbol{x}_i - \boldsymbol{\mu}_j^{(t+1)})^T. \tag{10.25}$$

Repeat the updates until the likelihood function satisfies the condition

$$\left|\sum_{i=1}^{n} \log\left\{\sum_{j=1}^{g} \pi_j^{(t+1)} f_j(\boldsymbol{x}_i|\boldsymbol{\mu}_j^{(t+1)}, \Sigma_j^{(t+1)})\right\} - \sum_{i=1}^{n} \log\left\{\sum_{j=1}^{g} \pi_j^{(t)} f_j(\boldsymbol{x}_i|\boldsymbol{\mu}_j^{(t)}, \Sigma_j^{(t)})\right\}\right| \leq c,$$

where $c > 0$ is preset to a sufficiently small value.

Thus far, we have considered estimation of the mixture model with a fixed value for g. Since g represents the number of model components, which is equivalent to the number of clusters, its appropriate selection is important. It is generally essential to estimate a number of candidate models, evaluate them, and select the one that best fits the probability structure of the data generation. Let us consider the evaluation and selection of mixture models estimated by the EM algorithm with $g = 2, 3, \cdots$, using the Akaike information criterion (AIC) while noting that the Bayesian information criterion (BIC) may also be used for this purpose.

If we denote by $f(\boldsymbol{x}|\boldsymbol{\theta}) = \sum_{j=1}^{g} \hat{\pi}_j f_j(\boldsymbol{x}|\hat{\boldsymbol{\mu}}_j, \hat{\Sigma}_j)$ the Gaussian mixture model estimated from n p-dimensional observed data, the AIC may then be given by

$$\text{AIC} = -2\sum_{i=1}^{n} \log\left\{\sum_{j=1}^{g} \hat{\pi}_j f_j(\boldsymbol{x}_i|\hat{\boldsymbol{\mu}}_j, \hat{\Sigma}_j)\right\} + 2(\text{no. of estimated parameters}).$$

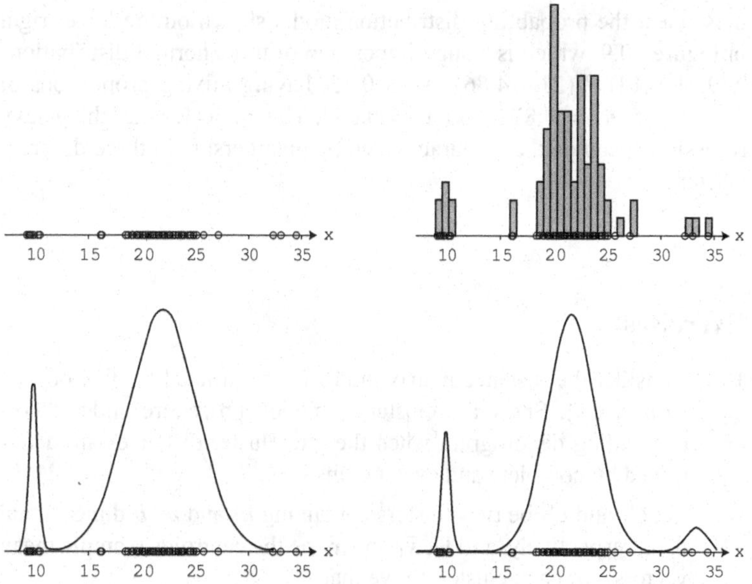

Figure 10.9 *Recession-speed data observed for 82 galaxies are shown on the upper left and in a histogram on the upper right. The lower left and lower right show the models obtained by fitting with two and three normal distributions, respectively.*

The number of estimated parameters in the model is $g - 1 + g\{p + p(p + 1)/2\}$, which is the sum of the number of mixing proportions $g - 1$, g times the number of component parameters in the p-dimensional mean vector, and g times the number of parameters $p(p + 1)/2$ in the variance-covariance matrix.

Example 10.2 (Galaxy recession-speed data) Our goal is to determine the number of clusters to be formed in the analysis of recession-speed data observed for 82 galaxies (Roeder, 1990) by fitting the one-dimensional Gaussian mixture model of (10.21) to the data. Figure 10.9 shows the 82 observed data on the upper left and in a histogram on the upper right. We used the EM algorithm to estimate the Gaussian mixture models with $g = 2$, 3, and 4 components. In Figure 10.9, the lower left and lower right show the models obtained by fitting with two and three normal distributions, respectively. The estimated probability distribution models were evaluated by the AIC, for which the number of estimated parameters with $g = 2$, 3, and 4 is 5, 8, and 11, respectively. We thus select the probability distribution model shown on the lower right in Figure 10.9, which is a superimposition of three normal distributions $N(9.71, 0.18)$, $N(21.4, 4.86)$, $N(33, 0.85)$ having mixing proportions of $\hat{\pi}_1 = 0.085$, $\hat{\pi}_2 = 0.878$, and $\hat{\pi}_1 = 0.037$. This indicates that the galaxy recession-speed data are characterized by membership in three different clusters.

Exercises

10.1 Consider the distance matrix in (10.7) constructed for five objects $\{a, b, c, d, e\}$. Show the cluster formation procedure and the corresponding dendrogram when the intercluster distances are determined by complete and average linkages.

10.2 Let C_1 and C_2 be two clusters containing n_1 and n_2 p-dimensional data, respectively, and let $\overline{\boldsymbol{x}}_1$ and $\overline{\boldsymbol{x}}_2$ be the centroids (sample mean vectors) for each cluster. Prove that

(a) $$\sum_{i \in C_1 \cup C_2} \left\| \boldsymbol{x}_i - \frac{n_1 \overline{\boldsymbol{x}}_1 + n_2 \overline{\boldsymbol{x}}_2}{n_1 + n_2} \right\|^2 - \left\{ \sum_{i \in C_1} \|\boldsymbol{x}_i - \overline{\boldsymbol{x}}_1\|^2 + \sum_{i \in C_2} \|\boldsymbol{x}_i - \overline{\boldsymbol{x}}_2\|^2 \right\}$$
$$= \sum_{j=1}^{2} n_j \left\| \overline{\boldsymbol{x}}_j - \frac{n_1 \overline{\boldsymbol{x}}_1 + n_2 \overline{\boldsymbol{x}}_2}{n_1 + n_2} \right\|^2.$$

(b) $$\sum_{j=1}^{2} n_j \left\| \overline{\boldsymbol{x}}_j - \frac{n_1 \overline{\boldsymbol{x}}_1 + n_2 \overline{\boldsymbol{x}}_2}{n_1 + n_2} \right\|^2 = \frac{n_1 n_2}{n_1 + n_2} \|\overline{\boldsymbol{x}}_1 - \overline{\boldsymbol{x}}_2\|^2,$$

where $\|\boldsymbol{x}\|^2 = \boldsymbol{x}^T \boldsymbol{x}$.

10.3 Let C_i, $i = 1, 2, 3$, be clusters with n_i p-dimensional observations and centroids $\overline{\boldsymbol{x}}_i$, respectively. Prove that

$$\left\| \overline{\boldsymbol{x}}_3 - \left(\frac{n_1 \overline{\boldsymbol{x}}_1 + n_2 \overline{\boldsymbol{x}}_2}{n_1 + n_2} \right) \right\|^2$$
$$= \frac{n_1}{n_1 + n_2} \|\overline{\boldsymbol{x}}_3 - \overline{\boldsymbol{x}}_1\|^2 + \frac{n_1}{n_1 + n_2} \|\overline{\boldsymbol{x}}_3 - \overline{\boldsymbol{x}}_2\|^2 - \frac{n_1 n_2}{(n_1 + n_2)^2} \|\overline{\boldsymbol{x}}_1 - \overline{\boldsymbol{x}}_2\|^2.$$

10.4 Let X be a random variable distributed according to

$$f(x|\boldsymbol{\theta}) = \sum_{i=1}^{2} \pi_i \frac{1}{\sqrt{2\pi\sigma_i^2}} \exp\left\{ -\frac{(x - \mu_i)^2}{2\sigma_i^2} \right\}$$

Show that the mean and variance of X are, respectively, given by

$$E[X] = \sum_{j=1}^{2} \pi_j \mu_j, \quad E[(X - E[X])^2] = \sum_{j=1}^{2} \pi_j(\sigma_j^2 + \mu_j^2) - \left(\sum_{j=1}^{2} \pi_j \mu_j \right)^2.$$

10.5 Let $f_j(\boldsymbol{x}|\boldsymbol{\theta}_j)$, $j = 1, 2, \cdots, g$, be p-dimensional density functions with mean vectors $\boldsymbol{\mu}_j$ and variance-covariance matrices Σ_j, respectively. Consider the mixture distribution

$$f(\boldsymbol{x}|\boldsymbol{\theta}) = \sum_{j=1}^{g} \pi_j f_j(\boldsymbol{x}|\boldsymbol{\theta}_j).$$

Show that the mean vector and variance-covariance matrix are, respectively, given by

$$E[\boldsymbol{X}] = \sum_{j=1}^{g} \pi_j \boldsymbol{\mu}_j,$$

$$E\left[(\boldsymbol{X} - E[\boldsymbol{X}])(\boldsymbol{X} - E[\boldsymbol{X}])^T\right]$$

$$= \sum_{j=1}^{g} \pi_j(\Sigma_j + \boldsymbol{\mu}_j \boldsymbol{\mu}_j^T) - \left(\sum_{j=1}^{g} \pi_j \boldsymbol{\mu}_j \right) \left(\sum_{j=1}^{g} \pi_j \boldsymbol{\mu}_j \right)^T.$$

Appendix A

Bootstrap Methods

The bootstrap methods proposed by Efron (1979) have received considerable interest due to their ability to provide effective solutions to problems that cannot be solved by analytic approaches based on theories or formulas. A salient feature of the bootstrap is that it uses massive iterative computer calculations rather than analytic expressions. This makes the bootstrap a flexible statistical technique for complex problems of inference under highly relaxed assumptions. The proposal led to a rapidly expanding range of research efforts on applications, and to its establishment as a practical computational technique.

Seminal literature on the bootstrap methods includes Efron and Tibshirani (1993) and Davison and Hinkley (1997) on practical aspects of its application to various statistical problems, and Efron (1982), Hall (1992), etc., on the theoretical aspects. In addition, Diaconis and Efron (1983), Efron and Gong (1983), and Efron and Tibshirani (1986) provide introductions to the basic concepts underlying the bootstrap methods.

A.1 Bootstrap Error Estimation

We introduce the basic concepts and procedures for the bootstrap methods through the evaluation of the bias and variance of an estimator, and also estimation of the sampling distribution of an estimator and percentile points. Let $\boldsymbol{X} = \{X_1, X_2, \cdots, X_n\}$ be a random sample of size n drawn from an unknown probability distribution $F(x)$. We estimate a parameter θ with respect to the probability distribution $F(x)$ by using an estimator $\hat{\theta} = \hat{\theta}(\boldsymbol{X})$. When observed data $\boldsymbol{x} = \{x_1, x_2, \cdots, x_n\}$ are obtained, critical statistical analysis tasks are estimating the parameter θ by the estimator $\hat{\theta} = \hat{\theta}(\boldsymbol{x})$ and evaluating the reliability of the estimation.

Parameter estimation by statistical analysis must generally be accompanied by effective extraction of information on the estimation errors included in the data and assessment of reliability. Two measures of estimation error that may be used for this purpose are the bias and variance of the estimator

$$b(F) = \mathrm{E}_F[\hat{\theta}] - \theta, \qquad \sigma^2(F) = \mathrm{E}_F\left[\left\{\hat{\theta} - \mathrm{E}_F[\hat{\theta}]\right\}^2\right], \tag{A.1}$$

where the expectation is taken with respect to the joint distribution $F(\boldsymbol{x})$ of the random variable $\boldsymbol{X}$. If we can obtain the probability distribution and percentile of the estimator, we can then construct the confidence interval. For this purpose we need the following distribution function $G(x)$ and 100α percentile c_α of $\hat{\theta} - \theta$. The bootstrap comprises an algorithm for numerical estimation of these quantities by computer iteration. Its execution is essentially as follows:

(1) *Iterative extraction of bootstrap samples* Designate as $\boldsymbol{x}^* = \{x_1^*, x_2^*, \cdots, x_n^*\}$ the set of data extracted by sampling with replacement (i.e., sampling in which the data extracted in a given iteration are returned to the data set for the next iteration), from the observed data $\boldsymbol{x} = \{x_1, x_2, \cdots, x_n\}$. A set of data extracted in a single iteration is referred to as a *bootstrap sample*. In B iterations of this observed data sampling with replacement, we thus obtain B sets of bootstrap samples $\{\boldsymbol{x}_i^*;\ i = 1, 2, \cdots, B\}$ of size n, and the corresponding B estimators are denoted as $\{\hat{\theta}^*(i);\ i = 1, 2, \cdots, B\}$.

(2) *Bootstrap error estimation* With estimate $\hat{\theta}$ based on the observed data and the B bootstrap estimates $\{\hat{\theta}^*(i);\ i = 1, 2, \cdots, B\}$, the bias and variance of the estimator $\hat{\theta}(\boldsymbol{X})$ are respectively estimated by

$$b(\hat{F}) \approx \frac{1}{B}\sum_{i=1}^{B}\hat{\theta}^*(i) - \hat{\theta}, \quad \sigma^2(\hat{F}) \approx \frac{1}{B-1}\sum_{i=1}^{B}\left\{\hat{\theta}^*(i) - \hat{\theta}^*(\cdot)\right\}^2, \tag{A.2}$$

where $\hat{\theta}^*(\cdot) = \sum_{i=1}^{B}\hat{\theta}^*(i)/B$.

(3) *Bootstrap estimates of probability distribution and percentile* The distribution and 100α percentile can respectively be approximated numerically as follows:

$$\hat{G}(x) \approx \frac{1}{B}\left\{\text{the number of } i \text{ such that } \hat{\theta}^*(i) - \hat{\theta} \leq x\right\},$$
$$\hat{G}^{-1}(\alpha) \approx \left\{B\alpha\text{-th largest value of } \{\hat{\theta}^*(i) - \hat{\theta};\ i = 1, 2, \cdots, B\}\right\}. \tag{A.3}$$

If $B\alpha$ is not an integer, then the selected size is that of the $[(B+1)\alpha]$-th value, where $[x]$ is the largest integer that does not exceed the real number x.

A.2 Regression Models

Suppose that we have n-sets of data $\{(y_i, \boldsymbol{x}_i);\ i = 1, 2, \cdots, n\}$ for response variable y and p predictor variables $\boldsymbol{x} = \{x_1, x_2, \cdots, x_p\}$. The value y_i at each point $\boldsymbol{x}_i$ is observed as $y_i = u(\boldsymbol{x}_i; \boldsymbol{\beta}) + \varepsilon_i \ (i = 1, 2, \cdots, n)$ with the error ε_i, where $\boldsymbol{\beta}$ is an unknown parameter vector. It is assumed that the errors are independently distributed according to the unknown probability distribution F and that $E[\varepsilon_i] = 0$ and $E[\varepsilon_i^2] = \sigma^2$. Suppose further that we estimate the model parameter $\boldsymbol{\beta}$ by, for example, least squares, thus obtaining the estimator $\hat{\boldsymbol{\beta}}$ and the estimated regression function $y = u(\boldsymbol{x}; \hat{\boldsymbol{\beta}})$. The bootstrap error estimation of $\hat{\boldsymbol{\beta}}$ is then executed essentially as follows:

(1) Find the residuals $\hat{\varepsilon}_i = y_i - u(\boldsymbol{x}_i; \hat{\boldsymbol{\beta}})\ (i = 1, 2, \cdots, n)$ at all points and their mean $\hat{\varepsilon}_{(\cdot)} = \sum_{i=1}^{n} \hat{\varepsilon}_i / n$, and set the mean-adjusted residuals as $e_i = \hat{\varepsilon}_i - \hat{\varepsilon}_{(\cdot)}$.

(2) For the sample $e_1^*, e_2^*, \cdots, e_n^*$ of size n from the mean-adjusted residuals $e_1, e_2, \cdots, e_n$ in step (1), construct the bootstrap sample $\{(y_i^*, \boldsymbol{x}_i);\ i = 1, 2, \cdots, n\}$ with $y_i^* = u(\boldsymbol{x}_i; \hat{\boldsymbol{\beta}}) + e_i^*\ (i = 1, 2, \cdots, n)$.

(3) Designate as $\hat{\boldsymbol{\beta}}^*_{(1)}$ the solution of $\min_{\boldsymbol{\beta}} \sum_{i=1}^{n} \{y_i^* - u(\boldsymbol{x}_i; \boldsymbol{\beta})\}^2$. Based on the values $\hat{\boldsymbol{\beta}}^*_{(1)}, \hat{\boldsymbol{\beta}}^*_{(2)}, \cdots, \hat{\boldsymbol{\beta}}^*_{(B)}$ obtained by repeating steps (2) and (3) B times, the bootstrap error estimates of $\hat{\boldsymbol{\beta}}$ can be assessed in the same way as in Section A.1.

For a regression model in which all of the n observed data $\{z_i = (y_i, \boldsymbol{x}_i);\ i = 1, 2, \cdots, n\}$ independently follow the same $(p + 1)$-dimensional probability distribution F, the bootstrap samples are obtained by sampling with replacement from $z_1, z_2, \cdots, z_n$ and the bootstrap error estimation can be performed on the basis of the values $\hat{\boldsymbol{\beta}}^*_{(1)}, \hat{\boldsymbol{\beta}}^*_{(2)}, \cdots, \hat{\boldsymbol{\beta}}^*_{(B)}$ obtained by iterative execution of step (3).

A.3 Bootstrap Model Selection Probability

In regression modeling, it is invariably necessary to construct models based on the observed data, evaluate their goodness of fit, and select an appropriate model. This poses the question of how to assess the reliability of models selected in a series of modeling processes. The assessment can be based on model selection probability by bootstrapping, essentially as follows:

(1) Let $\{(y_i, \boldsymbol{x}_i)\ ;\ i = 1, 2, \cdots, n\}$ be the set of n observed data with $\boldsymbol{x}_i$ as

multidimensional data. Here y_i is a response variable taking real values in regression modeling, a 0 and 1 binary variable in logistic modeling, and a class-indicator variable representing class membership in discriminant analysis. A bootstrap sample $\{(y_i^*, \boldsymbol{x}_i^*)\ ;\ i = 1, 2, \cdots, n\}$ is obtained by n repeated samples with replacement from these observed data.

(2) Construct the optimum model for the bootstrap sample $\{(y_i^*, \boldsymbol{x}_i^*)\ ;\ i = 1, 2, \cdots, n\}$. In regression modeling, for example, variable selection is performed and a single model is selected on the basis of the model evaluation criterion. In discriminant analysis, the selection is made on the basis of prediction error.

(3) Repeat steps (1) and (2) B times, and find the proportion of selected set of variables.

When bootstrapping is thus applied to model selection problems, the model selection probability can provide a quantitative measure of reliability or uncertainty for a selected model based on the observed data.

Appendix B

Lagrange Multipliers

The method of Lagrange multipliers is used to find the stationary points of a multivariate real-valued function with a constraint imposed on its variables. Let us consider the method of Lagrange multipliers in the context of solutions to the optimization problems under equality and inequality constraints used in the chapters on principal component analysis, support vector machines, and other techniques. For this purpose, we shall express the objective function as the p-variate real-valued function

$$f(x_1, x_2, \cdots, x_p) = f(\boldsymbol{x}), \tag{B.1}$$

and several p-variate real-valued functions that impose the equality constraints as

$$h_i(x_1, x_2, \cdots, x_p) = h_i(\boldsymbol{x}) = 0, \qquad i = 1, 2, \cdots, m. \tag{B.2}$$

Similarly, we shall express the p-variate real-valued functions imposing the inequality constraints as

$$g_i(x_1, x_2, \cdots, x_p) = g_i(\boldsymbol{x}) \leq 0, \qquad i = 1, 2, \cdots, n. \tag{B.3}$$

Application of Lagrange multipliers requires regularity conditions such as differentiability and convexity in both the objective function and the constraint functions, and we assume that the results considered here satisfy these conditions. For details, see Cristianini and Shawe-Taylor (2000), etc.

B.1 Equality-Constrained Optimization Problem

We first find the value that minimizes the objective function $f(\boldsymbol{x})$ under m equality constraints $h_i(\boldsymbol{x}) = 0$, $i = 1, 2, \cdots, m$. For the equality-constrained optimization problem, we define the following real-valued Lagrangian function containing both the objective function and the constraint equations as

$$L(\boldsymbol{x}, \boldsymbol{\lambda}) = f(\boldsymbol{x}) + \sum_{i=1}^{m} \lambda_i h_i(\boldsymbol{x}), \tag{B.4}$$

where $\boldsymbol{\lambda} = (\lambda_1, \lambda_2, \cdots, \lambda_m)^T$. The constraint equation coefficients λ_i are referred to as the *Lagrange multiplier*. The equality-constrained optimization problem then entails finding a solution that satisfies the following equations for the partial derivatives of the Lagrange function with respect to $\boldsymbol{x}$ and λ_i:

$$\begin{aligned} \frac{\partial L(\boldsymbol{x}, \boldsymbol{\lambda})}{\partial \boldsymbol{x}} &= \frac{\partial f(\boldsymbol{x})}{\partial \boldsymbol{x}} + \sum_{i=1}^{m} \lambda_i \frac{\partial h_i(\boldsymbol{x})}{\partial \boldsymbol{x}} = \mathbf{0}, \\ \frac{\partial L(\boldsymbol{x}, \boldsymbol{\lambda})}{\partial \lambda_i} &= h_i(\boldsymbol{x}) = 0, \qquad i = 1, 2, \cdots, m, \end{aligned} \tag{B.5}$$

where $\mathbf{0}$ is a p-dimensional vector with all components equal to 0.

Let us denote the solution to these equations by $(\hat{\boldsymbol{x}}, \hat{\boldsymbol{\lambda}})$. Since the constraint at this point is 0, we then have $L(\hat{\boldsymbol{x}}, \hat{\boldsymbol{\lambda}}) = f(\hat{\boldsymbol{x}})$ for the Lagrangian function and the objective function. It therefore follows that, to find the solution that minimizes the objective function $f(\boldsymbol{x})$ under the equality constraint, we simply find the stationary point with respect to $\boldsymbol{x}$ and $\boldsymbol{\lambda}$ of the Lagrangian function $L(\boldsymbol{x}, \boldsymbol{\lambda})$.

B.2 Inequality-Constrained Optimization Problem

To find the value of minimization of the objective function $f(\boldsymbol{x})$ under n inequality constraints $g_i(\boldsymbol{x}) \leq 0$, $i = 1, 2, \cdots, n$, we define the Lagrangian function containing the objective function and the inequality constraint expressions as

$$L(\boldsymbol{x}, \boldsymbol{\alpha}) = f(\boldsymbol{x}) + \sum_{i=1}^{n} \alpha_i g_i(\boldsymbol{x}) \tag{B.6}$$

with the Lagrange multipliers α_i as $\boldsymbol{\alpha} = (\alpha_1, \alpha_2, \cdots, \alpha_n)^T$. The solution to this inequality-constrained optimization problem can then be reduced to find the values ($\hat{\boldsymbol{x}}$ and $\hat{\boldsymbol{\alpha}}$) satisfying the following four conditions:

$$\begin{aligned} \frac{\partial L(\boldsymbol{x}, \boldsymbol{\alpha})}{\partial \boldsymbol{x}} &= \frac{\partial f(\boldsymbol{x})}{\partial \boldsymbol{x}} + \sum_{i=1}^{n} \alpha_i \frac{\partial g_i(\boldsymbol{x})}{\partial \boldsymbol{x}} = \mathbf{0}, \\ \frac{\partial L(\boldsymbol{x}, \boldsymbol{\alpha})}{\partial \alpha_i} &= g_i(\boldsymbol{x}) \leq 0, \qquad i = 1, 2, \cdots, n, \\ \alpha_i &\geq 0, \qquad i = 1, 2, \cdots, n, \\ \alpha_i g_i(\boldsymbol{x}) &= 0, \qquad i = 1, 2, \cdots, n. \end{aligned} \tag{B.7}$$

The equality- and inequality-constrained optimization problems described in Sections B.1 and B.2 can be generalized to optimization under both constraints, as next described.

B.3 Equality/Inequality-Constrained Optimization

Let us now consider the problem of finding the value of minimization of objective function $f(\boldsymbol{x})$ under the combination of m equality constraints $h_i(\boldsymbol{x}) = 0$, $i = 1, 2, \cdots, m$, and n inequality constraints $g_i(\boldsymbol{x}) \leq 0$, $i = 1, 2, \cdots, n$. For this purpose, we define the Lagrangian function that includes the objective function and both the equality and the inequality constraints as

$$L(\boldsymbol{x}, \boldsymbol{\alpha}, \boldsymbol{\beta}) = f(\boldsymbol{x}) + \sum_{i=1}^{n} \alpha_i g_i(\boldsymbol{x}) + \sum_{i=1}^{m} \beta_i h_i(\boldsymbol{x}), \tag{B.8}$$

where $\boldsymbol{\alpha} = (\alpha_1, \alpha_2, \cdots, \alpha_n)^T$ for the inequality constraint coefficients α_i, and $\boldsymbol{\beta} = (\beta_1, \beta_2, \cdots, \beta_m)^T$ for the equality constraint coefficients β_i. Solution of this general constrained optimization problem then consists of finding the values $(\hat{\boldsymbol{x}}, \hat{\boldsymbol{\alpha}}, \hat{\boldsymbol{\beta}})$ satisfying the following five conditions:

$$\begin{aligned}
&\frac{\partial L(\boldsymbol{x}, \boldsymbol{\alpha})}{\partial \boldsymbol{x}} = \frac{\partial f(\boldsymbol{x})}{\partial \boldsymbol{x}} + \sum_{i=1}^{n} \alpha_i \frac{\partial g_i(\boldsymbol{x})}{\partial \boldsymbol{x}} + \sum_{i=1}^{m} \beta_i \frac{\partial h_i(\boldsymbol{x})}{\partial \boldsymbol{x}} = \boldsymbol{0}, \\
&\frac{\partial L(\boldsymbol{x}, \boldsymbol{\alpha})}{\partial \alpha_i} = g_i(\boldsymbol{x}) \leq 0, \qquad i = 1, 2, \cdots, n, \\
&\frac{\partial L(\boldsymbol{x}, \boldsymbol{\lambda})}{\partial \beta_i} = h_i(\boldsymbol{x}) = 0, \qquad i = 1, 2, \cdots, m, \\
&\alpha_i \geq 0, \qquad i = 1, 2, \cdots, n, \\
&\alpha_i g_i(\boldsymbol{x}) = 0, \qquad i = 1, 2, \cdots, n,
\end{aligned} \tag{B.9}$$

which are referred to as the *Karush-Kuhn-Tucker conditions*.

For the Lagrangian function containing the objective function and the equality and inequality constraints defined in (B.8), we obtain the inferior limit with respect to the vector variable $\boldsymbol{x}$ given by

$$L_D(\boldsymbol{\alpha}, \boldsymbol{\beta}) = \inf_{\boldsymbol{x}} L(\boldsymbol{x}, \boldsymbol{\alpha}, \boldsymbol{\beta}). \tag{B.10}$$

The optimization problem of this section can then be replaced by the problem of optimizing the function $L_D(\boldsymbol{\alpha},\boldsymbol{\beta})$, which does not depend on the variable $\boldsymbol{x}$, as follows:

$$\max_{\boldsymbol{\alpha},\boldsymbol{\beta}} L_D(\boldsymbol{\alpha},\boldsymbol{\beta}) \quad \text{subject to} \quad \alpha_i \geq 0, \quad i = 1,2,\cdots,n. \tag{B.11}$$

With this expression, the primal problem of optimization in this section is converted to what is referred to as the *dual problem*.

What we may naturally ask is the relationship between the solution of this dual problem and the solution of the primal problem of constrained minimization of the objective function $f(\boldsymbol{x})$. The answer is based on the inequality

$$\begin{aligned} L_D(\boldsymbol{\alpha},\boldsymbol{\beta}) &= \inf_{\boldsymbol{x}} L(\boldsymbol{x},\boldsymbol{\alpha},\boldsymbol{\beta}) \\ &\leq L(\boldsymbol{x},\boldsymbol{\alpha},\boldsymbol{\beta}) = f(\boldsymbol{x}) + \sum_{i=1}^{n} \alpha_i g_i(\boldsymbol{x}) + \sum_{i=1}^{m} \beta_i h_i(\boldsymbol{x}) \\ &\leq f(\boldsymbol{x}). \end{aligned} \tag{B.12}$$

This inequality shows that $L_D(\hat{\boldsymbol{\alpha}},\hat{\boldsymbol{\beta}}) \leq f(\hat{\boldsymbol{x}})$ is generally established for the solution $(\hat{\boldsymbol{\alpha}}, \hat{\boldsymbol{\beta}})$ of the constrained optimization (maximization) of $L_D(\boldsymbol{\alpha},\boldsymbol{\beta})$ in (B.11) and the solution $\hat{\boldsymbol{x}}$ of the problem of optimization (minimization) of the objective function $f(\boldsymbol{x})$ under constraints of equality and inequality. In cases of equality, the solution of the dual problem is equivalent to the solution of the objective function under the equality and inequality constraints. That is, for the solution $(\hat{\boldsymbol{\alpha}},\hat{\boldsymbol{\beta}})$ of the equation

$$\max_{\boldsymbol{\alpha},\boldsymbol{\beta}} \left[\min_{\boldsymbol{x}} \{L(\boldsymbol{x},\boldsymbol{\alpha},\boldsymbol{\beta})\} \right], \qquad \alpha_i \geq 0, \qquad i = 1,2,\cdots,n, \tag{B.13}$$

it holds that $L_D(\hat{\boldsymbol{\alpha}},\hat{\boldsymbol{\beta}}) = f(\hat{\boldsymbol{x}})$. This also shows that the fifth Karush-Kuhn-Tucker condition in (B.9) is necessary for establishment of the equality in (B.12).

On this basis, the constrained optimal solution of an objective function satisfying the regularity conditions is obtained as follows:

(1) Take the partial derivative of the Lagrangian function $L(\boldsymbol{x},\boldsymbol{\alpha},\boldsymbol{\beta})$ with respect to $\boldsymbol{x}$ and set it equal to zero, and then obtain the solution $\hat{\boldsymbol{x}}(\boldsymbol{\alpha},\boldsymbol{\beta})$.

(2) Substitute the step (1) solution into the Lagrangian function and define

$$L_D(\boldsymbol{\alpha},\boldsymbol{\beta}) = L\left(\hat{\boldsymbol{x}}(\boldsymbol{\alpha},\boldsymbol{\beta}),\boldsymbol{\alpha},\boldsymbol{\beta}\right) \tag{B.14}$$

as the function that eliminates the dependence on $\boldsymbol{x}$.

(3) Solve the constrained optimization problem of (B.11) for the Lagrange multipliers.

The primal problem of constrained optimization is thus converted to a dual problem in order to find the optimum solution. The constrained optimization problem of a support vector machine is a quadratic programming that satisfies the regularity conditions, and its solution can be found through conversion of the primal problem into the dual problem.

Appendix C

EM Algorithm

The expectation-maximization (EM) algorithm was initially proposed by Dempster et al. (1977) as an iterative method of finding maximum likelihood estimates of model parameters from observed data that include missing data or are otherwise incomplete. In basic concept, it is a method for parsimonious utilization of likelihood maximization based on complete data, given incomplete data. Its initial proposal was followed by revisions and extensions of algorithms, and applications to a growing range of problems, and by widespread efforts for application of the basic concept to the solutions of complex statistical problems, such as estimation of mixture models (see McLachlan and Peel, 2000; McLachlan and Krishnan, 2008).

C.1 General EM Algorithm

A general description of the EM algorithm may be given essentially as follows. Suppose that we have n independent observations $\boldsymbol{x}_{(n)} = \{\boldsymbol{x}_1, \boldsymbol{x}_2, \cdots, \boldsymbol{x}_n\}$ from a probability distribution $f(\boldsymbol{x}|\boldsymbol{\theta})$, which are actually observed data. Denote by $\boldsymbol{y}_{(n)} = \{\boldsymbol{y}_1, \boldsymbol{y}_2, \cdots, \boldsymbol{y}_n\}$ the variables representing those data that are for some reason missing. The mixture model provides one example. If the group membership of each of the observed data $\boldsymbol{x}_i$ in the mixture model is not known, then $\boldsymbol{y}_i = (y_{i1}, y_{i2}, \cdots, y_{ig})^T$ may be regarded as latent variable vectors representing membership in groups from which data are missing. For those data that belong to group G_j, we may denote by $y_{ij} = 1$ just the j-th component and set all of the remaining components to 0.

For the complete data $\{\boldsymbol{x}_{(n)}, \boldsymbol{y}_{(n)}\} = \{(\boldsymbol{x}_1, \boldsymbol{y}_1), (\boldsymbol{x}_2, \boldsymbol{y}_2), \cdots, (\boldsymbol{x}_n, \boldsymbol{y}_n)\}$ thus comprising the observed data together with the data represented by the missing variables, we then have the joint density function $f(\boldsymbol{x}_{(n)}, \boldsymbol{y}_{(n)}|\boldsymbol{\theta})$. Given the n observed data $\boldsymbol{x}_{(n)}$, the conditional density function of the missing $\boldsymbol{y}_{(n)}$ may be denoted by $f(\boldsymbol{y}_{(n)}|\boldsymbol{x}_{(n)}; \boldsymbol{\theta})$. We then apply the EM algorithm to obtain the parameter estimates by iterative execution of the following E and M steps with solution updating, starting from an appropriate initial value $\boldsymbol{\theta}^{(0)}$:

EM algorithm

E step For the log-likelihood function $\log f(\boldsymbol{x}_{(n)}, \boldsymbol{y}_{(n)}|\boldsymbol{\theta})$ based on the complete data, compute the conditional expected value with respect to the conditional density function $f(\boldsymbol{y}_{(n)}|\boldsymbol{x}_{(n)}; \boldsymbol{\theta}^{(t)})$ having the parameter values $\boldsymbol{\theta}^{(t)}$ of the t-th step as

$$Q(\boldsymbol{\theta}, \boldsymbol{\theta}^{(t)}) = \mathrm{E}_{\boldsymbol{\theta}^{(t)}}[\log f(\boldsymbol{x}_{(n)}, \boldsymbol{y}_{(n)}|\boldsymbol{\theta})|\boldsymbol{x}_{(n)}] \tag{C.1}$$

$$= \begin{cases} \displaystyle\int f(\boldsymbol{y}_{(n)}|\boldsymbol{x}_{(n)}; \boldsymbol{\theta}^{(t)}) \log f(\boldsymbol{x}_{(n)}, \boldsymbol{y}_{(n)}|\boldsymbol{\theta}) d\boldsymbol{y}_{(n)}, & \text{continuous}, \\ \displaystyle\sum_{\boldsymbol{y}_{(n)}} f(\boldsymbol{y}_{(n)}|\boldsymbol{x}_n; \boldsymbol{\theta}^{(t)}) \log f(\boldsymbol{x}_{(n)}, \boldsymbol{y}_{(n)}|\boldsymbol{\theta}), & \text{discrete}, \end{cases}$$

where $\sum_{\boldsymbol{y}_{(n)}}$ is the sum over all discrete values taken by the missing variables.

M step Maximize $Q(\boldsymbol{\theta}, \boldsymbol{\theta}^{(t)})$ computed in the E step with respect to $\boldsymbol{\theta}$ and set the resulting value $\boldsymbol{\theta}^{(t+1)}$ for use in the next iteration.

Perform successive iterations of the E and M steps until the convergence condition is satisfied, which is the attainment of values less than sufficiently small specified values for terms such as

$$||\boldsymbol{\theta}^{(t+1)} - \boldsymbol{\theta}^{(t)}||, \ |Q(\boldsymbol{\theta}^{(t+1)}, \boldsymbol{\theta}^{(t)}) - Q(\boldsymbol{\theta}^{(t)}, \boldsymbol{\theta}^{(t)})|, \ \ell(\boldsymbol{\theta}^{(t+1)}|\boldsymbol{x}_{(n)}) - \ell(\boldsymbol{\theta}^{(t)}|\boldsymbol{x}_{(n)}),$$

where $||\cdot||$ is the Euclidian norm and $\ell(\boldsymbol{\theta}^{(t)}|\boldsymbol{x}_{(n)})$ is the log-likelihood of the t-th step based on the observed data.

C.2 EM Algorithm for Mixture Model

For estimation of the mixture model in (10.22) with the EM algorithm, it is required to obtain the E-step function $Q(\boldsymbol{\theta}, \boldsymbol{\theta}^{(t)})$. The log-likelihood function based on the observations $\boldsymbol{x}_1, \boldsymbol{x}_2, \cdots, \boldsymbol{x}_n$ drawn from the mixture model $f(\boldsymbol{x}|\boldsymbol{\theta})$ is given by

$$\ell(\boldsymbol{\theta}|\boldsymbol{x}_{(n)}) = \sum_{i=1}^{n} \log f(\boldsymbol{x}_i|\boldsymbol{\theta}) = \sum_{i=1}^{n} \log \left\{ \sum_{j=1}^{g} \pi_j f_j(\boldsymbol{x}_i|\boldsymbol{\theta}_j) \right\}. \tag{C.2}$$

As it is not known which groups have generated the data $\boldsymbol{x}_i$, we let the variable vectors $\boldsymbol{y}_i = (y_{i1}, y_{i2}, \cdots, y_{ig})^T$ represent the missing group membership. For data from group G_j, we designate only the j-th component as $y_{ij} = 1$ and set all of the remaining components to 0.

The probability of $y_{ij} = 1$ is the prior probability π_j of group G_j, and the distribution of $\boldsymbol{y}_i$ for data $\boldsymbol{x}_i$ is then a multinomial distribution

$$h(\boldsymbol{y}_i) = \prod_{j=1}^{g} \pi_j^{y_{ij}}, \qquad i = 1, 2, \cdots, n. \tag{C.3}$$

Given $y_{ij} = 1$, moreover, the conditional distribution of $\boldsymbol{x}_i$ is

$$h(\boldsymbol{x}_i|\boldsymbol{y}_i) = \prod_{j=1}^{g} \left\{f_j(\boldsymbol{x}_i|\boldsymbol{\theta}_j)\right\}^{y_{ij}}, \qquad i = 1, 2, \cdots, n, \tag{C.4}$$

since the data $\boldsymbol{x}_i$ is generated from $f_j(\boldsymbol{x}_i|\boldsymbol{\theta}_j)$. From the relation $h(\boldsymbol{x}_i, \boldsymbol{y}_i) = h(\boldsymbol{x}_i|\boldsymbol{y}_i)h(\boldsymbol{y}_i)$, it then follows that the joint density function of $(\boldsymbol{x}_i, \boldsymbol{y}_i)$ may be given as

$$f(\boldsymbol{x}_i, \boldsymbol{y}_i|\boldsymbol{\theta}) = \prod_{j=1}^{g} \left\{f_j(\boldsymbol{x}_i|\boldsymbol{\theta}_j)\right\}^{y_{ij}} \prod_{j=1}^{g} \pi_j^{y_{ij}}, \qquad i = 1, 2, \cdots, n. \tag{C.5}$$

The density function based on the complete data $(\boldsymbol{x}_{(n)}, \boldsymbol{y}_{(n)}) = \{(\boldsymbol{x}_1, \boldsymbol{y}_1), (\boldsymbol{x}_2, \boldsymbol{y}_2), \cdots, (\boldsymbol{x}_n, \boldsymbol{y}_n)\}$ is therefore

$$f(\boldsymbol{x}_{(n)}, \boldsymbol{y}_{(n)}|\boldsymbol{\theta}) = \prod_{i=1}^{n} \left[\prod_{j=1}^{g} \left\{f_j(\boldsymbol{x}_i|\boldsymbol{\theta}_j)\right\}^{y_{ij}} \prod_{j=1}^{g} \pi_j^{y_{ij}} \right]. \tag{C.6}$$

The log-likelihood function is accordingly given by

$$\begin{aligned} \ell(\boldsymbol{\theta}|\boldsymbol{x}_{(n)}, \boldsymbol{y}_{(n)}) &= \sum_{i=1}^{n} \log f(\boldsymbol{x}_i, \boldsymbol{y}_i|\boldsymbol{\theta}) \\ &= \sum_{i=1}^{n}\sum_{j=1}^{g} y_{ij} \log f_j(\boldsymbol{x}_i|\boldsymbol{\theta}_j) + \sum_{i=1}^{n}\sum_{j=1}^{g} y_{ij} \log \pi_j. \end{aligned} \tag{C.7}$$

In the E step of the EM algorithm, once we have obtained the parameter value $\boldsymbol{\theta}^{(t)}$ of the t-th update, we then obtain the expectation of the log-likelihood function of (C.7) with respect to the conditional density function of the missing variables given the observed data $\boldsymbol{x}_{(n)}$ as follows:

$$\begin{aligned} &E_{\boldsymbol{\theta}^{(t)}}\left[\ell(\boldsymbol{\theta}|\boldsymbol{x}_{(n)}, \boldsymbol{y}_{(n)})|\boldsymbol{x}_{(n)}\right] \qquad \text{(C.8)} \\ &= \sum_{i=1}^{n}\sum_{j=1}^{g} E_{\boldsymbol{\theta}^{(t)}}[y_{ij}|\boldsymbol{x}_{(n)}] \log f_j(\boldsymbol{x}_i|\boldsymbol{\theta}_j) + \sum_{i=1}^{n}\sum_{j=1}^{g} E_{\boldsymbol{\theta}^{(t)}}[y_{ij}|\boldsymbol{x}_{(n)}] \log \pi_j, \end{aligned}$$

and take this as $Q(\boldsymbol{\theta}, \boldsymbol{\theta}^{(t)})$. The expectation here corresponds to the case of discrete missing variables in (C.1).

Since $y_{ij} = 1$ when the data $\boldsymbol{x}_i$ is from group G_j, the conditional expectation is

$$E_{\boldsymbol{\theta}^{(t)}}[y_{ij}|\boldsymbol{x}_{(n)}] = \frac{\pi_j^{(t)} f_j(\boldsymbol{x}_i|\boldsymbol{\theta}_j^{(t)})}{\displaystyle\sum_{k=1}^{g} \pi_k^{(t)} f_k(\boldsymbol{x}_i|\boldsymbol{\theta}_k^{(t)})} \equiv p^{(t)}(j|\boldsymbol{x}_i), \tag{C.9}$$

which is equal to the posterior probability of (10.23). The function Q as obtained in the E step is accordingly given by

$$\begin{aligned} Q(\boldsymbol{\theta}, \boldsymbol{\theta}^{(t)}) &= E_{\boldsymbol{\theta}^{(t)}}\left[\ell(\boldsymbol{\theta}|\boldsymbol{x}_{(n)}, \boldsymbol{y}_{(n)})|\boldsymbol{x}_{(n)}\right] \qquad \text{(C.10)} \\ &= \sum_{i=1}^{n}\sum_{j=1}^{g} p^{(t)}(j|\boldsymbol{x}_i)\log f_j(\boldsymbol{x}_i|\boldsymbol{\theta}_j) + \sum_{i=1}^{n}\sum_{j=1}^{g} p^{(t)}(j|\boldsymbol{x}_i)\log \pi_j. \end{aligned}$$

In the M step which follows, we then maximize $Q(\boldsymbol{\theta}, \boldsymbol{\theta}^{(t)})$ with respect to $\boldsymbol{\theta}$. We first note that the partial derivative with respect to the mixture ratio π_j is

$$\frac{\partial Q(\boldsymbol{\theta}, \boldsymbol{\theta}^{(t)})}{\partial \pi_j} = \sum_{i=1}^{n}\left\{\frac{p^{(t)}(j|\boldsymbol{x}_i)}{\pi_j} - \frac{p^{(t)}(g|\boldsymbol{x}_i)}{\pi_g}\right\} = 0, \quad j = 1, 2, \cdots, g-1,$$

in which $p^{(t)}(j|\boldsymbol{x}_i)$ is the label variable of the t-th update, and thus satisfies

$$\sum_{j=1}^{g} p^{(t)}(j|\boldsymbol{x}_i) = 1, \qquad \sum_{i=1}^{n}\sum_{j=1}^{g} p^{(t)}(j|\boldsymbol{x}_i) = n. \tag{C.11}$$

Hence $\pi_j^{(t)}$ is updated as follows:

$$\pi_j^{(t+1)} = \frac{1}{n}\sum_{i=1}^{n} p^{(t)}(j|\boldsymbol{x}_i), \qquad j = 1, 2, \cdots, g-1. \tag{C.12}$$

In the $(t+1)$-th step, the updated parameter values for $\boldsymbol{\theta}_k$ are given by the solution of the equation

$$\frac{\partial Q(\boldsymbol{\theta}, \boldsymbol{\theta}^{(t)})}{\partial \boldsymbol{\theta}_k} = \sum_{i=1}^{n}\sum_{j=1}^{g} p^{(t)}(j|\boldsymbol{x}_i)\frac{\partial \log f_j(\boldsymbol{x}_i|\boldsymbol{\theta}_j)}{\partial \boldsymbol{\theta}_k} = \mathbf{0}, \qquad k = 1, 2, \cdots, g.$$

In the particular case of a Gaussian mixture model, the probability density function is given by

$$f_j(\boldsymbol{x}|\boldsymbol{\mu}_j, \Sigma_j) = \frac{1}{(2\pi)^{p/2}|\Sigma_j|^{1/2}} \exp\left\{-\frac{1}{2}(\boldsymbol{x} - \boldsymbol{\mu}_j)^T \Sigma_j^{-1} (\boldsymbol{x} - \boldsymbol{\mu}_j)\right\},$$

and the parameter values $\boldsymbol{\mu}_j^{(t)}$, $\Sigma_j^{(t)}$ of the t-th step may therefore be updated as

$$\boldsymbol{\mu}_j^{(t+1)} = \frac{1}{\sum_{i=1}^{n} p^{(t)}(j|\boldsymbol{x}_i)} \sum_{i=1}^{n} p^{(t)}(j|\boldsymbol{x}_i)\boldsymbol{x}_i,$$

$$\Sigma_j^{(t+1)} = \frac{1}{\sum_{i=1}^{n} p^{(t)}(j|\boldsymbol{x}_i)} \sum_{i=1}^{n} p^{(t)}(j|\boldsymbol{x}_i)(\boldsymbol{x}_i - \boldsymbol{\mu}_j^{(t+1)})(\boldsymbol{x}_i - \boldsymbol{\mu}_j^{(t+1)})^T.$$

Bibliography

[1] Akaike, H. (1969). Fitting autoregressive models for prediction. *Annals of the Institute of Statistical Mathematics* **21**, 243–247.

[2] Akaike, H. (1973). Information theory and an extension of the maximum likelihood principle. *2nd International Symposium on Information Theory* (Petrov, B.N. and Csaki, F., eds.), Akademiai Kiado, Budapest, pp. 267–281. (Reproduced in *Breakthroughs in Statistics,* Volume 1, S. Kotz and N. L. Johnson, eds., Springer Verlag, New York, (1992)).

[3] Akaike, H. (1974). A new look at the statistical model identification. *IEEE Transactions on Automatic Control* **AC-19**, 716–723.

[4] Akaike, H. (1978). A Bayesian analysis of the minimum AIC procedure. *Annals of the Institute of Statistical Mathematics* **30**, 9–14.

[5] Akaike, H. (1979). A Bayesian extension of the minimum AIC procedure of autoregressive model fitting. *Biometrika* **66**, 237–242.

[6] Akaike, H. and Kitagawa, G. (eds.) (1998). *The Practice of Time Series Analysis.* Springer-Verlag, New York.

[7] Anderson, T. W. (2003). *An Introduction to Multivariate Statistical Analysis* (3rd ed.). Wiley, New York.

[8] Ando, T., Konishi, S. and Imoto, S. (2008). Nonlinear regression modeling via regularized radial basis function networks. *Journal of Statistical Planning and Inference* **138**, 3616–3633.

[9] Andrews, D. F. and Herzberg, A. M. (1985). *DATA.* Springer-Verlag, New York.

[10] Barndorff-Nielsen, O. E. and Cox, D. R. (1989). *Asymptotic Techniques for Use in Statistics.* Chapman & Hall, New York.

[11] Bishop, C. M. (1995). *Neural Networks for Pattern Recognition.* Oxford University Press, Oxford.

[12] Bishop, C. M. (2006). *Pattern Recognition and Machine Learning.* Springer, New York.

[13] Bozdogan, H. (1987). Model selection and Akaike's information criterion (AIC): The general theory and its analytical extensions. *Psychometrika* **52**, 345–370.

[14] Breiman, L. (1996). Heuristics of instability and stabilization in model selection. *Annals of Statistics* **24**, 2350–2383.

[15] Breiman, L., Friedman, J., Olshen, R. and Stone, C. (1984). *Classification and Regression Trees*. Wadsworth, New York.

[16] Burnham, K. P. and Anderson D. R. (2002). *Model Selection and Multimodel Inference: A Practical Information-Theoretic Approach*. 2nd ed., Springer, New York.

[17] Cai, J., Fan, J., Li, R. and Zhou, H. (2005). Variable selection for multivariate failure time data. *Biometrika* **92**, 303–316.

[18] Chatterjee, S., Hadi, A. S. and Price, B. (1999). *Regression Analysis by Example*. 3rd ed., Wiley, New York.

[19] Clarke, B., Fokoué, E. and Zhang, H. H. (2009). *Principles and Theory for Data Mining and Machine Learning*. Springer, New York.

[20] Craven, P. and Wahba, G. (1979). Smoothing noisy data with spline functions: Estimating the correct degree of smoothing by the method of generalized cross-validation. *Numerische Mathematik* **31**, 377–403.

[21] Cristianini, N. and Shawe-Taylor, J. (2000). *An Introduction to Support Vector Machines*. Cambridge University Press, Cambridge.

[22] Davison, A.C. (1986). Approximate predictive likelihood. *Biometrika* **73**, 323–332.

[23] Davison, A. C. and Hinkley, D. V. (1997). *Bootstrap Methods and Their Application*. Cambridge University Press, Cambridge.

[24] de Boor, C. (2001). *A Practical Guie to Splines*. Springer, Berlin.

[25] Dempster, A. P., Laird, N. M. and Rubin, D. B. (1977). Maximum likelihood from incomplete data via the EM algorithm. *Journal of the Royal Statistical Society* B **39**, 1–37.

[26] Diaconis, P. and Efron, B. (1983). Computer-intensive methods in statistics. *Scientific American* **248**, 116–130.

[27] Donoho, D. L. and Johnstone, I. M. (1994). Ideal spatial adaptation by wavelet shrinkage. *Biometrika* **81**, 425–455.

[28] Draper, N. R. and Smith, H. (1998). *Applied Regression Analysis*. 3rd ed., John Wiley & Sons, New York.

[29] Efron, B. (1979). Bootstrap Methods: Another look at the jackknife. *Annals of Statistics* 7, 1–26.

[30] Efron, B. (1982). *The Jackknife, the Bootstrap and Other Resampling Plans*. CBMS-NSF 38, SIAM.

[31] Efron, B. (1983). Estimating the error rate of a prediction rule: improvement on cross-validation. *Journal of the American Statistical Association* **78**, 316–331.

[32] Efron, B. and Gong, G. (1983). A leisurely look at the bootstrap, the jackknife, and cross-validation. *The American Statistician* **37**, 36–48.

[33] Efron, B. and Tibshirani, R. (1986). Bootstrap methods for standard errors, confidence intervals, and other measures of statistical accuracy. *Statistical Science* **1**, 54–77.

[34] Efron, B. and Tibshirani, R. J. (1993). *An Introduction to the Bootstrap*. Chapman & Hall, New York.

[35] Efron, B. and Tibshirani, R. (1997). Improvements on cross-validation. The .632+ bootstrap method. *Journal of the American Statistical Association* **92**, 548–560.

[36] Efron, B., Hastie, T. Johnstone, I. and Tibshirani, R. (2004). Least angle regression. *Annals of Statistics* **32**, 407–499.

[37] Fan, J. and Li, R. (2001). Variable selection via nonconcave penalized likelihood and its oracle properties. *Journal of the American Statistical Association* **96**, 1348–1360.

[38] Fan, J. and Li, R. (2002). Variable selection for Cox's propotional hazards model and frailty model. *Annals of Statistics* **30**, 74–99.

[39] Fan, J. and Li, R. (2004). New estimation and model selection procedures for semiparametric modeling in longitudinal data analysis. *Journal of the American Statistical Association* **99**, 710–723. Fan, J. and Peng, H. (2004). On non-concave penalized likelihood with diverging number of parameters. *Annals of Statistics* **32**, 928–961.

[40] Frank, I. and Friedman, J. (1993). A statistical view of some chemometrics regression tools. *Technometrics*, **35**, 109–148.

[41] Friedman, J., Hastie, T. and Tibshirani, R. (2010). Regularization paths for generalized linear models via coordinate descent. *Journal of Statistical Software* **33**, 1–22.

[42] Friedman J., Hastie T. and Tibshirani R. (2013). *glmnet: Lasso and Elastic-Net Regularized Generalized Linear Models*. R package version 1.9-5, URL http://CRAN.R-project.org/package=glmnet.

[43] Friedman, J., Hastie, T., Höfling, H. and Tibshirani, R. (2007). Pathwise coordinate optimization. *Annals of Applied Statistics* **1**, 302–332.

[44] Fu, W. (1998). Penalized regressions: the bridge vs the lasso. *Journal of Computational and Graphical Statistics* **7**, 397–416.

[45] Green, P. J. and Silverman, B. W. (1994). *Nonparametric Regression and Generalized Linear Models*. Chapman and Hall, London.

[46] Hall, P. (1992). *The Bootstrap and Edgeworth Expansion*. Springer-Verlag, New York.

[47] Härdle, W. (1990). *Applied Nonparametric Regression*. Cambridge University Press, Cambridge.

[48] Harrison, D. and Rubinfeld, D. L. (1978). Hedonic housing prices and the demand for clean air. *Journal of Environmental Economics and Management* **5**, 81–102.

[49] Hastie, T. and Efron, B. (2013). *lars*: *Least Angle Regression, Lasso and Forward Stagewise*. R package version 1.2, URL http://CRAN.R-project.org/package=lars.

[50] Hastie, T. and Thibshirani, R.J. (1990). *Generalized Additive Models*. Chapman and Hall, London.

[51] Hastie, T., Tibshirani, R. and Friedman, J. (2009). *The Elements of Statistical Learning*. 2nd ed., Springer, New York.

[52] Hastie, T., Rosset, S., Tibshirani, R. and Zhu, J. (2004). The entire regularization path for the support vector machine. *Journal of Machine Learning Research* **5**, 1391–1415.

[53] Hoerl, A. E. and Kennard, R. W. (1970). Ridge regression: biased estimation for nonorthogonal problems. *Technometrics* **42**, 55–67.

[54] Hoeting, J. A., Madigan, D., Raftery, A. E. and Volinsky, C. T. (1999). Bayesian model averaging: a tutorial (with discussion). *Statistical Science* **14**, 382–417.

[55] Hurvich, C. and Tsai, C.-L. (1989). Regression and time series model selection in small samples. *Biometrika* **76**, 297–307.

[56] Hurvich, C. and Tsai, C.-L. (1991). Bias of the corrected AIC criterion for underfitted regression and time series models. *Biometrika* **78**, 99–509.

[57] Ishiguro, M., Sakamoto, Y. and Kitagawa, G. (1997). Bootstrapping log likelihood and EIC, an extension of AIC. *Annals of the Institute of Statistical Mathematics* **49**, 411–434.

[58] James, G., Witten, D., Hastie, T. and Tibshirani, R. (2013). *An Introduction to Statistical Learning: with Applications in R*. Springer, New York.

[59] Johnson, R. A. and Wichern, D. W. (2007). *Applied Multivariate Statistical Analysis*. Pearson Prentice Hall, Upper Saddle River, NJ.

[60] Kass, R. E. and Raftery, A. E. (1995). Bayesian factors. *Journal of the American Statistical Association* **90**, 773–795.

[61] Kass, R. E. and Wasserman, L. (1995). A reference Bayesian test for nested hypotheses and its relationship to the Schwarz criterion. *Journal of the American Statistical Association* **90**, 928–934.

[62] Kawano, S. and Konishi, S. (2007). Nonlinear regression modeling via regularized Gaussian basis functions. *Bulletin of Informatics and Cybernetics* **39**, 83–96.

[63] Kim, Y., Choi, H. and Oh, H. (2008). Smoothly clipped absolute deviation on high dimensions. *Journal of the American Statistical Association* **103**, 1665–1673.

[64] Kitagawa, G. (2010). *Introduction to Time Series Modeling*. Chapman & Hall, New York.

[65] Kitagawa, G. and Gersch, W. (1996). *Smoothness Priors Analysis of Time Series*. Lecture Notes in Statistics **116**, Springer-Verlag, New York.

[66] Kitagwa, G. and Konishi, S. (2010). Bias and variance reduction techniques for bootstrap information criteria. *Annals of the Institute of Statistical Mathematics* **62**, 209–234.

[67] Kohonen, T. (1990). The self-organizing map. Proceedings of the IEEE **78**, 1464–1479.

[68] Kohonen, T. (2001). *Self-Organizing Maps*. 3rd ed., Springer, New York.

[69] Konishi, S. (1999). Statistical model evaluation and information criteria. In *Multivariate Analysis, Design of Experiments and Survey Sampling*, S. Ghosh, ed., 369–399, Marcel Dekker, New York.

[70] Konishi, S. (2002). Theory for statistical modeling and information criteria –Functional approach–. *Sugaku Expositions* **15**, 89–106, American Mathematical Society.

[71] Konishi, S. and Kitagawa, G. (1996). Generalized information criteria in model selection. *Biometrika* **83**, 875–890.

[72] Konishi, S. and Kitagawa, G. (2008). *Information Criteria and Sta-*

tistical Modeling. Springer, New York.

[73] Konishi, S., Ando, T. and Imoto, S. (2004). Bayesian information criteria and smoothing parameter selection in radial basis function networks. *Biometrika* **91**, 27–43.

[74] Kullback, S. and Leibler, R. A. (1951). On information and sufficiency. *Annals of Mathematical Statistics* **22**, 79–86.

[75] Lanterman, A. D. (2001). Schwarz, Wallace, and Rissanen: Intertwining themes in theories of model selection. *International Statistical Review* **69**, 185–212.

[76] LeCun, Y., Bottou, L., Bengio, Y. and Haffner, P. (1998). Gradient-based learning applied to document recognition. *Proceedings of the IEEE* **86**, 2278–2324.

[77] Linhart, H. (1988). A test whether two AIC's differ significantly. *South African Statistical Journal* **22**, 153–161.

[78] Linhart, H. and Zucchini, W. (1986). *Model Selection*. Wiley, New York.

[79] Mallows, C. L. (1973). Some comments on C_p. *Technometrics* **15**, 661–675.

[80] McLachlan, G. (2004). *Discriminant Analysis and Statistical Pattern Recognition*. John Wiley & Sons, New York.

[81] McLachlan, G. and Krishnan, T. (2008). *The EM Algorithm and Extensions*. 2nd ed., John Wiley & Sons, New York.

[82] McLachlan, G. and Peel, D. (2000). *Finite Mixture Models*. John Wiley & Sons, New York.

[83] McQuarrie, A. D. R. and Tsai, C.-L. (1998). *Regression and Time Series Model Selection*. World Scientific, Singapore.

[84] Meinshausen, N. and Bühlmann, P. (2006). High-dimenional graphs and variable selection with the lasso. *Annals of Statistics* **34**, 1436–1462.

[85] Moody, J. and Darken, C. J. (1989). Fast learning in networks of locally-tuned processing units. *Neural Computation* **1**, 281–294.

[86] Neath, A. A. and Cavanaugh, J. E. (1997). Regression and time series model selection using variants of the Schwarz information criterion. *Communications in Statistics* A**26**, 559–580.

[87] Nelder, J. A. and Wedderburn, R. W. M. (1972). Generalized linear models. *Journal of the Royal Statistical Society* A**135**, 370–84.

[88] O'Hagan, A. (1995). Fractional Bayes factors for model compar-

ison (with Discussion). *Journal of the Royal Statistical Society* B **57**, 99–138.

[89] Park, M. Y. and Hastie, T. (2007). L_1 regularization path algorithm for generalized linear models. *Journal of the Royal Statistical Society* B **69**, 659–677.

[90] Park, M. Y. and Hastie, T. (2013). *glmpath: L_1 Regularization Path for Generalized Linear Models and Cox Proportional Hazards Model.* R package version 0.97, URL http://CRAN.R-project.org/package=glmpath.

[91] Pauler, D. (1998). The Schwarz criterion and related methods for normal linear models. *Biometrika* **85**, 13–27.

[92] Rao, C. R. (1973). *Linear Statistical Inference and Its Applications.* 2nd ed., Wiley, New York.

[93] Reaven, G. M. and Miller, R. G. (1979). An attempt to define the nature of chemical diabetes using a multidimensional analysis. *Diabetologia* **16**, 17–24.

[94] Ripley, R. D. (1996). *Pattern Recognition and Neural Networks.* Cambridge University Press, Cambridge.

[95] Roeder, K. (1990). Density estimation with confidence sets exemplified by superclusters and voids in the galaxies. *Journal of the American Statistical Association* **85**, 617–624.

[96] Rosset, S., Zhu, J. and Hastie, T. (2004). Boosting as a regularized path to a maximum margin classifier. *Journal of Machine Learning Research* **5**, 941–973.

[97] Ruppert, D., Wand, M. P. and Carroll, R. J. (2003). *Semiparametric Regression.* Cambridge University Press, Cambridge.

[98] Sakamoto, T., Ishiguro M. and Kitagawa, G. (1986). *Akaike Information Criterion Statistics.* D. Reidel, Dordreeht.

[99] Schwarz, G. (1978). Estimating the dimension of a model. *Annals of Statistics* **6**, 461–464.

[100] Seber, G. A. F. (1984). *Multivariate Observations.* John Wiley & Sons, New York.

[101] Sen, A. and Srivastava, M. (1990). *Regression Analysis: Theory, Methods, and Applications.* Springer-Verlag, New York.

[102] Shimodaira, H. (1997). Assessing the error probability of the model selection test. *Annals of the Institute of Statistical Mathematics* **49**, 395–410.

[103] Siotani, M., Hayakawa, T. and Fujikoshi, Y. (1985). *Modern Multivariate Statistical Analysis: A Graduate Course and Handbook.* American Sciences Press, Columbus, OH.

[104] Stone, C. J. (1974). Cross-validatory choice and assessment of statistical predictions (with discussion). *Journal of the Royal Statistical Society Series* B **36**, 111–147.

[105] Sugiura, N. (1978). Further analysis of the data by Akaike's information criterion and the finite corrections. *Communications in Statistics* A7, 13–26.

[106] Tibshirani, R. (1996). Regression shrinkage and selection via the lasso. *Journal of the Royal Statistical Society* B **58**, 267–288.

[107] Tibshirani, R. and Wang, P. (2008). Spatial smoothing and hot spot detection for CGH data using the fused lasso. *Biostatistics* **9**, 18–29.

[108] Tibshirani, R., Saunders, M., Rosset, S., Zhu, J. and Knight, K. (2005). Sparsity and smoothness via the fused lasso. *Journal of the Royal Statistical Society* B **67**, 91–108.

[109] Tierney, L. and Kadane, J.B. (1986). Accurate approximations for posterior moments and marginal densities. *Journal of the American Statistical Association* **81**, 82–86.

[110] Vapnik, V. (1996). *The Nature of Statistical Learning Theory.* Springer, New York.

[111] Vapnik, V. (1998). *Statistical Learning Theory.* Wiley, New York

[112] Wahba, G. (1990). *Spline Models for Observational Data.* Society for Industrial and Applied Mathematics, Philadelphia.

[113] Wasserman, L. (2000). Bayesian model selection and model averaging. *Journal of Mathematical Psychology* **44**, 92–107.

[114] Wu, T. T. and Lange, K. (2008). Coordinate descent algorithms for lasso penalized regression. *Annals of Applied Statistics* **2**, 224–244.

[115] Ye, J. (1998). On measuring and correcting the effects of data mining and model selection. *Journal of the American Statistical Association* **93**, 120–131.

[116] Yuan, M. and Lin, Y. (2006). Model selection and estimation in regression with grouped variables. *Journal of the Royal Statistical Society* B **68**, 49–67.

[117] Yuan, M. and Lin, Y. (2007). On the non-negative garrotte estimator. *Journal of the Royal Statistical Society* B **69**, 143–161.

[118] Zou, H. (2006). The adaptive lasso and its oracle properties. *Journal of the American Statistical Association* **101**, 1418–1429.

[119] Zou, H and Hastie, T. (2005). Regularization and variable selection via the elastic net. *Journal of the Royal Statistical Society* B **67**, 301–320.

[120] Zou H. and Hastie T. (2013). *elasticnet*: *Elastic-Net for Sparse Estimation and Sparse PCA*. R package version 1.1, URL http://CRAN.R-project.org/package=elasticnet.

[121] Zou, H. and Zhang, H. H. (2009). On the adaptive elastic-net with a diverging number of parameters. *Annals of Statistics* **37**, 1733–1751.

Index

adaptive lasso, 46
additive model, 71
AIC, 6, 31, 74, 83, 98, 102, 116, 125, 278
AIC for Gaussian linear regression model, 32
AIC minimization, 116
apparent error rate, 155
average linkage, 261

B-spline, 63
basis expansion, 67
basis function, 58
Bayes' theorem, 173
Bayesian classification, 8, 173
Bernoulli distribution, 95
best linear unbiased estimator, 27
between-class matrix, 166
between-class variance, 139
bias of the log-likelihood, 116, 121, 124
BIC, 7, 128, 130, 132
binary response data, 91, 94
biplot, 246
bootstrap error estimation, 283
bootstrap method, 283
bootstrap model selection probability, 285
bootstrap prediction error estimate, 156
bootstrap sample, 284
bootstrap selection probability, 128, 162
bridge regularization, 43

calcium oxalate crystals, 160, 182
canonical discriminant analysis, 164
centroid linkage, 262
chaining effect, 265
chemical substance data, 12
classification, 7
cluster analysis, 11
clustering, 11, 259
competitive layer, 273
complementary log-log model, 104
complete linkage, 261
conditional probability, 173
cross-validation, 84, 108, 156
cubic spline, 60

de Boor's algorithm, 63
dendrogram, 12, 259
dependent variable, 18
design matrix, 24
diagnosis of diabetes, 151
dimension reduction, 11, 234
dimension reduction by canonical discriminant analysis, 164
discriminant analysis, 7, 137
dispersion measure, 267, 268
distance matrix, 263
dual problem, 198, 200, 290

effective degrees of freedom, 82
Efron's .632 estimator, 158, 159
eigenvalue problem, 246

elastic net, 44
EM algorithm, 277, 293
EM algorithm for mixture model, 294
empirical distribution, 115
empirical distribution function, 155
equality- and inequality-constrained optimization, 289
equality-constrained optimization, 287
error term, 18
Euclidean distance, 260
expected log-likelihood, 115
explanatory variable, 18

final prediction error, 111
finite correction, 126
Fisher's linear discriminant analysis, 137, 144
Fisher's scoring algorithm, 97
fused lasso, 49
fusion distance, 263
fusion level, 263

Gauss-Markov theorem, 27
Gaussian basis function, 66
Gaussian distribution, 20
Gaussian linear regression model, 28, 118
Gaussian mixture model, 275
Gaussian nonlinear regression, 70
generalized cross-validation, 85, 109
generalized eigenvalue problem, 166
generalized variance, 268
geometric interpretation, 34
group lasso, 51
growth curve model, 56

hat matrix, 83, 109
heat evolution by cement, 32
hierarchical clustering, 12, 259

idempotent matrix, 26, 36
image compression and decompression, 239
independent variable, 18
inequality-constrained optimization, 288
information criterion, 112, 116, 121
information loss, 234
input layer, 273
intercluster distance, 261
interobject similarity, 260

K-fold cross-validation, 108, 156
k-means clustering, 271
Karush-Kuhn-Tucker conditions, 202, 289
kernel dual problem, 219, 220
kernel method, 216, 252
kernel principal component analysis, 246
kernel trick, 217
knot, 60
Kullback-Leibler information, 113, 114

L_1 distance, 260
L_1 norm regularization, 44
Lagrange multiplier, 287, 288
Lagrangian function, 199
Laplace approximation for integrals, 130, 132
Laplace distribution, 134
LARS, 42
lasso, 2, 40
law of total probability, 174
learning coefficient, 274
least squares, 18, 24, 69
least squares estimates, 19, 25
leave-one-out CV, 108

likelihood, 21, 175
likelihood equation, 28
likelihood for logistic regression, 96
likelihood function, 22, 28
linear discriminant function, 141, 178
linear logistic regression classifier, 179
linear model, 16
linear regression, 15, 22
linear regression model, 18, 24
linear separability, 193
linearly nonseparable, 203
local quadratic approximation, 48
log-likelihood for logistic regression, 96
log-likelihood function, 22, 28
logistic regression, 5, 87, 88
logistic regression for classification, 179
logit transformation, 89, 94

Mahalanobis distance, 146, 148, 150
Mallows' C_p, 110, 111
margin, 196
margin maximization, 196
marginal distribution, 129
marginal likelihood, 129
maximum likelihood method, 19, 28, 70
median linkage, 262
Mercer's theorem, 218
Minkowski distance, 260
mixture model, 275
mixture model-based clustering, 276
model averaging, 127, 132
model selection, 5, 31, 72
motorcycle impact data, 2
multiclass classification, 149
multiclass nonlinear logistic regression classifier, 187
multimodel inference, 7, 127, 132
multinomial distribution, 135
multiple linear regression, 24
multiple logistic regression, 94
multivariate normal distribution, 30

natural cubic spline, 61
neighborhood function, 273
nonhierarchical clustering, 12, 270
nonlinear classification via SVM, 218
nonlinear logistic regression, 98
nonlinear logistic regression classifier, 183
nonlinear logistic regression model, 100
nonlinear regression, 3, 55
normal distribution, 20
normal equation, 25

optimum separation hyperplane, 196
oracle property, 46

penalized least squares, 77
penalized log-likelihood function, 80
penalized maximum likelihood, 79
polynomial regression, 117
pooled sample variance-covariance matrix, 142
posterior distribution, 133
posterior probability, 174
posterior probability of the model, 129, 133
predicted value, 19, 26
prediction error, 105, 154, 155
prediction of housing price, 43

predictive distribution, 133
predictive mean squared error, 107
predictive sum of squares, 106
predictor variable, 18
primal problem, 199
primal problem for the soft margin, 208
principal component analysis, 11, 225
principal components, 225, 230
prior probability, 146, 173
probability distribution model, 20
probability of occurrence of kyphosis, 102
probability of the presence of calcium oxalate crystal, 97
probit model, 104
projection, 35
projection matrix, 35

quadratic discriminant function, 148, 178
quadratic programming, 198

radial basis function, 65
regression, 1
regression coefficients, 18
regression model, 15, 56, 285
regularization, 3, 36, 76
regularization parameter, 77
regularization path, 42, 43
regularization term, 77
regularized least squares, 77
regularized least squares estimator, 78
regularized log-likelihood function, 80
regularized maximum likelihood, 79
regularized maximum likelihood estimator, 80
residual, 19, 26
residual sum of squares, 106
response variable, 18
ridge estimator, 38
ridge regression, 2, 37
risk models, 4
risk prediction, 87

SCAD, 47
self-organizing map, 273
semiparametric model, 72
separating hyperplane, 193
single linkage, 261
singular value decomposition, 243
slack variable, 206
smoother matrix, 83, 109
smoothing parameter, 77
soft margin, 204
software packages in *R*, 37
spectral decomposition, 233
spline, 59
standardized data, 37, 39
standardized Euclidean distance, 260
statistical model, 113
stepwise procedure, 162
support vector, 201, 203, 212
support vector machines, 9, 193

total variance-covariance matrix, 166
tree diagram, 259

variable selection, 31, 154

Ward's method, 267, 269
weight decay, 38
weight in model averaging, 127
within-class matrix, 166
within-class variance, 139

Printed in the United States
by Baker & Taylor Publisher Services